Energy Technology Perspectives:

Conservation, Carbon Dioxide Reduction and Production from Alternative Sources

ENERGY TECHNOLOGY PERSPECTIVES

Related titles include:

- *Carbon Dioxide Reduction Metallurgy*
- *REWAS 2008: Global Symposium on Recycling, Waste Treatment, Minimization, and Clean Technology*
- *Processing Materials for Properties III*

HOW TO ORDER PUBLICATIONS

For a complete listing of TMS publications, contact TMS at (724) 776-9000 or (800) 759-4TMS or visit the TMS Knowledge Resource Center at http://knowledge.tms.org:

- Purchase publications conveniently online and download electronic publications instantly.
- View complete descriptions and tables of contents.
- Find award-winning landmark papers and webcasts.

MEMBER DISCOUNTS

TMS members receive a 30 percent discount on TMS publications. In addition, members receive a free subscription to the monthly technical journal *JOM* (both in print and online), free downloads from the Materials Technology@TMS digital resource center (www.materialstechnology.org), discounts on meeting registrations, and additional online resources to name a few of the benefits. To begin saving immediately on TMS publications, complete a membership application when placing your order or contact TMS:

Telephone: (724) 776-9000 / (800) 759-4TMS

E-mail: membership@tms.org or publications@tms.org

Web: www.tms.org

Energy Technology Perspectives:
Conservation, Carbon Dioxide Reduction and Production from Alternative Sources

Proceedings of symposia sponsored by the Light Metals Division
of The Minerals, Metals & Materials Society (TMS)

Held during TMS 2009 Annual Meeting & Exhibition
San Francisco, California, USA
February 15-19, 2009

Edited by:

Neale R. Neelameggham
Ramana G. Reddy
Cynthia K. Belt
Edgar E. Vidal

A Publication of

A Publication of **The Minerals, Metals & Materials Society (TMS)**
184 Thorn Hill Road
Warrendale, Pennsylvania 15086-7528
(724) 776-9000

Visit the TMS Web site at
http://www.tms.org

ISBN Number 978-0-87339-733-9

If you are interested in purchasing a copy of this book, or if you would like to receive the latest TMS Knowledge Resource Center catalog, please telephone (724) 776-9000, ext. 270, or (800) 759-4TMS.

TABLE OF CONTENTS
Energy Technology Perspectives

Preface.. xi
Editor Information xv

Carbon Dioxide Reduction Metallurgy 2009

Mechanisms and Electrolysis

Metal Cations in CO2 Assimilation and Conversion
by Plants...5
S. Shabala

Effect of Electrode Surface Modification on
Dendritic Deposition of Aluminum on Cu Substrate
Using Emic-Alcl3 Ionic Liquid Electrolytes 17
D. Pradhan, and R. Reddy

Room-Temperature Production of Ethylene from
Carbon Dioxide...25
K. Ogura

The Electrochemical Reduction of Carbon Dioxide
in Ionic Liquids..39
H. Lu, X. Zhang, and P. Wang

Silicon Dioxide as a Solid Store for CO2 Gas..............................49
V. Zavodinsky, and S. Rogov

Recent Developments in Carbon Dioxide Capture
Materials and Processes for Energy Industry...............................53
M. Goel

Ferrous and Titanium Metallurgy

Reduction of CO2 Emissions in Steel Industry Based on LCA Methodology 65
A. Losif, J. Birat, O. Mirgaux, and D. Ablitzer

Electrolytic Reduction of Ferric Oxide to Yield Iron and Oxygen 77
A. Cox, and D. Fray

Enhanced Energy Efficiency and Emission Reduction through Oxy-Fuel Technology in the Metals Industry 87
N. Bell

Suspension Ironmaking Technology with Greatly Reduced CO2 Emission and Energy Requirement 93
H. Sohn, M. Choi, Y. Zhang, and J. Ramos

Investigation of Carbonic Anhydrase Assisted Carbon Dioxide Sequestration Using Steelmaking Slag 103
C. Rawlins, S. Lekakh, K. Peaslee, and V. Richards

Novel Alkali Roasting of Titaniferous Minerals and Leaching for the Production of Synthetic Rutile 115
A. Jha, and A. Lahiri

Accelerated Electro-reduction of TiO2 to Metallic Ti in CaCl2 Bath using an Intermetallic Inert Anode 127
X. Yang, A. Lahiri, and A. Jha

Author Index 261

Subject Index 263

TABLE OF CONTENTS
Energy Technology Perspectives

Preface xi
Editor Information xv

Energy Conservation in Metals Extraction and Materials Processing II

Energy Conservation and Technology

Mechanism and Application of Catalytic
Combustion of Pulverized Coal 139
Z. Guo, Z. Lin, and C. Li

Improving Energy Efficiency in a Modern
Aluminum Casting Operation 149
C. Eckert, M. Osborne, and R. Peterson

Oxyfuel - Energy Efficient Melting 161
T. Niehoff, and D. Stoffel

Energy Savings and Productivity Increases at an
Aluminium Slug Plant Due to Bottom Gas Purging 169
K. Gamweger, and P. Bauer

Energy Conservation and Productivity Improvement
Measures in Electric Arc Furnaces 173
A. Jaiswal

Evaluating Aluminum Melting Furnace Transient
Energy Efficiency 181
E. Williams, D. Stewart, and K. Overfield

Billions of Dollars could be saved with Reliability Excellence....187
D. Wikoff, and S. Williams

Extraction Processes/Refractories/Modeling and Analysis

Energy and Sustainable Development in Hydrometallurgy - An Emerging Perspective....195
K. Rao

Copper Electrowinning using Noble Metal Oxide Coated Titanium - Based Bipolar Electrodes....201
K. Asokan, and K. Subramanian

Microbial Reduction of Lateritic Nickel Ore for Enhanced Recovery of Nickel and Cobalt through Bio-hydrometallurgical Route....213
L. Sukla, N. Pradhan, R. Mohapatra, B. Mohapatra, B. Nayak, and B. Mishra

Energy Saving Strategies for the Use of Refractory Materials in Molten Material Contact....225
J. Hemrick, K. Peters, J. Damiano, and J. Keiser

Energy Savings through Phosphate-bonded Refractory Materials....233
J. Decker

Advanced Ceramic Composites for Improved Thermal Management in Molten Aluminum Applications....241
K. Peters, R. Cravens, and J. Hemrick

Energy Efficient, Non-Wetting, Microporous Refractory Material for Molten Aluminum Contact Applications 249
K. McGowan

Author Index 261

Subject Index 263

Preface

With the initial symposia on CO_2 Reduction Metallurgy, and Energy Conservation held in March 2008 at the New Orleans Annual Meeting, TMS felt the need to enlarge the role of scientific contribution to the matter of energy use in metals and materials processing in addition to finding useful ways of reducing carbon dioxide. This resulted in the formation of the Energy Committee of TMS under the Light Metals Division and Extraction & Processing Division. The present book encompasses both as parts of the proceedings of Energy Technolgy Perspectives. There are two sections depicting each of the symposium sessions. Some of the noteworthy papers, which form part of the symposia presentations, are moved to the April 2009 JOM - special energy issue.

Carbon Dioxide Reduction Metallurgy 2009 Symposium:

This symposium continues to address both the chemical reduction of carbon dioxide into useful products like fuels using alternative energy as well as the physical reduction of carbon dioxide in metal production by both energy conservation and use of non-carbon reduction technologies. Here the extractive metallurgists apply their expertise in the basic understanding of ability to reduce any oxides into useful material and metals.

This symposium is divided into two sessions. Session one presentations concentrate on the mechanisms and electrolysis in carrying out the CO_2 reduction. Session two presentations encompass activities of CO_2 reduction in the ferrous metallurgy as well as titanium materials field including novel approaches of technology, which are just emerging.

The keynote address for the symposium is by Sergey Shabala, an expert in Plant nutrition who discusses the role of metal cations in the carbon dioxide assimilation by botanical matter from atmosphere. Understanding of some of these natural concepts by extractive metallurgists is expected to help develop newer approaches in this quest for carbon dioxide reduction worldwide. This paper will become part of the JOM - special energy issue.

The following topics which are not part of the proceedings will be covered during the symposium. (Readers are referred to the TMS Annual Meeting program for the abstracts.):

CO2 Mitigation in Metallurgical Processes Using Concentrated Solar Energy
Aldo Steinfeld; ETH Zurich

Aluminum Industry and Climate Change — Assessment and Responses
Subodh Das; John Green; Phinix LLC; Secat Inc.

Some the emerging technologies come from asking "Why Not" - in spite of the fact outwardly these techniques may not reduce the monetary aspect of producing the materials compared to conventional use of carbon as the only reductant possible. The conventional thinking is slowly changing with the possible advent of carbon tax type controls needed for cutting down the CO_2 emissions, which would make the use of carbon reductant as equally costly as alternative sources of energy. Thermal emission reductions, which will be the main method of global warming reduction, will continue to be addressed by becoming more energy efficient in our emerging technologies.

Approaches to carbon dioxide emission reductions in metal production by improved energy efficiency in life-cycle fuel use, reductions in carbonate based flux/raw material usage, and thermodynamically feasible reactions leading to lower emissions are all part of this program.

Energy Conservation in Metals Extraction and Materials Processing II Symposium:

Energy conservation is being addressed within the metals industry from two different viewpoints. The altruistic point-of-view says that results from energy conservation will be a reduction of greenhouse gas emissions which is important to the world's environment. The capitalistic point-of-view says that energy costs have greatly escalated and improved energy efficiency will keep overall production costs down. Whatever your view, or most likely a combination of both, we all understand that addressing energy conservation in our industry is increasingly essential.

The papers within the Energy Conservation in Metals Extraction and Materials Processing symposium are an effort to share ideas to improve energy efficiency within our industry. Now in its second year, this year's papers look at energy conservation from different ways. This year we look at extractive process work, energy savings through optimizing refractories, modeling efforts, technology improvements, and energy management. No one method will be the solution and we thank all our authors for their input and point-of-view. We hope the sharing of these papers both help conservation efforts and generate new ideas for future efforts in this area.

The following presentations during the symposium have been selected to be part of the special issue on energy in the JOM - April 2009. (Readers are referred to the TMS Annual Meeting program for the abstracts.):

Elements of an Energy Management Program
Ray Peterson; Cynthia Belt; Aleris International Inc.

Understanding and Evaluating Energy Saving Options
William Choate; Robert D. Naranjo; BCS

Overview of the Department of Energy's Industrial Technologies Program
Bob Gemmer; US Department of Energy

The following topics which are not part of the Proceedings will be covered during the symposium. (Readers are referred to the TMS Annual Meeting program for the abstracts.):

CFD Modeling for Optimization of Aluminum Melting Furnace Design Parameters
Mohamed Hassan Ali; Zhengdong Long; Shridas Ningileri; Subodh Das; University of Kentucky; Phinix LLC

A Study of Exergy Analysis for Combustion in Direct Fired Heater
Ahmed abd elrahman; Egyptalum

Waste Heat Recovery in an Aluminium Cast House
Jan Migchielsen; Tom Schmidt; Thermcon Ovens BV

We acknowledge the efforts by all the co-organizers: Neale Neelameggham, Ramana Reddy, Jean- Pierre Birat and Jiann-Yang Hwang (CO_2 Reduction Metallurgy Symposium) and Cindy Belt, Edgar Vidal, Marie Kistler, and Mark Cooksey (Energy Conservation Metals Extraction and Materials Processing). The co-editors also selected a few pertinent articles to go in the special JOM issue in April 2009 on energy related topics being organized by Xingbo Liu.

Our thanks go to both the TMS Light Metals and Extraction & Processing divisions and staff who helped promote this symposium and proceedings to be of value. We appreciate the efforts by the participating authors and session chairs.

Neale R. Neelameggham, Ramana G. Reddy, Cynthia Belt and Edgar Vidal
(Editors)

Editors

Neale R. Neelameggham is the Technical Development Scientist for US Magnesium LLC. He has 36 years of expertise in magnesium production technology having been with the plant from its startup company NL Magnesium. Dr. Neelameggham's expertise includes all aspects of the magnesium process, from solar ponds through the cast house including solvent extraction, spray drying, molten salt chlorination, electrolytic cell and furnace designs, lithium ion battery chemicals and by-product chemical processing. In addition, he has an in-depth and detailed knowledge of alloy development as well as all competing technologies of magnesium production, both electrolytic and thermal processes worldwide. Dr. Neelameggham holds 13 patents and has several technical papers to his credit. As a member of TMS, AIChE, and a former member of American Ceramics Society he is well versed in energy engineering, biofuels and related processes. Dr. Neelameggham has served in the Magnesium Committee of LMD since its inception in 2000, chaired it in 2005, and has been a co-organizer of the magnesium Committee since 2004. In 2007 he was made a Permanent Co-organizer for magnesium Committee. He has been a member of Reactive Metals Committee, Recycling Committee and Programming Committee Representative of LMD. In 2008 LMD and EPD created the Energy Committee following the symposium on CO2 Reduction Metallurgy symposium initiated by him. Dr. Neelameggham was selected as the inaugural Chair for the Energy Committee with a two-year term. He is also a member of LMD council. Dr. Neelameggham's doctorate in extractive metallurgy from the University of Utah.

Ramana G. Reddy is an ACIPCO Professor and Head, Department of Metallurgical and Materials Engineering at The University of Alabama, Tuscaloosa, Alabama. He has 27 years of teaching and research experience in the field of chemical and materials engineering. Professor Reddy has conducted projects involving thermodynamics and kinetics of metallurgical reactions; plasma processing of materials, molten salts and ionic liquids electrolysis, and fuel cells.

He is the recipient of four U.S. Patents and has published more than 300 research papers and 21 books including an undergraduate student textbook in thermodynamics. Professor Reddy has also delivered more than 192 invited lectures and research presentations in 23 nations. He is the associate editor of the Journal of Phase Equilibria and Diffusion and editor-in-chief of the International Journal for Manufacturing Sciences & Production. Professor Reddy has organized 23 national and international symposia including the National Science Foundation 2003 Design Service and Manufacturing Grantees and Research Conference. He was appointed the University of Alabama's co-coordinator for the National Space Science and Technology Center. and NASA, and served as council member of the Alabama State Committee for the Department of Defense EPSCoR Programs. The recipient of many honors and awards, Professor Reddy received the 2008 Distinguished Member Award from the Society for Mining, Metallurgy & Exploration and the TMS Extraction & Processing Division 2002 Distinguished Lecturer Award. He is a Fellow of American Society for Metals International. He obtained his doctorate from the University of Utah.

Cynthia K. Belt is Energy Engineer for Aleris International working in energy conservation along with optimization of throughput, recovery, and refractory throughout the corporation.

Cindy earned her Bachelor of Science, Mechanical Engineering degree from Ohio Northern University with additional graduate work in materials at Case Western Reserve University and Akron University. Cindy is trained as a Black Belt in Six Sigma. She is Vice-Chair of the Energy Committee within TMS, an active member of the Recycling & Environmental Technologies Committee, and a member on the Process Heating System Assessment Standard Committee for ASME. During March 2008, when the Energy Committee of LMD/EPD was formed, Cindy was selected as the Vice Chair for the first two years of this inagural committee.

Dr. Edgar E. Vidal is a Senior Technologist in Beryllium Products R&D at Brush Wellman, Inc. He obtained his PhD in Metallurgy from the University of Idaho and a MS and BS in Materials Engineering from Simon Bolivar University in Caracas, Venezuela. Dr. Vidal's most recent work has been directed towards the chemical and physical processing of beryllium and its potential use for increasing the efficiency of nuclear reactors and energy consumption of semiconductors. As a member of the Extraction Processing Division (EPD) of TMS, Dr. Vidal has been actively involved in promoting more energy efficient methods and processes for the extraction of metals and valuables from the earth. Dr. Vidal is also a member of the American Nuclear Society (ANS) and ASM International, and enjoys his free time piloting airplanes.

Energy Technology Perspectives:

Conservation, Carbon Dioxide Reduction and Production from Alternative Sources

Carbon Dioxide Reduction Metallurgy 2009

Organizers

Neale Neelameggham
Ramana Reddy

Energy Technology Perspectives:

Conservation, Carbon Dioxide Reduction and Production from Alternative Sources

Mechanisms and Electrolysis

Session Chairs

Ramana Reddy
Jiann-Yang Hwang

Energy Technology Perspectives
Edited by: Neale R. Neelameggham, Ramana G. Reddy, Cynthia K. Belt, and Edgar E. Vidal
TMS (The Minerals, Metals & Materials Society), 2009

METAL CATIONS IN CO_2 ASSIMILATION AND CONVERSION BY PLANTS

Sergey Shabala

School of Agricultural Science, University of Tasmania
Private Bag 54, Hobart, Tas 7001, Australia

Keywords: photosynthetic light conversion; carbon dioxide assimilation; light; metal cations; artificial photosynthesis

Abstract

Green leaf tissues convert solar energy into the energy of chemical bonds of sugar molecules during the process of photosynthesis. The efficiency of this conversion can be much higher than efficiency of any of the currently known silicon-based solar panels. Importantly, vast amounts of CO_2 are assimilated during this process. The efficiency of photosynthesis is critically dependent on the availability of a large number of nutrients, among which metal cations such as K, Ca, Mg, Cu, Zn, Fe, Mn, and Ni play a key role. In this paper I summarize basic requirements and major functions for each of these essential nutrients in plant photosynthesis, both at the whole-plant and molecular level. I will talk about how these requirements may be affected by the global climate trends and discuss the prospects of creating artificial photosynthetic "bioreactors" for efficient energy conversion and CO_2 assimilation.

Introduction: Atmospheric CO_2 and Global Warming

Human activities have resulted in a gradual increase in the atmospheric CO_2 concentration from 280 ppm in pre-industrial times to ~ 380 ppm at present and potentially to 700 ppm towards the end of the century (Luo 2007). As a consequence of the greenhouse gas buildup, Earth's surface temperature has increased by 0.74°C since 1850 and is expected to increase by another 1.1 to 6.4 °C by 2100 (Forster et al. 2007). CO_2 contributes to ~ 60% of this buildup, and the atmospheric CO_2 content has been increasing at the rate of 1.9 ppm per year (Dalal and Allen 2008). The major sources of CO_2 are the use of fossil fuels, biomass burning and industrial processes, as well as biological respiration. Counterbalancing these processes is biological CO_2 fixation (by terrestrial plants and phytoplankton) and carbonate formation. Among biological ecosystems, forests store about 45% of terrestrial carbon and can sequester large amounts of carbon annually (at least 1/3 of the total anthropogenic emission; Denman et al. 2007). However, the capacity of forests (as well as other terrestrial and marine ecosystems) to efficiently sequester CO_2 is strongly affected by numerous factors. As a result, the existing non-linear interaction between plant ecosystems and atmosphere can either dampen or amplify anthropogenic climate change (Bonan 2008). As such, plant ecosystems can reduce global warming directly, by evapotranspiration, and indirectly, by removing CO_2 from the atmosphere. At the same time, elevated ambient temperatures may suppress the rate of CO_2 fixation in some plant types, providing a positive feedback between terrestrial carbon cycles and global warming. It is becoming increasingly evident that the existing coupled carbon-climate models reported in the literature do not describe all the diversity of the possible ecosystem responses (Luo 2007). Of specific interest is the link between plant nutrition and efficiency of CO_2 sequestration. It is well known that the efficiency of photosynthesis is critically dependent on the availability of a large number of nutrients, among which metal cations such as K, Ca, Mg, Cu, Zn, Fe, Mn, and Ni play a key role. At the same time, changing climate will (and already does) place some significant constraints on the availability of these nutrients to plants. In this review, I summarize basic requirements and major functions for each of these essential nutrients in plant photosynthesis, both at the whole-plant and molecular level. I also discuss how these requirements may be affected by the global climate trends and discuss the prospects of creating artificial photosynthetic "bioreactors" for efficient energy conversion and CO_2 assimilation.

Photosynthesis: an Overview

Photosynthesis is a process of conversion of solar energy into the energy of chemical bonds of organic molecules, and can be summarized by a simple equation (Buchanan et al. 2000):

$$CO_2 + 2H_2A + Light \rightarrow (CH_2O) + 2A + H_2O$$

In this redox process, CO_2 is the electron acceptor, and H_2A is any reduced compound that can serve as the electron donor. With the relatively small exception of prokaryotic organisms (e.g. some genera of bacteria), most organisms on Earth use water as a reductant. Water is oxidized and the electrons released are energized and transferred to CO_2, yielding oxygen and carbohydrates

$$CO_2 + 2H_2O + Light \rightarrow (CH_2O) + 2O_2 + H_2O$$

This reaction is light-driven and enderogenic, with a free energy change ($\Delta G^{o'}$) of 2840 kJ/mol of hexose formed (Buchanan et al. 2000). Without any doubt, this is the most important chemical reaction on the planet which supplies us with oxygen to breathe, carbohydrates to fuel all the metabolic processes in our bodies, and which removes CO_2 from the atmosphere at the same time. It is the latter that will be the focus of this review.

Being relatively simple in chemical terms, this reaction is mediated by a large number of specific components and is found in nature almost exclusively in highly specialized organelles (the so-called chloroplasts) in the cells of green plant tissues. Two distinct phases of photosynthesis are known – the so-called light (or energy-transduction) and dark (or carbon-fixation) reactions. These are briefly summarized below, along with their major nutritional requirements.

Light Reactions of Photosynthesis

Light reactions take place in the thylakoid membranes of chloroplasts. Between 250 and 400 pigment molecules (chlorophylls and carotenoids) are embedded in the thylakoids to form discrete units called photosystems (Fig. 1). Two different types of photosystems (PSII and PSI) are known; the main difference between these is in their maximum absorption peak (680 and 700 nm, respectively). Each photosystem consists of two components: (i) an antenna complex (the major function of which is to trap the light) and (ii) a reaction center. The light absorbed by the antenna pigments is passed to the reaction center in a process of resonant energy transfer. The energized electrons are then transferred to a primary acceptor molecule and then through a chain of secondary acceptors, down the energy gradient. As the electrons are passed through cytochrome b/f complex, a sharp proton gradient across the thylakoid membrane is generated which drives the synthesis of ATP (a universal cell "currency") from ADP and Pi in a process called photophosphorylation. The electrons lost in the energy transfer by PSII are replaced by splitting the water into oxygen gas and hydrogen atoms (electrons and protons) in the interior of the thylakoid. This light-dependent oxidative splitting of water molecules is called photolysis. The electrons are ultimately accepted by $NADP^+$ and H^+, producing NADPH. PSI and PSII are linked together by an electron transport chain, forming the so-called Z-scheme of photosynthesis. ATP can also be produced solely by PSI via cyclic electron flow, in a process called cyclic phosphorylation (not show in Fig. 1).

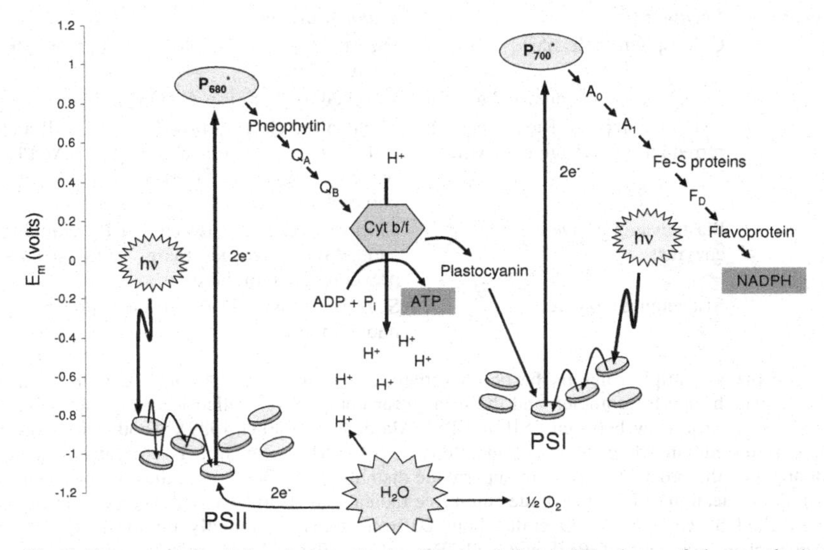

Fig. 1. Energy transfer during photosynthesis, phosphorylation and Z-scheme of photosynthesis. Q_A – plastoquinone A; Q_B – plastoquinone B; Cyt b/f – cytochrome b/f complex; A_O – modified chlorophyll a; A_1 – phylloquinone; F_D – ferredoxin. At least three Fe-S proteins act as electron acceptors in PSI. Based on Raven et al. (1999) and Atwell et al. (1999).

The above process of electron transport in thylakoid membranes of chloroplasts is critically dependent on the availability of a large number of essential nutrients. Most of these are metal cations. Their essentiality and major functions are summarized below in Table 1.

A major function of Mg is its role as the central atom of the chlorophyll molecule. In Mg^{2+} deficient plants, over 50% of total Mg is associated with chlorophyll (Marschner 1995). Mg^{2+} availability affects the size, structure and function of chloroplasts, including electron transfer in PSII (Lavon and Goldsmith 1999). There are also a large number of photosynthetic-relevant enzymes and processes controlled by Mg^{2+}; these are discussed in the following sections.

Manganese-containing enzyme complex, associated with PSII, is responsible for the process of photolysis (required to "inject" electrons into the electron transport chain) and oxygen evolution. The details of this process remain elusive. It is assumed that Mn clusters can act as a device for the storing energy prior to the oxidation of two H_2O molecules, or as a binding site for H_2O molecules which are oxidized (Rutherford 1989). Manganese deficiency is usually associated with a dramatic decline in O_2 evolution and correlates with the loss of functional PSII units in the stacked areas of thylakoid membranes (Kriedemann et al. 1985). In addition, a special class of Mn^{2+}-activated enzymes, the so-called superoxide dismutases (SOD), plays a crucial role in scavenging free oxygen radicals (O_2^{-}) (Bowler et al. 1991). These radicals are produced in leaves under stress conditions and are deleterious to cell metabolism.

Table 1. Essentiality and major functions of metal cations in light reactions of photosynthesis (based on Marschner 1995).

Element	*Location*	*Major function*
Mg	Chlorophyll molecule	Directly absorb photons initiating the electron flow
Mn	Enzyme complex attached to PSII	Involved in photolysis (H_2O splitting)
Fe	Cyt b/f complex; Rieske protein; ferredoxin; scavenging enzymes	Mediate the electron flow between PSII and PSI; acts as e^- transmitter to NADPH; scavenge toxic ROS species preventing photoinhibition
Cu	Plastocyanin; scavenging enzymes	Electron acceptor between Cyt b/f complex and PSI; scavenge toxic ROS species preventing photoinhibition
Zn	Scavenging enzymes	Scavenge toxic ROS species preventing photoinhibition

Iron plays multiple roles in the photosynthetic electron transfer. Being a central atom in both the cytochrome b/f complex and the iron-sulfur complex (so-called Rieske protein), Fe is crucial for electron flow between PSII and PSI (Marder and Barber 1989). Another iron-sulfur protein is ferredoxin which transmits electrons to $NADPH^+$. Also, iron is present as a metal component of the prosthetic group in superoxide dismutases (FeSOD) and, thus, plays a central role in detoxification of the superoxide anion free radicals. It has to be added that this role can be also fulfilled by Cu and Zn. Overall, about 20 iron atoms are directly involved in electron transport chain per unit of PSII and PSI (Rutherford 1989). This structural and functional dependence of the thylakoid membrane on Fe explains the sensitivity of chloroplasts (and photosynthesis in general) to Fe deficiency.

Copper is a constituent of plastocyanin and thus is also crucial for the electron transport between PSII and PSI. In general, more than 50% of the copper localized in chloroplasts is bound to this protein (Marschner 1995). It should also be noted that the total amount of plastocyanin comprises ~ 0.4% of the overall chlorophyll content.

Dark reactions of Photosynthesis

In the second series of photosynthetic reactions, the ATP and NADPH generated by the light reactions are used to fix and reduce carbon and to synthesize simple sugars. While light reactions of photosynthesis are very similar for all plant species, dark reactions are strikingly different between different groups of plants. Four major pathways of photosynthetic CO_2 fixation are known, called C_3, C_4, CAM (crassulacean acid metabolism) and SAM (submerged aquatic macrophytes) photosynthesis. Despite the differences, all these pathways focus upon a single enzyme which is the most abundant protein on earth, namely ribulose 1,5-bisphosphate (RuBP) carboxylase/ oxygenase (or Rubisco). Localised in the stroma of chloroplasts, this enzyme enables the primary catalytic step in photosynthetic carbon reduction in all green plants and algae. Approximately 85% of terrestrial plant species are C_3 plants; 10% are CAM plants, and only 5% belong to C_4 type of photosynthesis. As the efficiency of different types of photosynthesis is crucially dependent on such factors as light intensity, ambient temperature, and water availability, these species usually belong to very different habitats.

Most crops are C_3-plants that use series of reactions called the Calvin cycle (Fig. 2). The key compound of this cycle is a five-carbon sugar with two phosphate groups known as ribulose 1,5-biphosphate (RuBP). Initial reactions in the Calvin cycle are catalyzed by the Rubisco enzyme. Carbon fixation in C_3 species occurs in three stages: fixation, reduction, and regeneration of acceptor. At each full turnover of the cycle, one molecule of glyceraldehyde 3-phosphate (C_3 – compound) is formed and later used to synthesize a sugar molecule. In C_3 species, CO_2 fixation occurs exclusively in the stroma.

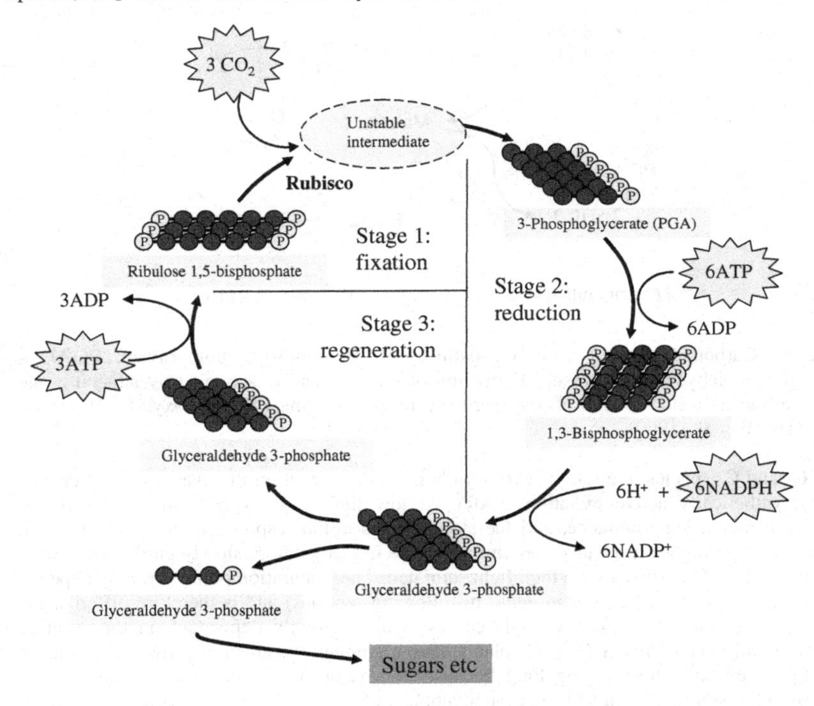

Fig. 2. Carbon fixation in C_3-plants (the Calvin cycle).

Under high CO_2 conditions, the thermodynamic efficiency of the Calvin cycle is ~ 90% (Raven et al. 1999). However, Rubisco's specificity for CO_2 as a substrate is not absolute, and it can also catalyze the condensation of oxygen with ribulose 1,5-bisphosphate. No carbon is fixed during this process (called photorespiration), and some extra energy should be spent to salvage the carbon from the phosphoglycolate molecule formed during this oxygenation process. From this point of view, the observed global trends in atmospheric CO_2 rise should result in a higher affinity of Rubisco towards CO_2, reducing photorespiratory losses and increasing the efficiency of photosynthesis in C_3 species.

The above problem of Rubisco's affinity towards O_2 is partially solved in the so-called C_4 plants (examples: maize, sugar cane or sorghum). Here, biochemical CO_2 fixation is spatially separated between two compartments (Fig. 3). One part of the process occurs in the mesophyll cells (similar to C_3 species); however, the final product of the biochemical cycle is a four (not 3)-carbon molecule. This molecule (either malate or asparate) then moves in the special bundle-sheath cells surrounding the vascular bundles (where O_2 concentration is low) to be decarboxylised. Such wreath-like anatomy of the C_4 species is usually referred to as Kranz anatomy.

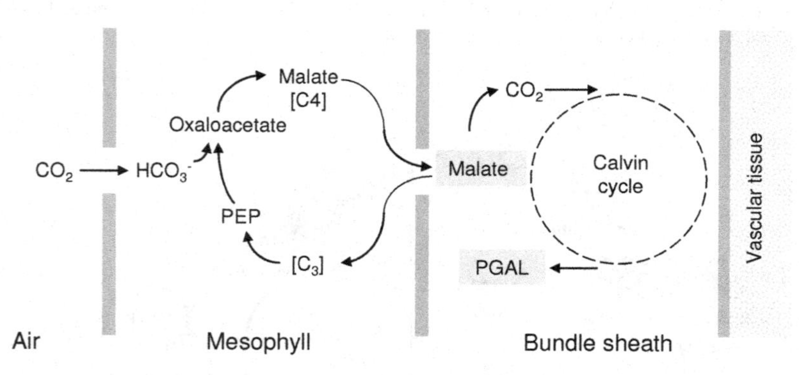

Fig. 3. Carbon fixation by the C_4 pathway. PEP - phosphoenolpyruvate; PGAL – glyceraldehyde -3-phosphate. Formation of oxaloacetate is achieved by attaching one carbon molecule to phosphoenolpyruvate by the enzyme PEP carboxylase. Based on Atwell et al. (1999).

In both C_3 and C_4 species, the rate of photosynthesis is dependent on the available light energy (or photosynthetically active radiation, PAR). At low light CO_2 assimilation rate increases linearly with increasing irradiance, and the slope of this initial response represents maximum quantum yield. At higher irradiances, assimilation rate increases more slowly until eventually a plateau is reached. Chloroplasts are then light saturated. The "saturation point" strongly depends on the Rubisco activity (discussed in detail in following sections) and is therefore different for different crop species. Saturation of light-curves is also strongly dependent on the ambient temperature and availability of CO_2. C_4 plants have a competitive advantage over C_3 plants at high temperature and under strong light because of a reduction in photorespiration plus an increase in the absolute rates of CO_2 fixation at ambient CO_2.

Crassulacean acid metabolism (CAM) occurs in many succulent plants. In CAM plants, the fixation of CO_2 to phosphoenolpyruvate to form oxaloacetate occurs at night, when the stomata are open. The oxaloacetate is rapidly converted to malate, which is stored overnight in the vacuole as malic acid. During the daytime, when the stomata are closed, the malic acid is recovered from the vacuole and the fixed CO_2 is transferred to RuBP of the Calvin cycle. The C_4 pathway and the Calvin cycle occur within the same cells in CAM plants; hence, these two pathways, which are spatially separated in C_4 plants, are temporally separated in CAM plants. In general, CAM plants are much less efficient at CO_2 assimilation (about 30% of C_3 plants and only 20% of C_4 plants; Atwell et al. 1999) and hence are not considered in further sections.

Metal Cations and Their Control over Plant Photosynthesis

Altogether, 14 inorganic elements are considered to be *essential* in higher plants (Marschner 1995). Of these, nine (K, Ca, Mg, Fe, Mn, Zn, Cu, Mo and Ni) are metals. The role of metal cations is not limited by their regulation of electron transport flow in thylakoid membranes of chloroplasts (discussed above). Even more important is their role in regulating the efficiency of CO_2 assimilation either directly (by controlling some key biochemical reactions of carbon fixation in stroma during dark reactions) or indirectly (by regulating a large number of associated physiological processes at the tissue and whole-plant level). Some of these are discussed below.

Potassium. Potassium is the second most abundant inorganic nutrient in plants comprising ~ 6 % of a plant's dry weight. Potassium is involved in numerous functions such as osmo- and turgor regulation, charge balance, and control of stomata and organ movement (Shabala 2003). In addition, K^+ activates over 50 various enzymes critical for numerous metabolic processes, including photosynthesis, oxidative metabolism and protein synthesis (Suelter 1970). Of special importance is Rubisco synthesis, where K^+ is involved in several steps of the translational processes, including the binding of tRNA to ribosomes (Wyn Jones et al. 1979). Other important enzymes affected by K^+ are those related to carbohydrate metabolism (e.g. pyruvate kinase, phosphofructokinase, starch synthase; Lauchli and Pflugger 1978). Under K^+ deficient conditions, a drop in the activity of these enzymes may cause over-accumulation of sugars in chloroplast stroma and result in decreased CO_2 assimilation due to substrate inhibition. Also crucial is K^+ activation of membrane-bound H^+-ATPases. Not only does the H^+-ATPase activity provide a driving force for energized uptake of many essential nutrients required for cell photosynthesis, but also cell turgor pressure and control of stomata aperture will be directly affected by the rate of H^+-ATPase pumping. Light-induced activation of the plasma membrane H^+-ATPase in stomata guard cells is considered to be the first step in light-induced stomata opening (Shimazaki et al. 2007) and subsequent CO_2 assimilation. Lower H^+-ATPase activity under K^+ deficiency condition may significantly impair this process. In addition, K^+ is considered to be a major osmoticum in plant cells, responsible for water movement from subsidiary into guard cells during stomatal opening (Shabala 2003). The importance of potassium to crop production has long been recognised, with potassium fertilizers, traditionally derived from livestock manure and wood ashes, being supplied for over 250 years. Today, farmers apply over 20 million tones of potassium fertilizer annually (FAOSTAT 2003). However, it should be noted that K^+ presence in the soil does not automatically make it available for plant uptake. Potassium delivery to the root surface is limited primarily by the rate of its diffusion though the soil. This rate is determined by several factors, of which the most important is the soil moisture content. Thus, climate change trends and reduced levels of precipitation might significantly affect K^+ availability to plants.

Magnesium. In addition to being the central atom of the chlorophyll molecule, Mg^{2+} directly affects many metabolic processes. This includes enzymatic reactions of photosynthesis, cellular pH control, RNA polymerization, and ATP synthesis (Marschner 1995; Laing et al. 2000). Magnesium is known to affect numerous enzymatic reactions of photosynthesis including RuBP carboxylase (Shaul 2002) and PEP carboxylase (Wedding and Black 1988) activity. It appears that binding Mg^{2+} to Rubisco increases both its affinity (K_m) for the substrate CO_2 and the turnover rate V_{max} (Marschner 1995). Also, chlrolophyll biosynthesis in green plant tissues is controlled by Mg chelatase enzyme (Walker and Weinstein 1991). Mg^{2+} also shifts the pH optimum of the reaction towards the physiological range. Another example is the ABI1 gene product, which is used to regulate various responses such as stomatal closure, maintenance of seed dormancy and inhibition of plant growth. In conjunction with abscisic acid (ABA) mediation, ABI1 activity is highly affected by magnesium ion concentration (Leube et al. 1998).

As a bridging element for the aggregation of ribosome subunits, Mg^{2+} availability has a strong impact on the rate of protein biosynthesis (Marschner 1995). Under Mg^{2+} deficiency conditions, Mg^{2+} binding to the ribosome is reduced, causing a drop in protein synthesis (Sperrazza and Spremulli 1983). Mg^{2+} also plays a key role in carbohydrate partitioning (Cakmak et al. 1994ab) and export of photosynthates from source leaves. Under Mg^{2+} deficiency, starch concentrations are high in source leaves (Fischer and Bremer 1993) and low in sink organs (Beringer and Forster 1981). Variation in Mg^{2+} concentration in leaves causes disturbance to phloem loading, affecting carbohydrate and amino acid supply from the source leaves (Cakmak et al. 1994ab, Fisher et al. 1998).

Mg^{2+} deficiency-induced depression of photosynthesis is a widely reported phenomenon observed in a wide range of species (Bottrill et al. 1970; Cakmak et al. 1994; Troyanos et al. 1997; Fischer 1997; Ridolfi and Garrec 2000). In addition to its direct effect on dark reactions of photosynthesis, the observed reduction in net CO_2 assimilation under Mg^{2+} deficiency conditions is also associated with reduced stomatal conductance (Fisher and Bremer 1993; Fischer 1997; Sun et al. 2001).

Calcium. It appears that Ca^{2+} effects on CO_2 assimilation and plant photosynthesis are indirect and mediated by at least two principal mechanisms: (1) the role of Ca^{2+} as a second messenger in intracellular signaling cascades and (2) by impact of Ca^{2+} nutrition on uptake and compartmentation of other (critical to photosynthesis) ions. It is becoming increasingly evident that elevation in the cytosolic free Ca^{2+} levels is the first and most essential step in plant cell responses to literally every known environmental stimulus (Sanders et al. 1999). These changes are then translated into conformational or catalytic activity of "sensor" proteins such as calmodulin, calcineurin B-like proteins, and Ca^{2+}-dependent protein kinases (White and Broadley 2003). The latter changes, in turn, affect all aspects of cell metabolism, including efficiency of electron transport in thylakoid membranes and CO_2 fixation in the stroma. However, it should be commented that cytosolic Ca^{2+} content in plant cells is extremely low (~ 100 nM) and thus does not require large amount of Ca^{2+} in the external media. Hence, most detrimental effects of Ca^{2+} deficiency on leaf photosynthesis are explained by the second mechanisms. In natural environment, plant acquisition of essential nutrients required for photosynthesis (e.g. K^+ or Mg^{2+}) is strongly affected by adverse soil conditions such as salinity or acidity (Marschner 1995; Shabala and Cuin 2008). Addition of supplemental Ca^{2+} to the growth media is known to ameliorate these detrimental effects on plant nutrition (Shabala et al. 2003, 2006) and improve availability of these essential metal cations in photosynthesizing leaf tissues.

Microelements

In addition to their direct involvement in the regulation of electron flow in the light reactions of photosynthesis, metal microelements (such as Fe, Mn, Cu, Zn and Ni) also have a significant impact on the activity of numerous enzymes, associated with various aspects of plant photosynthesis and CO_2 fixation. This role can be either catalytic or structural. One of the obvious examples is their role in superoxide dismutase (SOD) activity (see above). The isoenzymes of SOD differ in their metal component which might be either iron (FeSOD), manganese (MnSOD) or copper + zinc (CuZnSOD). In addition to this, a wide range of other enzymes are regulated by these metal cations. For example, Mn can exist in six oxidation states and acts as a co-factor for 35 different enzymes (Marschner 1995) catalyzing various oxidation-reduction, decarboxylation and hydrolytic reactions. Zn also has catalytic, co-catalytic or structural roles in a large number of enzymes (e.g. catalytic - carbonic anhydrase, carbopepsidase; structural - alcohol dehydrogenase, PNA polymerase, alkaline phosphatase etc). As a result, Zn tightly controls both the CO_2 fixation in chloroplast stroma (especially in C_4 species; Hatch and Burnell 1990) and carbohydrate metabolism. Examples of Fe-containing enzymes are various catalases and peroxidases, both of which play a crucial role in the dismutation of H_2O_2 to H_2O and O_2. Some other (less characterized) Fe-dependent enzymes are known. Copper is a transition element and thus shares similarities with Fe. Nearly 99% of total plant copper is present in a complex form, forming three types of proteins: (i) blue proteins without oxidase activity (e.g. plastocyanin); (ii) non-blue proteins (e.g. peroxidazes), and (iii) multicopper proteins (e.g. ascorbate oxidase)(Sandmann and Boger 1983).

Despite Mo and Ni contents in plants being extremely low (~ 0.1 ppm), these are still classified as essential nutrients and thus are important for normal photosynthetic performance. Mo has structural and catalytic function in enzymes such as nitrogenase, nitrate reductase, or xanthine oxidase, while Ni is a constituent of urease (N metabolism) and hydrogenases (redox reactions) (Marschner 1995).

Improving Efficiency of Photosynthetic Energy Conversion

Can the efficiency of CO_2 assimilation be improved? To what extent? And how much will it cost us? These questions are indeed critical, given current global climate trends.

The answer to the first question is "yes indeed". The bottleneck of the entire problem (e.g. the reasons for the limitation of the maximum rate of light-saturated photosynthesis) is the efficiency by which Rubisco catalyses CO_2 fixation in the Calvin cycle (Woodrow and Berry 1988). Rubisco has a low rate of catalysis, and the net rate of carboxilation *in vivo* is further reduced by its poor affinity for CO_2 and the competing reaction for O_2. This inefficiency is partially compensated by the large amount of Rubisco protein in the leaf (reaching up to 25% of total leaf protein; Evans 1989) and by the high concentration of ribulose-1,5-bisphosphate maintained in the leaf during photosynthesis (Stitt and Schulze 1994). However, this strategy is far too expensive to be really energy cost-efficient. As a result, plants have developed a complex combination of both biochemical and anatomical specialization (reflected in C_4 pathway) which provides an elevation of the CO_2 at the site of Rubisco and makes the entire process more efficient. As a result, 11 out of the 12 most productive plant species on the planet are C_4 plants (von Caemmerer and Furbank 2003). But this is only the tip of the iceberg. A real possibility exists, using rDNA technology, to modify the Rubisco structure and to finish up with a much smaller, more efficient molecule. It was recently shown that some non-green algae can discriminate twice as effectively against O_2 while maintaining higher carboxylation efficiencies (Andrews and Whitney 2003; Collings and Critchley 2005), opening prospects for creating transgenic crops with much more efficient Rubisco function. Even within higher plant species, a significant diversity in the Rubisco kinetics exists. The existing chloroplast transformation technology makes it possible to transplant altered or foreign Rubiscos into model plant species and assess the biochemical and physiological consequences of the engineered changes (Andrews and Whitney 2003). And, without any doubt, this is only the first step in this direction. In very plain terms, Rubisco represents a single Mg^{2+} ion in its binding site surrounded by a "globe of grease" comprised of tens of thousands of ions (Collings 2004). The physiological and chemical rationale for such a structure is not clear. Given that natural photosynthesis has emerged from the ocean, such "waterproof structure" might have had some advantages several million years ago, but not any more for terrestrial plants. As a result, the Rubisco efficiency is only 1.5% (Collings 2004). Just imagine what the tremendous benefits would be of, say, doubling it!

Artificial Photosynthesis

In addition to optimizing plant photosynthetic performance by better agricultural practices (e.g. plant nutrition) or by transgenic means, another exciting opportunity exists to mimic natural plant photosynthesis by technological means, e.g. by creating artificial photosynthesis. Both the advance in our knowledge of the molecular mechanisms underlying the photosynthetic process and recent technological developments make it a real opportunity these days (Collings and Critchley 2004). Several years ago, the Australian Artificial Photosynthesis Network (AAPN) was created, including (at that time) 40 scientists from eight Australian universities and four divisions of the CSIRO. So far, this is the world's largest coordinated research effort to develop effective artificial photosynthesis technology. In 2002, the AAPN submitted a proposal to the Australian Government requesting to give this initiative a national research priority status and to provide some dedicated funding for the Artificial Photosynthesis program. The concept of this program is summarized in Fig. 4 below.

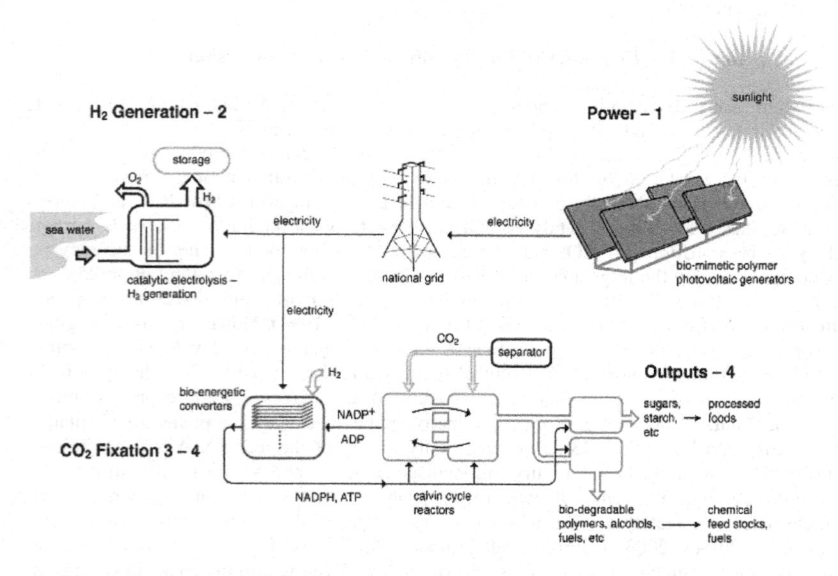

Fig. 4. Artificial photosynthesis concept. Reproduced from Pace (2002), with the permission of the author.

The major objectives of this program are three-fold (NRP submission 49, 2002):

- to extract CO_2 from the atmosphere and to photoreduce it to synthesise useful products, foods, fuels and fibres, without the excessive use of water;
- to devise artificial photosynthetic devices to produce electricity;

- to diminish CO_2 emissions through fossil fuels by the photoproduction of hydrogen as an alternative energy source.

Within this concept, a bio-mimetric polymeric photovoltaic generators are made, using the existing knowledge of molecular processes of photosynthesis and modern nanotechnology tools. At the first stage, this is done by copying the actual natural process by using materials analogous to chlorophyll. The already existing photovoltaic generators currently have 8% efficiency (Collings 2004) but can theoretically well exceed the best silicon-based panels which are capped at 33% of the maximum possible efficiency in converting sun into electricity. The suggested technology is highly cost-efficient, as literally paint-thin layers are used in these photovoltaic cells, creating a sandwich-like structure of high power capacity. The generated energy can be used for H_2 generation (in catalytic electrolysis) as well as in bio-energetic converters to assimilate CO_2 and produce carbohydrates then used to synthesise foods, fuels and fibres. Crucially, this conversion will require little (if any) water, leading to the so-called "Dry Agriculture" concept. In natural systems, only a tiny fraction of water supplied to the plant is incorporated in the final product. Depending on the type of photosynthesis, between 450 and 700 water molecules should be evaporated from plant leaves to fix one CO_2 molecule in the process of photosynthesis. For example, 4700 L of water is needed to make 1 kg of cotton (Collings 2004). Using the enzyme bed reactor systems to fix CO_2 from the air (as in the "Dry Agriculture" program), the water usage is reduced to the absolute chemical minimum (e.g. ~0.9 water molecules spent per one CO_2 molecule assimilated). This nearly 1000-fold increase in water use efficiency is of paramount importance in the light of current global climate trends and reduced water availability for agricultural production. Also, the threat of wide-spread secondary-induced soil salinisation (a direct consequence of the irrigation practice) will be eliminated or at least significantly reduced.

From my point of view, there is little doubt that the artificial photosynthesis approach offers a realistic and viable economic strategy for moving towards the use of renewable energy resources. Unfortunately, this view is not shared by everyone (and – most importantly – by the Government and funding bodies). As stated by Dr Tony Collings (one of the initiators and coordinators of the AAPN), "the actual overall project and the vision of artificial photosynthesis as such is .... "overlooked" would be a nice way of putting it". Let's hope things will change for the better in the future. Given the current global climate trends, there is not much time left to hesitate.

References

Andrews TJ, Whitney SM (2003) *Arch Biochem Biophys* **414**: 159-169

Atwell BJ, Kriedeman PE, Turnbull CGN (1999) Plants in Action. MacMillan, Melbourne.664 p

Beringer H, Forster H (1981) *Z Pflanzerenahr Bodenk* **114**: 8-15

Bonan GB (2008) *Science* **320**: 1444-1449

Bottrill DE, Possingham JV, Kriedemann PE (1970) *Plant Soil* **32**: 424-438

Bowler et al. (1991) *EMBO J* **10**: 1723-1732

Buchanan BB, Gruissem W, Jones RL (2000) Biochemistry and Molecular Biology of Plants. ASPP, Rockville. 1365 p

Cakmak I, Hengeler C, Marschener H (1994a) *J Exp Bot* **45**:1245-1250

Cakmak I, Hengeler C, Marschener H (1994b) *J Exp Bot* **45**:1251-1257

Collings T (2004). Artificial photosynthesis. In: ABC Radio National: the buzz (Sept 4). http://www.abc.net.au/rn/science/buzz/stories/s1191870/htm

Collings AF, Critchley C. (2005). Artificial Photosynthesis: from basic biology to industrial applications. Wiley-VCH. 313 p

Dalal RC, Allen DE (2008) *Austral J Bot* **56**: 369-407
Denman et al. (2007) In: *Climate Change 2007*. Univ. Press, Cambridge. pp 499-587
Evans JR (1989) *Oecologia* **78**: 9-19
FAOSTAT (2003) Fertilizer usage. http://apps.fao.org/lim500/nph-wrap.pl? Fertilizers& Domain=LUI&servlet=1
Fischer ES (1997) *Photosynthetica* **33**:385-390
Fischer ES, Bremer E (1993) *Physiol Plantar* **89**:271-276
Fischer ES, Lohaus G, Heineke D, Heldt HW (1998) *Physiol Plantar* **102**:16-20
Forster et al. (2007) In: *Climate Change 2007*. Univ. Press, Cambridge. pp 129-234
Hatch MD, Burnell JN (1990) *Plant Physiol* **93**: 825-828
Kriedemann Pe, Graham RD, Wiskich JT (1985) *Aust J Plant Physiol* **11**: 287-301
Laing W, Greer D, Sun O, Beets P, Lowe A, Payn T (2000) *New Phytol* **146**:47-57
Lasa et al. (2000) *Plant Soil* **225**:167-174
Lauchli A, Pflugger R (1978) In: *Proc 11th Int Congr Int Potash Inst Bern*, 111-163
Lavon R, Goldschmidt EE (1999) *J Amer Soc Hort Sci* **124**: 158-162
Leube MP, Grill E, Amrhein N (1998) *FEBS Lett* **424**:100-104
Luo Y (2007) *Annu Rev Ecol Evol Syst* **38**: 683-712
Marder JB, Barber J (1989) *Plant Cell Environ* 12: 595-614
Marschner H (1995) Mineral Nutrition of Higher Plants. Academic Press, San Diego. 889 p
NRP submission 49 (2002) http://www.dest.gov.au/NR/rdonlyres/180718C2-EFDA-4D44-B195-730BF851E304/3195/pdf49p.pdf
Pace RJ (2002) In: *Proceed 26th Annu Conf Austral Soc Biophys*. Melbourne, Nov 29-30
Raven PH, Evert RF, Eichhorn SE (1999). Biology of Plants. WH Freeman and Company, New York. 944 p
Ridolfi M, Garrec JP (2000) *Ann Forestry Sci* **57**: 209-218
Rutherford AW (1989) *Trends Biochim Sci* **14**: 227-232
Sanders D, Brownlee C, and Harper JF (1999) *Plant Cell* **11**: 691–706
Sandmann G, Boger P (1983) In: *Encyclopedia of Plant Physiol*. Vol **15A**: 563-596
Shabala S (2003) *Ann Bot* **92**: 627-634
Shabala S, Cuin TA (2008) *Phisiol Plantar* **133**: 651-669
Shabala S, Demidchik V, Shabala L, Cuin TA, Smith SJ, Millar AJ, Davies JM, Newman IA (2006) *Plant Physiol* **141**: 1653-1665
Shabala S, Shabala L, Van Volkengurgh E (2003) *Funct Plant Biol* **30**: 507-514
Shaul O (2002) *BioMetals* **15:** 309–323
Shimazaki KI, Doi M, Assmann SM, Kinoshita T (2007) *Annu Rev Plant Biol* **58**: 219-247
Sperrazza JM, Spremulli LL (1983) *Nucleic Acids Res* **11**: 2665-2679
Stitt M, Schulze D (1994) *Plant Cell Environ* **17**: 465-487
Suelter CH (1970) *Science* **168**: 789-795
Sun OJ, Gielen GJHP, Sands R, Smith CT, Thorn AJ (2001) *Trees* **15**: 335-340
Troyanos YE, Hipps NA, Moorby J, Ridout MS (1997) *Plant Soil* **197**: 25-33
von Caemmerer S, Furbank RT (2003) *Photosysthesis Res* **77**: 191-203
Walker CJ, Weinstein JD (1991) *Proc Natl Acad Sci USA* **88:** 5789-5793
Wedding RT, Black MK (1988) *Plant Physiol* **87**: 443-446
White PJ, Broadley MR (2003) *Ann Bot* **92**: 487-511
Woodrow IE, Berry JA (1988) *Annu Rev Plant Physiol Plant Mol Biol* **39**: 533-594
Wyn Jones RG, Brady CJ, Speirs J (1979) In: *Recent Advances in the Biochemistry of Cereals*. pp 63-103. Acad. Press, London

Energy Technology Perspectives
Edited by: Neale R. Neelameggham, Ramana G. Reddy, Cynthia K. Belt, and Edgar E. Vidal
TMS (The Minerals, Metals & Materials Society), 2009

EFFECT OF ELECTRODE SURFACE MODIFICATION ON DENDRITIC DEPOSITION OF ALUMINUM ON Cu SUBSTRATE USING EMIC-$AlCl_3$ IONIC LIQUID ELECTROLYTES

D. Pradhan[1] *and R. G. Reddy*[2]

[1]Graduate Student, [2]ACIPCO Professor and Head
[1, 2]Department of Metallurgical and Materials Engineering
The University of Alabama, Tuscaloosa, AL 35487, USA
E-mail: rreddy@eng.ua.edu

Keywords: Ionic liquid, Electrodeposition, 1-Ethyl-3-methyl-imidazolium chloride, Dendrite.

Abstract

Electrorefining of aluminum scrap was investigated from $AlCl_3$ –1-Ethyl-3-methyl-imidazolium chloride (EMIC) (Molar ratio 1.65) chloroaluminate electrolyte using copper cathodes at temperature of 90±3°C and cell voltage of 1.5 V. The deposits were characterized using scanning electron microscope (SEM), energy dispersive spectroscopy (EDS) and X-ray diffraction (XRD). The study was focused to determine the effect of electrode surface modification as well as deposition time on dendritic depositions of aluminum. Their effect on current density was investigated also. It was shown that the surface modification of electrodes reduced the dendritic depositions of aluminum. Pure aluminum (>99%) was deposited for all experiments with the current efficiency of 94-97%. The use of ionic liquids not only reduced the energy consumption but also minimized pollutant emissions (CO, CO_2 etc.) compared to the commercial high temperature aluminum production.

1. Introduction

Aluminum electrorefining/recycling from aluminum scrap using ionic liquid electrolyte is an attractive process today because of the unique chemical and physical properties such as wide temperature range for the liquid phase, high thermal stability, negligible vapor pressure, low melting point and wide electrochemical window. But in conventional practice, melting of aluminum scrap and high temperature molten salt electrolysis makes this process more expensive and less energy intensive (average 13 kWh/ kilogram of aluminum) [1] . Although recycling scrap aluminum requires only 5% of the energy used to make new aluminum [1-2], but the green house emissions like CO_2, CO and HF have to be counted for environmental reason. Approximately, 5.31 kilogram of CO_2 emitted per kilogram of aluminum production in most of the US industries.

For this reason, extraction and refining of aluminum from low temperature or room temperature ionic liquid electrolytes gaining interest. Many researchers have studied electrodeposition of aluminum using different ionic liquid melts [3-9]. But bulk electrodeposition and commercialization was not established. In recent years, Reddy and co-workers [10-14] have reported bulk electrowinning and electrorefining of aluminum using low temperature chloroaluminate ionic liquid electrolytes, yield high purity aluminum deposits with zero pollutant emissions. But optimization of dendritic deposition of aluminum from ionic liquid electrolytes was not fully investigated. In electrorefining cell, coarse grained, rough and adhesive deposits are required. As dendritic deposition is not firm enough to endure handling before melting and casting into desirable shapes for further processing, they add additional cost for handling and processing. Not only that, some times dendrites caused short circuiting between anodes and cathodes. Although there are number of experimental parameters such as electrolyte concentration, surface roughness of electrodes, addition of additives, impurities, stirring, temperature and deposition time to prevent the dendritic deposition, but current density and cathode overpotential plays the major role. Dendrites are formed after a critical potential is reached.

There are only few investigations reported concerning aluminum deposition on copper [4, 10-14] and aluminum [15, 16]. Aluminum deposition on aluminum or copper substrates was divided into two steps: (a) a thin layer is deposited on the substrate, (b) dendrites attach to this thin layer. Adherence depends directly on the reactivity between the aluminum and cathode material. There is strong reactivity when intermetallic compounds like Al_2Cu are formed by interdiffusion. The presence of an underlayer is correlated with good adherence of the dendrites. In the case of aluminum cathode, there is a low reactivity between aluminum and the substrate. At the beginning of the growth of the layer, just after nucleation: Budevski et al., have proposed models taking into account the binding energy of the electrodeposited metal (M), either on substrate (S), E_{MS} or on the pre-deposited metal E_{MM} [17, 18]. When $E_{MS} \leq E_{MM}$ (Volmer-Weber model), M becomes deposited on M rather than on S, leading to three-dimensional islands. This initial process favors the rapid formation of axial dendrites on the cathode surface. With copper, a surface alloy is formed by metal inter-diffusion with the underlayers of aluminum and dendrite formation occurs when the intermetallic layer is saturated; at this time E_{MM} is higher than $E_{M\text{-alloy}}$ and once again, M forms 3D islands. During deposition growth, orthogonal needles result from diffusional phenomena favoring current at the top of the islands. The heterogeneity of the electrolyte layer at the interface with the cathode material causes a particular fluid motion which readily influences the morphology of the deposition. Fleury et al. [19, 20] demonstrated that the depletion of cations near the cathode generates strong local electric fields. According to these authors, the electrolyte motion follows contrarotative vortices at the tip of each needle, producing ramification at the edges.

In this present work, we have investigated the effect of electrode surface modification on deposit morphology of aluminum using $AlCl_3$ – EMIC (Molar ratio 1.65) electrolyte at 90 ± 3°C.

2. Experimental Procedure

2.1 Chemicals

The chloroaluminate electrolytes of $AlCl_3$–1-Ethyl-3-methyl-imidazolium chloride (EMIC) (Molar ratio 1.65) were used without further processing.

2.2 Experimental Setup

The electrorefining experiments were conducted in a 40 ml Pyrex® glass beaker fitted with teflon cap. The schematic diagram and actual electrolytic cell are shown in Figure 1 (a) and (b), respectively. The anodes were prepared by cutting the aluminum alloy ingots material into 45×20×5 mm^3 plate. Table 1 shows the compositional analysis of aluminum alloy scrap. Copper (99.999 % pure, Alfa Aesar®) sheets with dimensions of 45×20×1 mm^3 were used as cathodes. Pure aluminum wire (2 mm diameter, 99.999 % pure, Alfa Aesar®) was used as reference electrode.

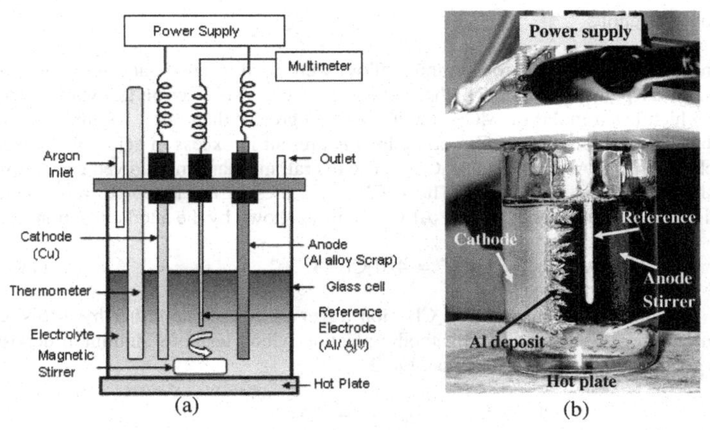

Figure 1 (a) Schematic diagram of the experimental setup (b) actual electrorefining cell.

All electrodes were polished with a 320 grit SiC emery paper to expose fresh metal surface before electrorefining. Some of the anode and cathode surfaces were "modified" with different techniques to investigate the effect on deposit morphology. The cathode area of deposition and anode working area were calculated from the portion of electrodes immersed into the electrolyte. The purpose of reference electrode was only to measure the electrode voltage of anode and cathode individually, using a multimeter (Keithley® 2000 Multimeter). The constant voltage power source (Kepco® ABC Programmable Power Supply) supplied the required voltage across anode and cathode. The electrolyte was stirred at a constant speed (60 rpm) using a magnetic stirrer and the temperature of were controlled by placing a hot plate below the electrorefining cell. A thermometer was inserted through the teflon cap to monitor the temperature. The electrolyte was heated stepwise to required temperature under control atmosphere (argon purging). Then the electrolyte was allowed to stand for at least 30 minutes at final temperature for temperature stabilization before starting the experiment. All experiments were conducted in a ventilated hood, under argon atmosphere at 90±3°C with the applied cell voltage between anode and cathode of 1.5 V. All of the experiments were done for duration of 5 hours.

Table 1 Elemental analysis of aluminum alloy scrap

Elements	Al	Si	Fe	Mn	Mg	Cu	Ti	+Cr	Ni	Zn
Wt %	74.25	25.07	0.08	0.008	0.345	0.02	0.13	0.002	0.007	0.03

2.3 Characterization Instruments

The qualitative and quantitative analysis of the electrodeposited aluminum samples were carried out using Philips® XL30 scanning electron microscopy (SEM) with Energy Dispersive Spectroscopy (EDS). The X-ray diffraction (XRD) analysis was performed using Philips® PW3830 X-ray diffractometer which uses monochromatic CuK_{α} radiation (λ = 1.5406 Å).

3. Results and Discussion

3.1 Electrode Reactions

In electrorefining process, impure aluminum alloys were used as anode and were electrorefined and deposited as pure aluminum on the cathode surface. The electrolyte exhibit Lewis acid properties which has a molar of $AlCl_3$: EMIC (1.65:1) greater than one. Aluminum deposition proceeds by reduction of only $Al_2Cl_7^-$ ion which is present in excess in acidic melts with other anions such as $AlCl_4^-$, $Al_3Cl_{10}^-$ and $Al_4Cl_{13}^-$. The overall mechanism involved can be explained by the following electrode reactions. The $AlCl_4^-$ ions present in electrolyte react with anode metal to dissolve aluminum and produce $Al_2Cl_7^-$ ions as shown by the anodic reaction, Eq. 1.

$$Al_{(anode)} + 7AlCl_4^- = 4Al_2Cl_7^- + 3e^- \quad (1)$$

Dissolved aluminum in the form of $Al_2Cl_7^-$ ion traverses to cathode either by diffusion or by convection and gets discharged. The cathodic reaction which leads to aluminum deposition at cathode is given by the following reaction, Eq. 2.

$$4Al_2Cl_7^- + 3e^- = Al_{(cathode)} + 7AlCl_4^- \quad (2)$$

The energy consumption for Al electrorefining varies from 4.79-5.05 kWh/ kg of Al alloy, which is well below the commercial and molten salt electrolysis of Al (13 kWh/ kg of Al). Also, the efficiency of electrorefining is high (94.10-96.55%).

It is also noticeable that electrorefining of Al provided zero CO, CO_2 and CF_4 emission, thus reduce the pollutant emission compared to conventional high temperature Al electrolysis process.

3.2 Characterization of Electrodeposits

Structural and compositional characterizations were done for deposits using SEM-EDS and X-Ray diffraction (XRD). The XRD pattern of the electrodeposit is shown in Fig. 2. It shows only the diffraction peaks of Al, indicating that the deposits have a high purity.

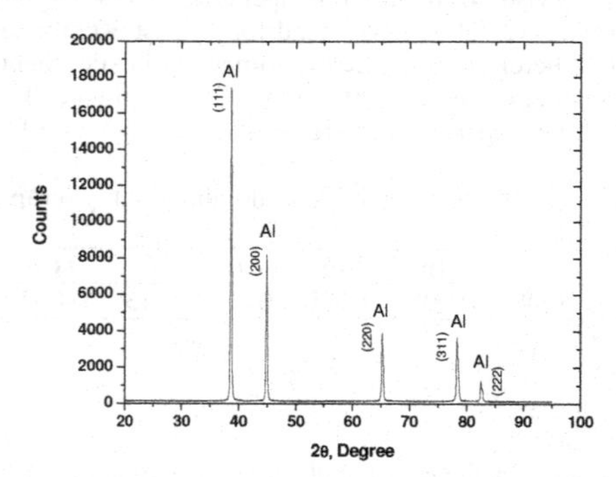

Figure 2 X-ray diffraction pattern of electrodeposited aluminum using non-modified electrodes.

Figure 3 (a)-(c) shows the SEM microstructure of the aluminum dendrites obtained after 5 hours of experiment using non-modified electrodes and the corresponding compositional analysis by EDS. The EDS analysis shows that the deposit is composed of pure aluminum. The SEM micrograph of the electrodeposit obtained by using modified electrodes shows granular deposit rather than dendritic as shown in Figure 4 (a)-(c).

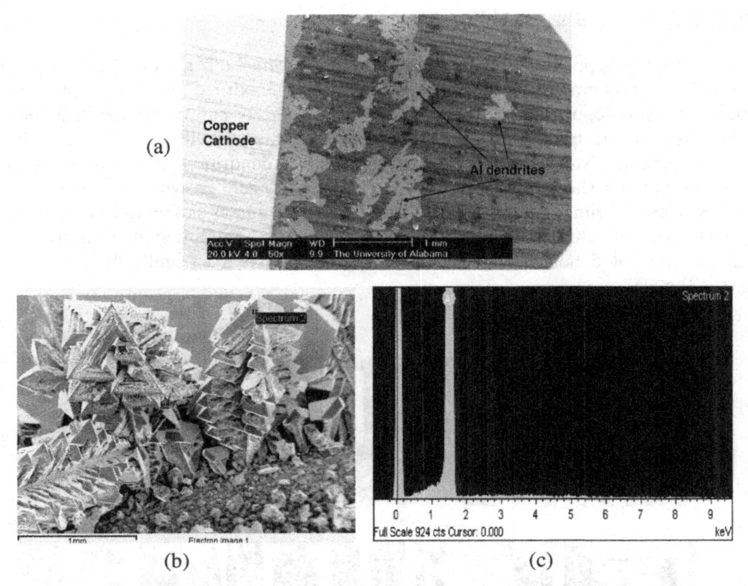

Figure 3 (a) Cross-sectional SEM image of aluminum deposit, (b)-(c) SEM micrographs and EDS analysis of aluminum dendrites obtained by using “non-modified” electrodes.

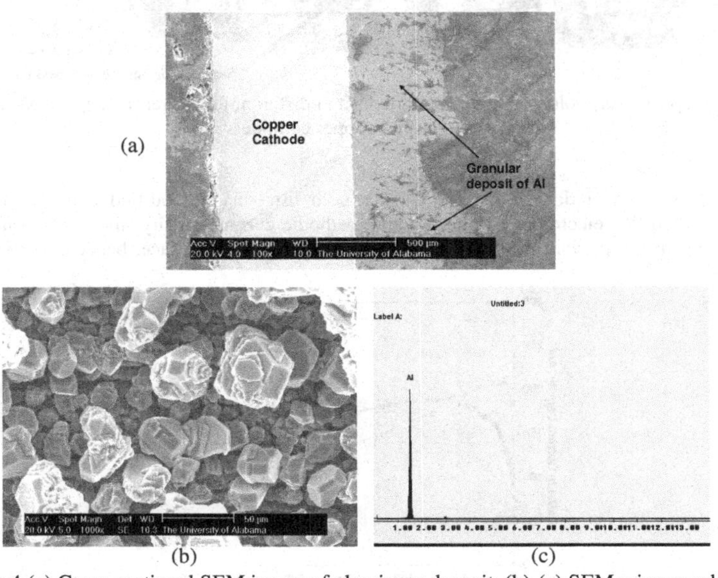

Figure 4 (a) Cross-sectional SEM image of aluminum deposit, (b)-(c) SEM micrographs and EDS analysis of granular aluminum deposits obtained by using “modified” electrodes.

3.3 Effect of electrode surface modification

Figure 5 shows the deposit morphology of aluminum using both modified and non-modified electrodes. The photographs clearly show that surface modified electrodes produced non-dendritic or granular deposition of aluminum on the cathode for entire duration of 5 hours. After 1 hour both type of electrodes show bright deposit of aluminum which is clearly visible at the edges of the cathodes. Non-modified electrode shows fine dendritic deposition of aluminum on the edges and on the surfaces. As time proceeded, dendrites were grown more and more. After 5 hours of experiment, primary dendritic arms grow up to 4-6 mm as shown in SEM micrograph in Fig. 3. But modified electrodes did not produce dendritic aluminum deposit for the entire experimental time of 5 hours. The deposit was granular, compact and adherent with copper surface (see Fig. 4).

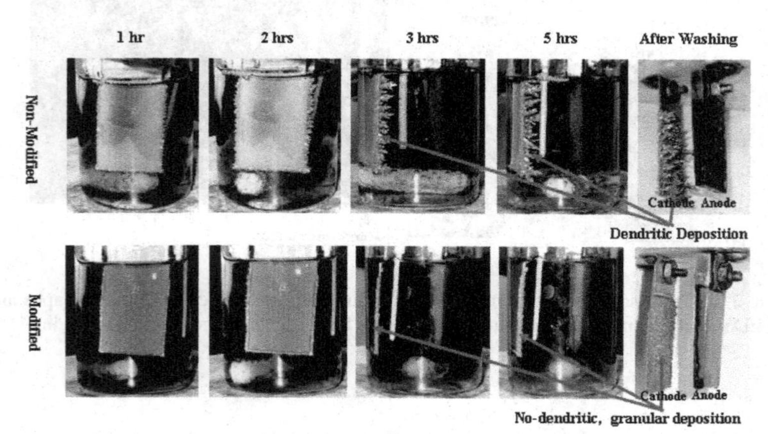

Figure 5 Deposit morphology of aluminum deposit at different time interval using modified and non-modified copper cathodes.

Fig. 6 shows the variation of current density with time using modified and non-modified electrodes. Modified electrodes exhibit a higher cathodic current density (max. 207 A/m^2). The modified electrodes prevent the loss of current density on the back face, hence current density increases.

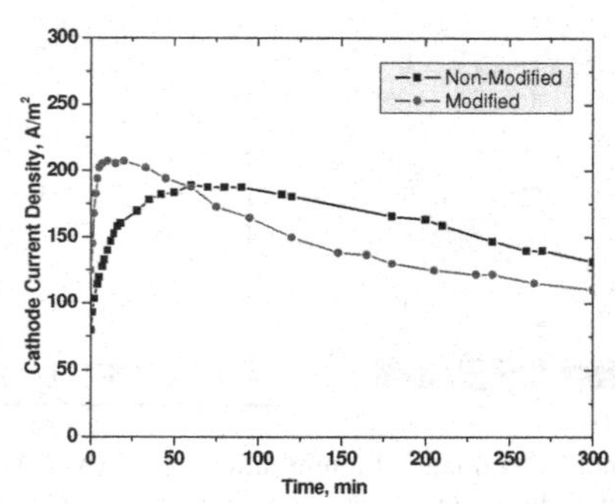

Figure 6 Effect of electrode surface modification on cathode current density.

The morphology of metal deposit is mainly dependent on kinetic parameters and cathodic overpotential or current density. All experiments using modified electrodes produced non-dendritic and granular structure and it can be explained from the cathodic over potential data as shown in Fig. 7. Modified cathodes showed lower over potential compared to non-modified electrodes. The maximum overpotential obtained with modified electrode is 0.32 V and that of non-modified is 0.52 V. The maximum current densities obtained in both above experiments are nearly same but measured overpotentials are almost a factor of two lower for modified electrodes. This indicates that overpotentials are plying a significant role in the morphology of the deposits and lower overpotentials prevent the dendrite formation.

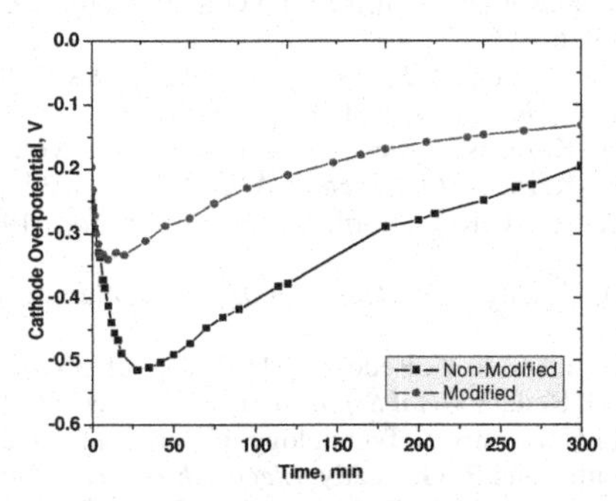

Figure 7 Cathodic overpotential as a function of time using non-modified and modified electrode.

However, further investigations are undergoing to find out the critical cathode overpotential/current density for dendritic deposition of aluminum using different fixed cathode overpotential between cathode and reference electrode.

4. Conclusions

The electrorefining experiments were carried out using aluminum anode alloys and copper cathode materials from $AlCl_3$ – EMIC (Molar ratio 1.65) mixture at 1.5 volts, 90 ± 3°C, 60 rpm for 5 hours. The deposit was analyzed using SEM and XRD techniques. The analysis of the deposits showed pure aluminum. The electrorefining process has high current efficiency (94-97%) and is very energy efficient (4.58-5.40 kWh/kg of Al). The process also provides zero emission of pollutant (CO, CO_2 etc.), thus makes them environmentally benign. The surface modification of electrodes not only improves the cathode current density but also eliminates the dendritic deposition of aluminum. It was observed that cathodic overpotentials obtained from experiments using modified electrodes are significantly lower than the overpotentials using non-modified electrode. It can be concluded that the overpotential is the key in preventing dendrite formation. Lower overpotential (<0.4 V) are recommended for prevention of dendrite formation on copper cathode.

Acknowledgements

The authors gratefully acknowledge the financial support from NSF, ACIPCO and The University of Alabama.

References

1. International Aluminum Institute: Statistics, Electrical Power Used in Primary Aluminum Production, 2006.Retrieved from- http://www.world- aluminium.org.
2. USGS Publications and Data Products, 2007. Retrieved from- http://minerals.usgs.gov/minerals/pubs/commodity/aluminum/mcs-2008-alumi.pdf
3. R. T. Carlin, W. Crawford, and M. Bersch, *J. Electrochem. Soc.*, 1992, vol.139, pp. 2720-2727.
4. Q. Liao, W.R. Pitner, G. Stewart, C.L. Hussey and G.R. Stafford, *J. Electrochem. Soc.*, 1997, vol.144, pp. 936-943.
5. J. Robinson and R. A. Osteryoung, *J. Electrochem. Soc.*, 1980, vol. 127, pp.122.
6. C.-C. Yang, *Mater. Chem. Phys.*, 1994, vol. 37, pp. 355.
7. P. K. Lai and M. Skyllas-Kazacos, *J. Electroanal. Chem.*, 1988, vol. 248, pp. 431.
8. P. K. Lai and M. Skyllas-Kazacos, *Electrochim. Acta*, 1987, vol. 32, pp. 1443.
9. W.R. Pitner, C.L. Hussey and G.R. Stafford, *J. Electrochem. Soc.,* 1997, vol. 143, pp. 130-138.
10. V. Kamavaram and R. G. Reddy: *Light Metals 2005*, TMS, Warrendale, PA, 2005, pp. 501-05.
11. M. Zhang, V. Kamavaram, and R. G. Reddy: *JOM*, 2003, vol. 55 (11), pp. 54-57.
12. V. Kamavaram and R. G. Reddy, *Metal Separation Technologies III*, R. E. Aune and M. Kekkonen, ed., Helsinki University of Technology, Espoo, Finland, 2004, pp. 143-51.
13. V. Kamavaram, D. Mantha, and R. G. Reddy, *Electrochim. Acta*, 2005, vol. 50, pp. 3286-95.
14. V. Kamavaram, D. Mantha, and R. G. Reddy, *J. Min. Met.*, 2003, vol. 39 (1-2) B, pp. 43-58.
15. A. P. Abbott, C. A. Eardly, N. R. S. Faley, G. A. Griffith and A. Pratt, *J. Appl. Electrochem.*, 2001, vol. 31, pp.1345.
16. T. P. Moffat, *J. Electrochem. Soc.*, 1994, vol. 141, pp. 3059.
17. W.J. Lorenz, G. Staikov, *Surf. Sci.*, 1995, vol. 335, pp. 32.
18. E. Budevski, G. Staikov, W.J. Lorenz, Electrochemical Phase Formation and Growth, VCH Publications, New York, 1996.
19. V. Fleury, J.N. Chazalviel, M. Rosso, *Phys. Rev.*, 1993, vol. E 48 (2), pp.1279.
20. V. Fleury, J.H. Kaufman, D.B. Hibbert, *Nature,* 1994, vol. 367 (3), pp. 435.

Energy Technology Perspectives
Edited by: Neale R. Neelameggham, Ramana G. Reddy, Cynthia K. Belt, and Edgar E. Vidal
TMS (The Minerals, Metals & Materials Society), 2009

Room-Temperature Production of Ethylene from Carbon Dioxide

Kotaro Ogura

Department of Applied Chemistry, Yamaguchi University, Tokiwadai 1-17-30
Ube 755-0097, Japan. E-mail: ko.ogura@nifty.com

Abstract

Ethylene has been produced in aqueous solution from CO_2 by the electrochemical reduction driven by a natural energy. This process is useful for storing a large amount of the natural energy. In the closed system, the conversion efficiency of CO_2 is almost 100%, and the maximum selectivity for the formation of ethylene is more than 70%. On the other hand, the current efficiency for the competitive reduction of water is less than 10%. The electrolysis is practicable under such special conditions as three-phase interface consisting of gas, solution and metal, concentrated solution of potassium halide, low pH and copper or copper halide-confined metal electrode. These requirements are thoroughly examined, and the grounds to reply upon are revealed. A series of chemical apparatuses including an electrolytic cell in a large scale are designed for the ethylene production, which allow us continuously to supply raw CO_2 and to extract the product.

Introduction

In connection with the depletion of the petroleum reserve, much attention has been paid to the utilization of natural energies such as sunlight, wind force and waves. However, these energies are always changeable, and some means of storing them are necessary for the common use. In general, natural energies are transformed to the electric energy at the time of utilization, but the electricity is not very storable. This is one of the reasons why the application of natural energy is not so expanded. On the other hand, the storing of natural energy by plants is very skilful. In the photosynthesis, solar energy is transformed to chemical energy by fixing carbon dioxide.

$$CO_2 + H_2O \rightarrow \frac{1}{6}C_6H_{12}O_6 + O_2 \quad (1)$$

$$\Delta G^0 = 479.9 \text{ kJ/mol}$$

The energy of 479.9 kJ/mol is stored when one mol of CO_2 is inverted to sugar. The energy stored in this way is stationary, and sugar can be assumed to be the natural energy of different form. The chemical fixation of natural energy requires an uphill reaction as reaction (1). It is desirable that abundant substances like CO_2 and H_2O and useful compounds such as fuels and commodity chemicals are involved as reactants and products, respectively, in the uphill reaction, because the natural energy to be stored is enormous and the production of needless compounds is meaningless.

We have developed the electrolytic process in which CO_2 is selectively fixed to ethylene at cathode with accompanying the evolution of oxygen at anode by solar energy and water [1-3]. The overall reaction is given by

$$CO_2 + H_2O \rightarrow \frac{1}{2}C_2H_4 + \frac{3}{2}O_2 \quad (2)$$

$$\Delta G^0 = 665.7 \text{ kJ/mol}$$

This reaction is uphill and the energy of 665.7 kJ/mol is accumulated. The reactants are abundant and the products useful, suggesting that reaction (2) is qualified as a good means for the storage of natural energy. Ethylene is an important feedstock in the chemical industry for making fuels and petrochemicals involving polyethylene and vinyl chloride. As a rule, ethylene is produced by cracking naphtha, ethane or oil gas at around 800°C. The pyrolysis is attended with the formation of various hydrocarbons, and the separation and refinement are needed to obtain the desired product. So, the production of ethylene is achieved by the consumption of a great energy. The fixation of CO_2 to ethylene under the assistance of natural energy permits us to substitute CO_2 for fuels and chemicals derived from petroleum and natural gas, which means that the unsteady natural energy is transformed to the stationary form. Furthermore, the artificial fixation of CO_2 contributes to the mitigation of the global warming, which is better than just capturing CO_2 and squeezing it into the ground. That is, CO_2 is not disused but converted to valuable substances.

Our electrolysis is practicable under rather special conditions such as three-phase (gas/solution/metal) interface, concentrated solution of potassium halide, low pH and copper or copper halide-confined metal electrode [1]. This is attributed to the existence of kinetic barriers involving the capture of CO_2 on the electrode, the electron transfer at the metal/solution interface, and the hydrogenation of activated intermediates. The object of the present work is to disclose the grounds for requiring these conditions and to design a series of chemical apparatuses including an electrolytic cell in a large scale.

Electrolytic Solution and the Capture of CO_2 at Electrode

It is natural that the electrochemical reduction of CO_2 in aqueous solution is more favorable at lower pH, but acidic solution is unfavorable for obtaining high concentration of CO_2 owing to the solubility limit. These situations are antinomic, and the electrolytic reduction of CO_2 is generally carried out in neutral or weakly alkaline solution. To achieve the efficient electrochemical reduction of CO_2, both the concentrations of CO_2 and protons are necessary to be high enough on the electrode. These requirements can be fulfilled by performing the electrolysis at the three-phase interface [4, 5]. In the electrolysis, a metal-mesh electrode is partially immersed in solution, and the reaction takes place most efficiently at the gas/solution/metal interface. Carbon dioxide is always supplied from gas phase, and the concentration at the reaction zone is independent of the solubility, which allows us to use an acidic solution as the electrolyte. Thus, the concentrations of both CO_2 and protons can be maintained high on the electrode during the electrolysis. In acidic aqueous solution, however, there is another problem to be overcome; i.e., the electroreduction of CO_2 is considerably decelerated by the competitive reduction of water. We have succeeded in restraining the competitive reaction by promoting the specific adsorption of halide ions [1, 2]. In Table 1, the results are exhibited which were obtained by the galvanostatic reduction of CO_2 at the three-phase interface on a copper-mesh electrode in a 4 M KBr solution of pH 3. In the closed system, the conversion efficiency of CO_2 reached approximately 100%, while the current efficiency for the hydrogen evolution was controlled to less than 10%. The suppression of the reduction of water is related to the structure of the electric double-layer formed at the electrode/solution interface. It is known that an anion like halide ion specifically adsorbs on metal electrode. The center of the anion is located at the inner Helmholtz plane. The specific adsorption is attributable not to the electrostatic attraction but to a strong chemical affinity of the anion with metal [6]. On the other hand, a cation has not such an adsorption specificity since it is surrounded by many water molecules. Hence, potassium ion cannot approach the electrode beyond

Table 1 Conversion efficiency of CO_2 and current efficiency of H_2 on a CuBr-confined Cu-mesh electrode by the galvanostatic reduction at −19.6 mA/cm² [a)]

Electrolysis time (min)	Electrolysis Potential (V vs Ag/AgCl)	Cell Voltage (V)	pH [b)] Catholyte	Anolyte	Conversion efficiency of CO_2 (%)	Current efficiency of H_2 (%)
60	−1.80∼−1.93	4.83∼4.85	2.5	0.0	16.0	21.9
120	−1.78∼−1.83	4.83∼4.87	3.4	−0.4	41.0	23.0
180	−1.83∼−1.90	4.85∼4.91	3.1	−0.4	55.6	22.0
240	−1.78∼−1.88	4.81∼4.88	2.1	0.1	65.3	19.1
300	−1.78∼−1.93	4.80∼4.88	2.9	−0.3	79.7	14.1
360	−1.80∼−1.90	4.85∼4.91	2.3	−0.5	98.0	9.6

a) Catholyte was a 4 M KBr solution, and the surface area of cathode substrate 10.2 cm². Anolyte, 1 M $KHSO_4$; anode, Pt net; initial gas constitution, 50 % CO_2 + 50 % N_2.

b) pH value of the catholyte was adjusted below 3.5 by adding a drop of concentrated H_2SO_4 solution to the anode compartment during the electrolysis.

the outer Helmholtz plane of closest approach. High concentration of halide ion renders the specific adsorption favorable at the inner Helmholtz plane, and protons and water molecules are kept away from the electrode. Thus, the hydrogen overpotential is considerably enhanced in a concentrated solution of potassium halide, leading to the decline in current efficiency of hydrogen evolution.

It is said that CO_2 is very stable and chemically inert. As described above, however, the reactivity of CO_2 is fairly large if the electrolytic process is properly tailored. It is interesting to discuss how CO_2 is activated in this electrolysis. The ionization potential (I_P) and electron affinity (E_A) of CO_2 are 13.79 eV and 3.7 eV, respectively. As compared with the values for other common compounds; e.g., I_P(eV): 12.06(O_2); 15.57(N_2), E_A(eV): 0.45(O_2); 2.38(Cl_2); 2.55(Br_2) [8,9], it turns out that the electron-accepting ability of CO_2 is large although the oxidation is very difficult. In fact, the reaction of CO_2 with a nucleophilic reagent such as amine is considerably active.

$$-NH_2 + CO_2 \rightarrow -N(H)COOH \qquad (3)$$

Amine solutions are widely used as the absorbent of CO_2. In the Grignard reaction, CO_2 is inserted into the linkage between negatively charged alkyl group ($R^{\delta-}$) and positively charged magnesium ($Mg^{\delta+}$) (reaction (4)). The combination of $R^{\delta-}$ with CO_2 is induced by the flow of electrons from the alkyl group to the vacant orbital of CO_2. The resultant adduct easily reacts with water and decomposes to carboxylic acid and hydroxide. A chain of these reactions are downhill and exothermic. The halide ion

$$R^{\delta-}Mg^{2\delta+}X^{\delta-} + CO_2 \longrightarrow R-C(=O)-O^{\delta-}\cdots Mg^{2\delta+}X^{\delta-} \qquad (4)$$

$$\xrightarrow{+H_2O} RCOOH + Mg(OH)X$$

specifically adsorbed on the electrode may act as a nucleophilic reagent for CO_2 in the same manner as the above-mentioned reactions. The electron transfer from adsorbed halide ion (X^-_{ads}) to the vacant orbital of CO_2 leads to the formation of X^-_{ads}–CO_2 linkage, and CO_2 is restricted to the electrode as represented by reaction (5).

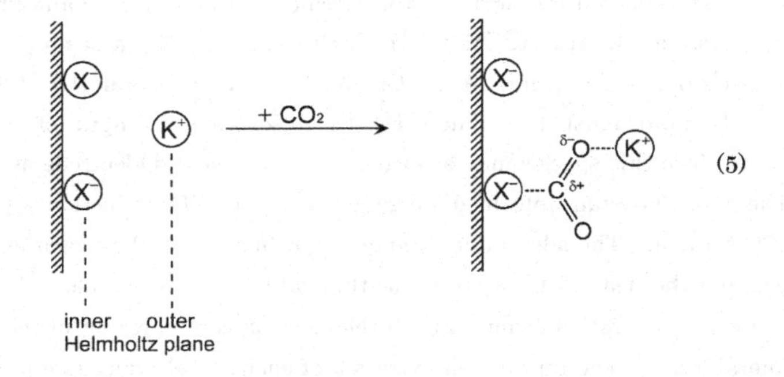

This reaction is exothermic and corresponds to the capture of CO_2 by the electric double-layer [10]. The occurrence of X^-_{ads}–C bond results in the weakening of C–O bond and so the activation of CO_2. The restriction of CO_2 to the electrode, which is very important in the succeeding electrochemical reduction, depends on the readiness of specific adsorption of anion. If there is no anion capable of adsorbing specifically, the approach of CO_2 to the electrode may be much disturbed by water molecules existing abundantly on the electrode, rendering the electrochemical reduction of CO_2 difficult. Halide ions give the strongest specific adsorption among the common anions [6]: $I^- > Br^- > Cl^- > NO_3^- > CO_3^{2-} > SO_4^{2-}$, and CO_2 may be bound on the electrode most effectively in the presence of halide ion. Therefore, it is understandable that the reduction of CO_2 at the three-phase interface is efficient in a concentrated solution of halide ion [7]. In the formation of X^-_{ads}–CO_2 bond, the contribution of potassium ion located at the outer Helmholtz plane is also important, because the conversion efficiency of CO_2 to ethylene was extremely low in the presence of cation (e.g., Na^+ and Li^+) other than K^+ [10]. The role of cation is to compensate the charge of X^{-}_{ads}–CO_2 occurring in the inner Helmholtz plane. The dependence of the reduction efficiency on the cation is presumably caused by the difference in size of cation. The radius of K^+ (1.33 Å) is larger than that of Li^+ (0.6 Å) or Na^+ (0.96 Å), and the degree of hydration of K^+ should be less than that of the other cations. Hence, the water molecules around K^+ ion are labile to be detached, and the direct interaction of K^+ with the nucleophilic oxygen atom of CO_2 becomes possible (reaction 5). The CO_2 thus stabilized in the electric double-layer is ready to accept electrons from the electrode.

Cathode Material

It is worthy of special attention that the reduction products from CO_2 are dependent on the cathode material. Formic acid is mainly formed at In, Sn, Hg, Tl and Pb (Metal I), carbon monoxide at Ag and Au (Metal II), hydrocarbon at Cu, and no product (only hydrogen evolution from water) at Ti, Cr, Fe, Ni and Pt (Metal III) [10]. Such a specificity of metal must be related to the adsorption strength of a reduction intermediate since this species may adsorb directly onto metal electrode as mentioned later. The most important intermediate generated in the CO_2 reduction is the formate radical, COOH [11]. The adsorption strength of formate radical on each metal can be estimated from the heat of adsorption, but this value is not available. So, let's infer qualitatively the adsorption strength from the electron configuration in the outermost shell of metal atom. The typical configuration of each metal group is represented as: $nd^{10}(n+1)s^2(n+1)p^{0-2}$ (Metal I); $nd^{10}(n+1)s^1$ (Metal II and Cu); $nd^m(n+1)s^{1-2}$ (Metal III). The d orbital of Metal I is saturated with electrons, and it is suggested that the interaction between the formate radical and Metal I is weak and the heat of adsorption is small. On the other hand, some vacancies are present in the d orbital of Metal III, and the formate radical may adsorb strongly onto those metals. In the case of Metal II or Cu, an electron in the d orbital is easily promoted to the s orbital, and the activated d orbital can interact with the formate radical, giving the medium strength of adsorption. Owing to the weak adsorption on Metal I, the formate radical may be desorbed from the electrode immediately after the combination with a hydrogen atom, resulting in the formation of formic acid. The formate radical strongly adsorbed on Metal III must be thoroughly reduced to carbon atom, and the metal surface be carbonated. The hydrogen overpotential would be considerably diminished on the carbonated metal, and the hydrogen evolution is dominant over the reduction reaction. On the other hand, the hydrogenation of formate radical on Metal II and Cu may advance to some extent, but the reduction species desorb in the form of CO from Metal II or combine with each other to become a more complicated form like ethylene on Cu.

As is mentioned above, silver electrode is known as the unique substrate that allows the selective conversion of CO_2 to CO. However, the surface of Ag electrode can be transmuted into a Cu-like substrate capable of converting CO_2 to ethylene by coating Cu(I) chloride [11]. The covering of Ag electrode with CuCl is performed by the electrochemical deposition (reaction (6)) at a more negative potential than 0.56 V vs NHE in an acidic KCl solution containing cupric ion.

$$Cu^{2+} + Cl^- + e^- \rightarrow CuCl \qquad (6)$$

$$E^0 = 0.559 \text{ V vs NHE}$$

The thickness of the deposited CuCl is proportional to the electrolysis time to the extent of a limited thickness [7]. The current efficiencies of the products obtained in the

Table 2 Effects of thickness of CuCl film on the current efficiencies for the products obtained in the electrochemical reduction of CO_2 on a CuCl-confined Ag-mesh electrode [a)]

Thickness[b)] (nm)	Faradaic efficiency / %									Conversion (%)	Selectivity (%)	Q (C)	h (%)
	Ethylene	Methane	CO	Ethane	EtOH	Formic	Acetic	Lactic	H_2				
0	1.3	4.9	79.6	0.7	0.4	2.4	0.3	1.4	7.4	11.6	0.5	153	98.4
3.4	16.5	1.8	59.7	1.1	2.6	1.3	1.4	0.5	9.2	13.0	0.6	215	94.1
7.9	28.8	1.8	48.9	0.7	2.7	1.3	1.8	0.8	12.7	11.7	15.3	212	99.5
8.9	42.8	4.0	31.8	1.0	2.3	1.0	2.0	1.5	14.5	9.3	28.0	206	101
10.2	64.0	3.8	9.3	1.8	2.0	2.0	1.0	1.1	13.0	10.4	59.7	323	98.0
12.9	60.4	2.7	14.4	0.9	0.4	2.0	0.5	0.4	13.5	10.7	52.9	307	95.2
15.3	60.2	4.5	15.8	0.8	1.8	2.6	0.6	0.3	9.2	10.1	49.0	278	95.8

a) Electrolyte, 3M KCl; initial pH, 3.0; the surface area of a silver-mesh substrate, 10.2 cm^2; electrolysis potential, -1.8V vs Ag/AgCl; electrolysis time, 30 min; initial concentration of CO_2, 100 %; volume, 480 cm^3.
b) Thickness of coated CuCl.

reduction of CO_2 on a CuCl-confined Ag-mesh electrode are exhibited as a function of thickness of CuCl film in Table 2. On a pure Ag electrode, the current efficiency for the formation of CO was about 80% without accompanying the production of C_2H_4. With increasing the thickness of deposited CuCl, however, the yields of CO and C_2H_4 decreased and increased, respectively. At the thickness of 10.2μm, the current efficiency of CO dropped to 9.3%, while that of C_2H_4 raised to 64%. Hence, the place where the electroreduction of CO_2 occurred is suggested to be on the copper metal newly formed from CuCl. This is comprehensible from the thermodynamics. The standard potential of reduction of CuCl is 0.121 V vs NHE (= −0.101 V vs Ag/AgCl).

$$CuCl + e^- \rightarrow Cu + Cl^- \qquad (7)$$
$$E^0 = 0.121 \text{ V vs NHE}$$

The electrode potential applied in Table 2 was −1.8 V vs Ag/AgCl. This potential is highly negative to E^0, indicating the favorable reduction of CuCl to Cu. So, the CuCl-confined Ag may behave like a Cu electrode in the cathodic electrolysis. Similar results are obtained with stainless steel and Monel metal (Ni + Co: 65%, Cu: 33%, Fe: 2%) electrodes [11]. No reduction reaction was observed on these electrodes except the hydrogen evolution, but CuCl-confined stainless steel and Monel metal electrodes yielded ethylene with the current efficiencies of 11.4% and 42.9%, respectively. Therefore, it is again emphasized that the adsorption strength of the intermediate is important in the CO_2 reduction and the medium strength given by copper is necessary for the formation of ethylene.

Electrocatalysis

As is noted above, CO_2 is first captured by the electric double-layer on the cathode (reaction (5)), and electrons are injected to CO_2 in the subsequent electrochemical step. The electron injection is achieved through the X^-_{ads}–carbon bond. The oxygen atom rich in electrons attracts a proton, and the formate radical is formed. The radical may adsorb directly onto the electrode, because it has a stronger affinity for metal than halide (reaction(8)). This reaction corresponds to the inner-sphere mechanism in which electrons are smoothly transferred via a bridge ion (X^-) without accompanying an additional energy. In such a reaction path, the overpotential necessary for the CO_2 reduction is largely mitigated as observed in our electrolysis [11]. On the other hand, in the absence of a bridge ion like in the usual electroreduction of CO_2, the injection of

$$X^- \cdots C^{\delta+}(=O)-O^{\delta-}\cdots K^+ \xrightarrow{+H^+ \ +e^-} -C(=O)OH \quad K^+ \qquad (8)$$

electrons follows the outer-sphere mechanism in which electrons need to hop from the electrode surface to CO_2 with an activation energy, and hence a highly negative overpotential is required. The reductive coupling of formate radicals on Cu electrode results in the generation of intermediates such as $-CH_2C(O)-$ and $CH_2=CH-$, and ethylene is finally produced [10,11].

$$2COOH + 10H^+ + 10e^- \rightarrow C_2H_4 + 4H_2O \qquad (9)$$

It is described above that the role of deposited copper(I) halide is to offer new sites of copper metal for the reduction of CO_2. In reality, however, copper(I) halide has another important role; i.e., to facilitate the specific adsorption of halide ion at the inner Helmholtz plane. Although the cuprous ion is reduced to Cu during the cathodic electrolysis, the halide ion must remain on the electrode, so that the specific adsorption is intensified. In fact, the considerable enhancement of the ethylene formation has been observed by increasing the concentration of halide ion [7].

The electrochemical deposition of Cu(I) halide can be performed beforehand in a separate cell, but this process is troublesome in a practical use. So, the *in situ* method of anodic deposition of Cu(I) halide has been proposed, which is achieved in the same solution as that used for the CO_2 reduction [10]. That is, a bare Cu-mesh electrode is polarized to the anodic side for a given period of time before the start of cathodic electrolysis of CO_2. In result, C(I) halide is anodically deposited by the reverse process of reaction (7) or (10).

$$CuBr + e^- \rightarrow Cu + Br^- \qquad (10)$$

$$E^0 = 0.033 \text{ V vs NHE}$$

In the anodic deposition of Cu(I) halide, the electrochemical parameters such as anodizing time, current density and electrode potential should be properly fixed to avoid the deposition of CuX_2 and $Cu(OH)_2$ which operate against the reduction of CO_2 [10]. In Table 3, the conversion efficiency of CO_2 and current efficiency of H_2, which were obtained by the galvanostatic reduction of CO_2 with an *in situ* CuBr-deposited Cu-mesh electrode, are exhibited as a function of the time spent in the anodic deposition. The results are compared with that (41.0%) obtained with the previously prepared electrode (i.e., without *in situ* anodizing). The conversion efficiency of CO_2 increased with an increase of the anodizing time, and attained to 56.4% at 180s, implying that the

Table 3 Conversion efficiency of CO_2 and current efficiency of H_2 obtained on a CuBr-confined Cu-mesh electrode versus the anodizing time spent for the *in situ* deposition of CuBr at a current density of +19.6 mA/cm^2 in a 4 M KBr solution of pH 3 [a]

Anodizing time (s)	Electrolysis Potential (V)	pH [b]		Conversion efficiency of CO_2 (%)	Current efficiency of H_2 (%)
		Catholyte	Anolyte		
0	−1.78∼−1.83	3.4	−0.4	41.0	23.0
1	−1.85∼−1.98	2.3	−0.4	38.0	25.0
30	−1.83∼−1.93	2.5	−0.4	43.2	22.3
90	−1.80∼−1.93	3.6	−0.3	45.8	21.0
120	−1.83∼−2.03	3.5	−0.3	49.0	20.1
180	−1.78∼−1.91	3.8	−0.3	56.4	23.1
300	−1.78∼−1.89	3.1	−0.4	40.7	23.3

a) The electrolysis of CO_2 was performed for 120 min with a CuBr-confined Cu-mesh electrode at a current density of −19.6 mA/cm^2 in the same solution as that used for the anodic deposition of CuBr. Anolyte was a 1 M $KHSO_4$ solution of pH 0. Initial gas constitution was 50 % CO_2 + 50 % N_2.

b) pH value of the catholyte was adjusted below 3.8 by adding a drop of concentrated H_2SO_4 to the anode compartment during the electrolysis.

in situ deposited Cu(I) chloride operated more positively upon the reduction of CO_2 than the previously immobilized one. On the other hand, the current efficiency of hydrogen evolution was almost independent of the deposition method. It is therefore confirmed that the *in situ* immobilization of Cu(I) halide is useful for the pretreatment of the working electrode. The catalytic activity of thus prepared electrode is not deteriorated at all during the electrolysis.

In the galvanostatic reduction of CO_2, it is important to examine the effect of current density applied. The conversion efficiency of CO_2 and current efficiency of H_2 are shown in Figs. 1 and 2, respectively, as a function of the electric charge passed in the electrolysis with various current densities. It is indicated from Fig. 1 that the conversion efficiency depends on both the electric charge and current density. This suggests that the electrochemical reduction of CO_2 is regulated by both the electron transfer and chemical reactions. In the application of higher current density (24.5 mA/cm^2), the chemical reactions involving the adsorption of halide ion, the formation of X^-_{ads}–CO_2 linkage and the desorption of products are unable to follow the rate of supply of electrons, and the excessive electrons may be consumed to reduce protons, leading to the increase in current efficiency of H_2 (Fig. 2). On the other hand, in the case of lower current density (14.7 and 9.8 mA/cm^2), the supply of electrons is insufficient and unable to match with the chemical reactions, and the conversion efficiency would show a tendency to approach a constant value. The conversion efficiency of about 100% was attained with a medium current density of 19.6 mA/cm^2, and in this case, the rate of supply of electrons may be well-balanced with the chemical

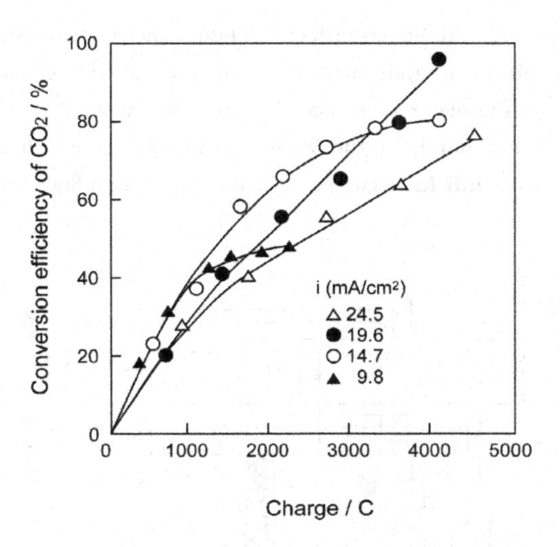

Fig.1 Conversion efficiency of CO_2 versus the electric charge passed in the galvanostatic reduction at different current densities on a CuBr-confined Cu-mesh electrode in a 4 M KBr solution of pH 3.

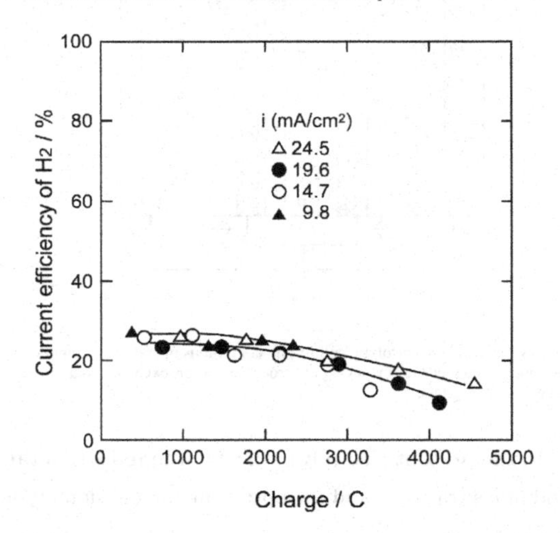

Fig.2, Current efficiency of H_2 versus the electric charge passed in the galvanostatic reduction under the same condition as that given in Fig.1.

reactions. The existence of an optimum current density is often observed in the industrial electroextraction of metals; e.g., copper and zinc are extracted from aqueous solutions containing various additives with current densities of about 20 and 50 mA/cm^2, respectively. This obviously indicates that the rate of the electroextraction is controlled not only by the electron transfer but also by chemical reactions involving catalysis. On the other hand, in the water electrolysis of which rate is mainly regulated by the electron transfer, a current density higher than 300 mA/cm^2 is

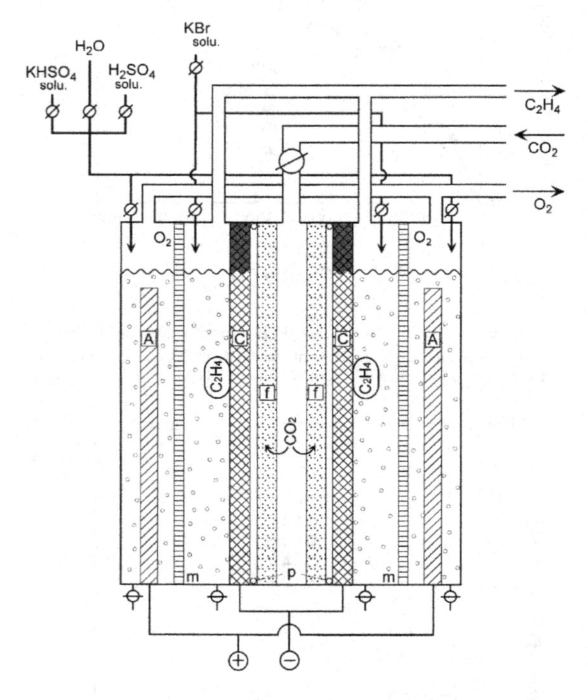

Fig.3 Schematic of electrolytic cell for the continuous operation. A, anode electrode; C, copper-mesh working electrode; m, cation-exchange membrane; f, glass filter; p, packing

generally applied. In the present electrolysis of CO_2, the reduction rate is regulated by both the electron and mass transfer, and the optimum current density is concluded to be around 20 mA/cm^2.

Continuous Electrolysis at a Large Scale

In the practical use of the electrolysis, it is important that the process is feasible in the mode of consecutive operation, because a batch-type electrolysis is unsuitable to treat a volume of CO_2. Fig. 3 shows the schematic diagram of the electrolytic cell which permits us continuously to supply CO_2 and to extract ethylene. Carbon dioxide is first bubbled on a copper-mesh electrode (C) by passing it through a glass filter (f) to form the interface consisting of gas, solution and metal. The three-phase interface extended on the entire surface can be constructed by adjusting the rate of gas flow, the gradient dimensions of pores of glass filter and/or the size of a packing (p). An anode electrode (A) is separated from the cathode compartment by a cation-exchange membrane (m). A platinum or platinized titanium plate is suitable as the anode [10]. A concentrated solution of KBr (4 M, pH 3) and a $KHSO_4$ solution (1 M, pH 0) are introduced to the cathode and anode compartments, respectively. The pH of the catholyte is elevated as

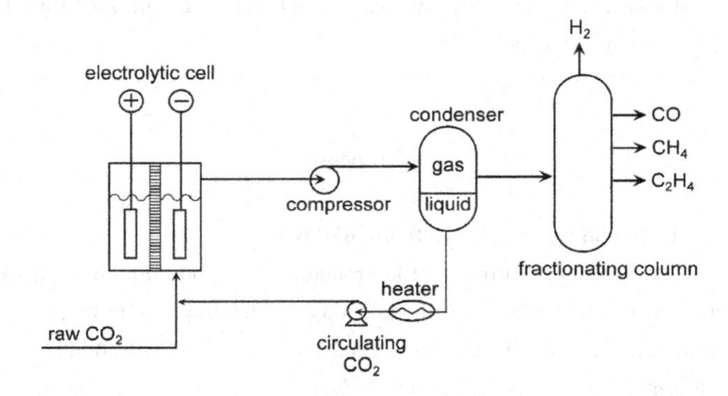

Fig.4 Flow schematic of apparatuses for the proposed manufacturing process of ethylene from CO_2

the electrolysis advances, and the pH value is needed to be kept below 3.5 by adding a concentrated solution of H_2SO_4 to the anode compartment. Such an indirect regulation of pH of the catholyte is necessary for the efficient reduction of CO_2 [10]. A quantity of fresh water is added to the anode compartment to compensate the volume consumed by the anodic decomposition. Gases from the cathode compartment contain ethylene, unreacted CO_2, hydrogen and byproducts (small quantities of CO and CH_4), while the gas from the anode compartment is almost pure oxygen. The gases from the cathode compartment are fractionated, and the unreacted CO_2 is put back to the electrolytic system. In Fig. 4, the flow schematic of apparatuses for the ethylene production from CO_2 is shown, consisting of electrolytic, circulating and fractionating branches. A portion of gases is extracted from the cathode compartment and condensed to liquefy only the unreacted CO_2. The CO_2 is again gasified, and given back to the electrolytic cell. Other gases are separated with a fractionating column. The volume of gases discharged from the cathode compartment is always filled up by introducing raw CO_2, and thus the electrolytic reduction of CO_2 is continuously carried out.

The energy required for the electrolysis can be supplied by various sources of energy. The application of a solar cell may accomplish the artificial photosynthesis in which CO_2 can be fixed to the organic substance with accompanying the evolution of oxygen by solar energy and water. This signifies that CO_2 can be substituted for fuels and chemicals derived from petroleum by using solar energy and water. In other words, the unsteady natural energy is transformable to the stationary form by means of abundant substances. Accordingly, we can construct a green sustainable energy system independent of fossil fuel.

References

1. K. Ogura, Electrochemistry, 71(2003), pp. 676-680.
2. K. Ogura, H. Yano, and F. Shirai, J. Electrochem. Soc., 150(2003), pp. D163-D168.
3. K. Ogura, H. Yano, T. Tanaka, Catalysis Today, 98(2004), pp. 515-521.
4. K. Ogura and N. Endo, J. Electrochem. Soc., 146(1999), pp. 3736-3740.
5. H. Yano, F. Shirai, M. Nakayama, and K. Ogura, J. Electroanal. Chem., 519(2002), pp. 93-100; 533(2002), pp. 113-118.
6. P. Delahay, "Double Layer and Electrode Kinetics," John Wiley & Sons, New York, pp. 53-63, 1965.
7. H. Yano, T. Tanaka, M. Nakayama, and K. Ogura, J. Electroanal. Chem., 565(2004), pp. 287-293.
8. W. A. Chupka, J. Berkowitz and D. Gutman, J. Chem. Phys., 55(1971), pp. 2724-2733.
9. D. W. Turner, C. Baker, A. D. Baker, and C. R. Brundle, "Molecular Photoelectron Spectroscopy," John Wiley & Sons, New York, 1970.
10. K. Ogura, "Carbon Dioxide Reduction Metallurgy," N. R. Neelameggham and R. G. Reddy, ed., TMS, pp. 147-160, 2008.
11. K. Ogura, R. Ohara, and Y. Kudo, J. Electrochem. Soc., 152(2005), pp. D213-D219.

Energy Technology Perspectives
Edited by: Neale R. Neelameggham, Ramana G. Reddy, Cynthia K. Belt, and Edgar E. Vidal
TMS (The Minerals, Metals & Materials Society), 2009

THE ELECTROCHEMICAL REDUCTION OF CARBON DIOXIDE IN IONIC LIQUIDS

Huimin Lu[1], Xiaoxiang Zhang[1], Pengkai Wang[1]

[1]Beijing Univ. of Aeronautics & Astronautics, School of Material Sci. & Eng.;
37 Xueyuan Road, Haidian District, Beijing 100191, China
* Corresponding author: lhm0862002@yahoo.com.cn

Keywords: carbon dioxide, electrochemical reduction, ionic liquid, 1-n-butyl-3-methylimidazolium hexafluorophosphate ([Bmim] PF_6)

Abstract

In this paper, it was studied an electrochemical reduction process of carbon dioxide in ionic liquids such as 1-n-butyl-3-methylimidazolium hexafluorophosphate ([Bmim] PF_6) as the electrolyte. The electrolysis experiments were carried out under current and potential controls. The cathode products contained carbon nanotubes, carbon nanofibers, nanographites, and amorphous carbon. To establish the actual current and potential ranges, the electroreduction of carbon dioxide dissolved in the ionic liquid was studied by cyclic voltammetry on glass-carbon (GC) electrode at room temperature. The electrochemical mechanism of carbon dioxide electroreduction was studied for explanation of all obtained results. Carbon dioxide in ionic liquids was electroreduced, similar to metals oxides being electroreduced in molten salts.

Introduction

Global warming caused by the increase of carbon dioxide in the atmosphere has become one of the most concerning environment problems, so the solution to this greenhouse effect is very urgent. However, the carbon dioxide is also a cheap and abundant raw material that represents a potential source for the production of chemicals and fuels. Therefore, the development of the technologies that achieve simultaneous reduction and conversion of carbon dioxide to fuels is important. The electrochemical reduction of carbon dioxide is one of the effective approaches to transform CO_2. Many investigations have been carried out on the electrochemical reduction of CO_2 in both aqueous [1] and non-aqueous [2] electrolyte solutions with different electrodes. The solubility of CO_2 in water is very low even under high pressure of CO_2, and some investigators have brought focus into the electrochemical reduction of carbon dioxide in organic solvents, because they dissolved much more CO_2 than water did. However, the hazardous or toxic properties of organic solvents hinder their use.

Room temperature ionic liquids (RTILs) are described as compounds composed entirely of ions, generally a bulky organic cation and an inorganic anion, existing in the liquid state at 298 K. Wide electrochemical windows, negligible volatility, high thermal stability, and high conductivity are the unique properties of ILs, making them widely useful in many fields. Additionally, the ability to change the cation or anion provides tunable solvents for specific purposes. RTILs have been successfully used as solvents for organic and inorganic synthesis [3-5], as well as for extractions [6], and also have been used as electrolytes in lithium batteries [7], solar cells [8], fuel cells [9], and electroplating [10]. With the further studies about the characters of CO_2 dissolved in RTILs, more and more attention was concentrated on their applications in electrochemistry. Yang [11] and coworkers studied the electrochemical activation of CO_2 for the synthesis of cyclic carbonates through cycloaddition of CO_2 to epoxides in the ILs under mild conditions, and found Al^{3+} or Mg^{2+} in the ILs after electrolysis, because Al or Mg sacrificed anode was used. It also has been described that the products containing carbon monoxide, hydrogen, and small amount of formic acid were produced by the electrochemical reduction of CO_2 in ionic liquid [Bmin]PF_6 under supercritical reaction condition. In this work, the most current efficiency could reach 91.9%, and no hydrocarbon was detected in the gas products [12].

Recently, Novoselova and co-workers[13]using a novel electrolytic method synthesized carbon nanotubes (CNTs) from CO_2 dissolved in molten salt, found that the platinum electrode surface eroded and its mass reduced after electrolysis, and gave the electrochemical-chemical-electrochemical mechanism of CO_2 electroreducation.

First stage: reduction of CO_2 to CO_2^{2-} radical

$$CO_2 + 2e^- = CO_2^{2-} \tag{1}$$

Second stage: chemical formation of carbon monoxide

$$CO_2^{2-} = CO + O^{2-} \tag{2}$$

Third stage: irreversible electroreduction of CO to elementary carbon

$$CO + 2e^- = C + O^{2-} \tag{3}$$

At the same time, carbon dioxide acts as acceptor of oxide anions, and the overall cathode reaction may be represented as

$$3CO_2 + 4e^- = C + 2CO_3^{2-} \tag{4}$$

Carbonate-ion discharge takes place at the anode to produce carbon dioxide and oxygen

$$2CO_3^{2-} = CO_2 + O_2 + 4e^- \tag{5}$$

The overall electrode reaction is as follows

$$CO_2 = C + O_2 \tag{6}$$

This method faces some problems, such as high reaction temperature and low solubility of CO_2 in molten salts.

In this paper, using ionic liquid 1-n-butyl-3-methylimidazolium hexafluorophosphate ([Bmim]PF_6) as electrolyte, the electrochemical reduction process of carbon dioxide was studied at room temperature at different pressures. And the characterization of produced carbon materials was studied by SEM. To establish the actual current and potential ranges, the electroreduction of carbon dioxide dissolved in the ionic liquid was studied by cyclic voltammetry on glass-carbon (GC) electrode at room temperature. The electrochemical mechanism of carbon dioxide electroreduction was studied for explanation of all obtained results.

Experimental Section

Materials

The ionic liquid (IL) 1-n-butyl-3methylimidazolium hexafluorophosphate ([Bmin] PF_6, - high purity) was used as received from Henan Lihua Pharmaceutical Co. Ltd., China. CO_2 gas (>99.995%) stored up in a steel cylinder was purchased from Beijing Qianxi Jingcheng Gas Marketing Centre.

Apparatus

The self-made high-pressure electrolysis cell of 100ml was composed of a Teflon-lined stainless steel container (Φ110/80mm ×H110/80mm), a Pt wire (>99.99%, Φ1mm) anode , glass carbon rod (GC, Φ2mm) cathode, and an Ag/AgCl reference electrode [from Tianjin Aidahengsheng Technology Co. Ltd]. The LPS DC Power Supply -5A/30V electric commutator, made by Beijing Tradex Electronic Technology Corp., China, was used supply direct current for the electrochemical reduction experiments; heating of the cell was using a S2CL-2 Digital Smart-Temperature Magnetic Stirrer (Henan Yuhua Instrument Co. Ltd); Cyclic voltammetry was performed with a CHI 604C Electrochemical System made in Shanghai Chenhua Inc, Shanghai, China, controlled by a hand-held computer. The CO_2 gas electrochemical reduction apparatus consisted mainly of these instruments described above as shown in Figure1.

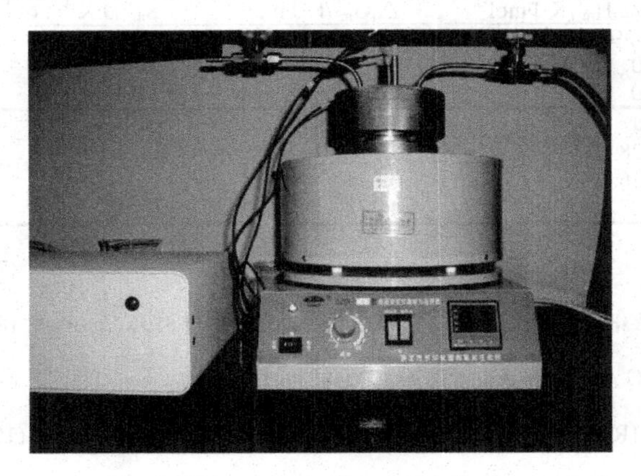

Fig. 1 CO_2 gas electrochemical reduction apparatus

Experimental Procedures

In the cyclic voltammetry and electrolysis experiments, the three-electrode high-pressure electrolysis cell was used. The platinum electrode was mechanically polished with 0.05μm Al_2O_3 slurry, sonicated successively in ethanol and redistilled water for 10 min, and then pretreated by electrolyzing at 2.5V in concentrated H_3PO_4 solution. Pt wire (Φ1mm×10mm) and GC rod (Φ2mm×10mm) worked as cathode and anode respectively, and an Ag/AgCl electrode was used as the reference electrode. Both the Pt cathode and GC anode had geometric area of $2.0cm^2$. They were arranged in parallel and the distance between them was 3cm, while a reference electrode was kept farther from the two electrodes. 100g IL [Bmim]PF_6 was charged into the electrolysis cell. The cell was sealed and, with the three electrodes were inserted into [Bmin]PF_6 electrolyte simultaneously through the cover of the cell. CO_2 gas was charged into the cell up to 0.5MPa, and then the CO_2 was released to atmosphere. This was repeated three times so that the air in the cell was replaced by CO_2. The electrolysis cell was controlled by S2CL-2 Digital Smart-Temperature Magnetic Stirrer, and CO_2 of desired pressure was fed into the electrolysis cell and the stirrer was started. The system was equilibrated at desired pressure for 4hour. The cyclic voltammetry experiments were performed with the CHI 604C Electrochemical System. The electrolysis was conducted galvanostatically at room temperature in the same electrolysis cell. After the electrolysis, the solid products were collected by a high vacuum micro-materials colatorium and analyzed by SEM. The IL left under the colatorium could be reused.

Results and Discussion

CO_2 Decomposition Voltage

Table 1 presents the standard thermodynamic data for CO_2 decomposition voltage computation. Through these thermodynamic equation (7) (8) and Nernst equation (9) (10), CO_2 decomposition voltage can be computation.

Table 1 CO_2 decomposing thredyne data ($CO_2 = C + O_2$ [25°C, P = 100KPa]) [14, 15]

Name	$\triangle_f H_m^\theta$/K·J·mol^{-1}	$\triangle_f G_m^\theta$/K·J·mol^{-1}	S_m^θ /J·K^{-1}·mol^{-1}
CO_2(g)	-393.509	-394.359	273.74
O_2(g)	0	0	205.138
C	0	0	5.740
$C_{p,m}$/ J·K^{-1}·mol^{-1}	a	10^3b	c
CO_2(g)	28.66	35.702	0
O_2(g)	36.162	0.835	0
C	17.15	4.27	0

$$\triangle rC_{p,m}{}^\theta = \triangle a + \triangle bT + \triangle cT^2 \quad (7)$$

$$\triangle G^\theta = -\triangle_f G_m{}^\theta(CO_2) + \triangle_f G_m{}^\theta(O_2) + \triangle_f G_m{}^\theta(C) = 394.359 KJ/mol \quad (8)$$

$$E^\theta = -\triangle G^\theta/nF = 394.359 \times 10^3/4 \times 96482 = -1.02V \quad (9)$$

$$E = E^\theta - \{RT/4F\} \times \ln\{p(O_2)/ p^\theta\}/\{p(CO_2)/ p^\theta\} \quad (10)$$

From the equation (10), it can be seen that CO_2 actual decomposition voltage only has a lot to do with the actual CO_2 pressure, and nothing to do with CO_2 solubility in ILs.

The computation formula of $\triangle G^\theta$ at different temperatures:

$$\triangle H_m{}^\theta(T) = \triangle_r H_m{}^\theta(298K) + \int_{298}^{T} \triangle_f C_{p,m}{}^\theta dT \quad (11)$$

$$\triangle S_m{}^\theta(T) = \triangle_r S_m{}^\theta(298K) + \int_{298}^{T} \triangle_f C_{p,m}{}^\theta / T\, dT \quad (12)$$

$$\triangle G_m{}^\theta = \triangle H_m{}^\theta(T) - T\triangle S_m{}^\theta(T) \quad (13)$$

$$\triangle_r C_{p,m}{}^{\theta} = \triangle a + \triangle bT + \triangle cT^2 = 24.652 - 30.597 \times 10^{-3}\, T^2 \quad (14)$$

$$\triangle G_m{}^{\theta} = \triangle H_m{}^{\theta}(T) - T\triangle S_m{}^{\theta}(T) \quad (15)$$
$$= 5.0997 \times 10^{-3} T^3 - 1192.486T - 24.652T\ln T + 0.65606$$

Because of φ^0 (standard electrode potential) of the Ag/AgCl electrode relative to standard H electrode is 0.224V, the maximum value of the decomposition voltage of CO_2 in ILs is lower than 3V. Therefore, when CO_2 is electrolyzed, the cell voltage should not be higher than 3V excepting for the resistive voltage which would go up with increased amperage.

Cyclic Voltammograms of CO_2 in [Bmim]PF_6 IL Electrolyte

One of the keys of this work is that CO_2 can be converted, while the IL remains stable during electrolysis. To verify this, it was first determined the voltammograms on the cathode electrode (Pt), anode electrode (GC) in the IL + CO_2 (0.1MPa, 0.2MPa, 0.3MPa, 0.4MPa and 0.5MPa) at 298K. Figure 2 a, b, c, d, and e are the voltammograms of CO_2 in [Bmim] PF_6 IL electrolyte. The data in the figures confirm that the IL can be used to electrolyze CO_2. This is in agreement with the conclusion that the IL has an electrochemical window of larger than 6V [16]. For all the experiments in this work, the cell voltages were less than 3V.

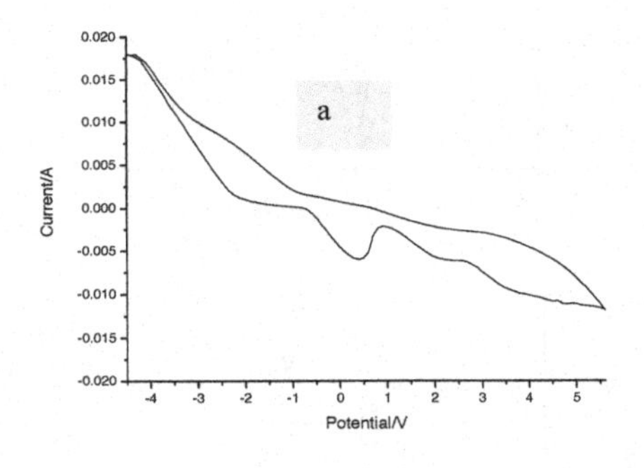

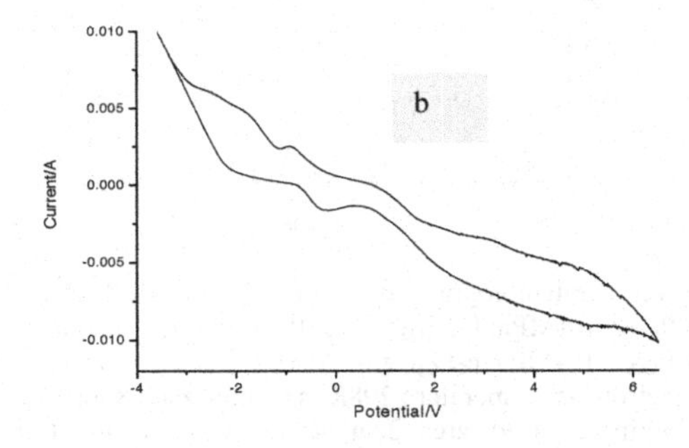

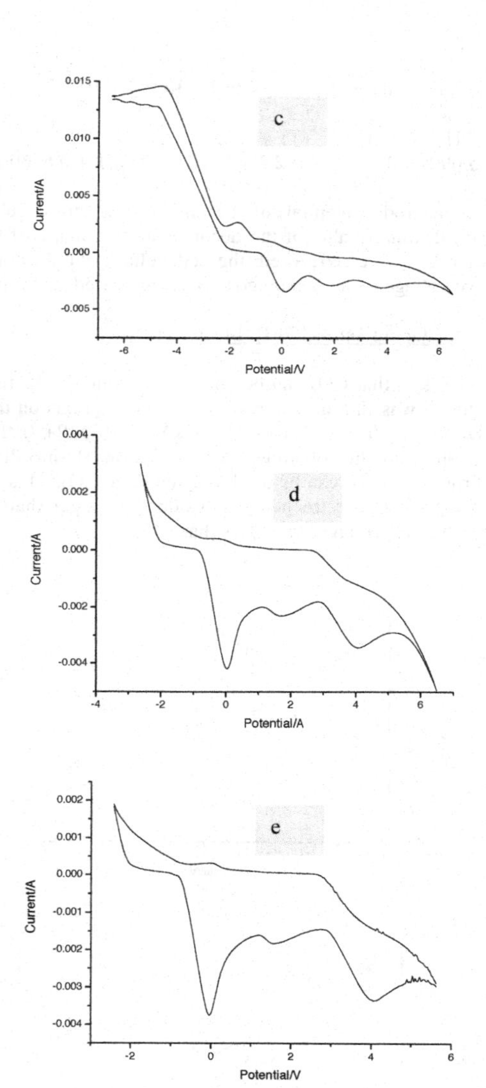

Fig.2 Cyclic voltammograms of CO_2 in [Bmim]PF_6 IL electrolyte
a-P_{CO2} = 0.1MPa; b-P_{CO2} = 0.2MPa; c-P_{CO2} = 0.3MPa;
d-P_{CO2} = 0.4MPa; e-P_{CO2} = 0.5Mpa.
Electrolysis temperature 298K; scanning rate: 80mV/s;
Working electrode area: $2cm^2$ at same voltage range from -6V to +6V.

Products of CO_2 Electrolysis in [Bmim]PF_6 IL Electrolyte

The CO_2 electrolysis experiments were conducted in the high-pressure electrolysis cell. The experimental conditions were as follows: CO_2 pressures 0.1MPa, 0.2MPa, 0.3MPa, 0.4MPa and 0.5Mpa; electrolysis temperature 298K, 313K, 353K and 373K; the cell voltages 2V, 2.5V and 3V. In the electrolysis process, the current was in about 5~250mA. The Faraday efficiency was in the range of 85 ~ 91%. These experiments showed that the Faraday efficiency has a lot to do with CO_2 pressure, the larger the CO_2 pressure, the higher the Faraday efficiency; the electrolysis temperature has little effect on the Faraday efficiency. The products were same such as carbon nanotubes, carbon nanofibers, nanographites, and amorphous carbon by SEM analyses. Figure 3 is the SEM of the CO_2 electrolysis products. The SEM analyses showed that the carbon nanotubes were about 33nm in length. The purity of the products was 88.75%, and some iron oxides in the products. It was thought that the iron oxides came from the stainless steel clip which held the GC electrode during electrolysis. The clip could be immersed in the IL while electrolysis. An attention will be paid for this in further experiments. In the experiment, the quantities of carbon nanotubes accounting for the overall carbon nanomaterials had not been detected.

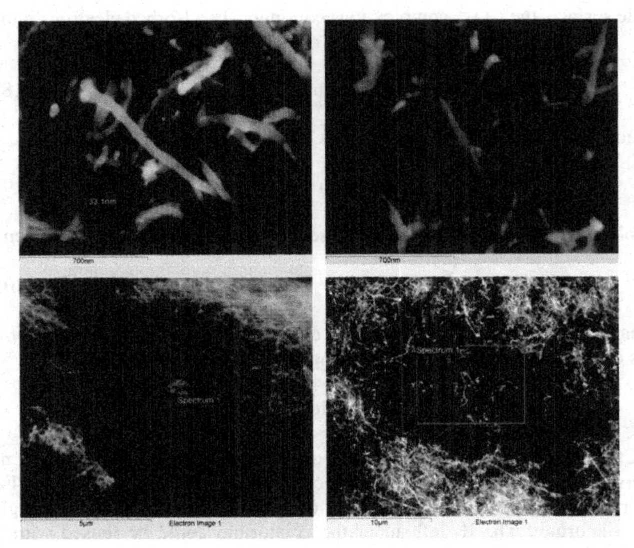

Fig.3 SEM of the CO_2 electrolysis products

Table 2 The composition of the CO_2 electrolysis products in [Bmim]PF_6 electrolyte

Elements	Weight%	Atomic%
C	88.75	91.69
O	10.51	8.15
	0.74	0.16
Totals	100.00	

Reaction Mechanism

The mechanism of the electrochemical reduction of high pressure CO_2 in [Bmim]PF_6 was investigated for Pt electrode. When the electrolysis was conducted mainly carbon nanomaterials was found as products. These experimental data suggest that the reaction pathway, by which carbon nanomaterials as carbon nanotubes, carbon nanofibers, nanographites, and amorphous carbon can be formed, similar to those in the electrochemical reduction of CO_2 in molten salts as NaCl –KCl –CsCl melt in 823K[13]. The reaction mechanism is perceived as follows:

First stage, the CO_2 gas in high pressure went into the IL, CO_2 (g) changed into CO_2 (ads) by IL.

$$CO_2\,(g) = CO_2\,(ads) \qquad (16)$$

Then on the Pt cathode, the CO_2 (ads) adsorbed electrons into CO_2^{2-}(surf).

$$CO_2\,(ads) + 2e^- = CO_2^{2-}(surf) \qquad (17)$$

On the Pt cathode surface, the CO_2^{2-} complex ions continuously adsorbed electrons into CO_2^{4-} ions.

$$CO_2^{2-}(surf) + 2e^- = CO_2^{4-}\,(surf) \qquad (18)$$

And then CO_2^{4-} ions forming complex compounds $C * 2O_2$.

$$CO_2^{4-}(surf) = C * 2O_2\,(surf) \qquad (19)$$

At last, the complex compounds $C * 2O_2$ decomposed into carbon products and oxygen gas.

$$C * 2O_2\,(surf) = C + 2O_2 \qquad (20)$$

From this mechanism, it is seen that the carbon dioxide in ionic liquids was electro-reduced similar to metals oxides such as Al_2O_3 or MgO being electro-reduced in molten salts.

Conclusions

In conclusion, CO_2 can be electrolyzed into carbon nanomaterials such as carbon nanotubes, carbon nanofibers, nanographites, and amorphous carbon and so on using [Bmim] PF_6 IL as an electrolyte. After electrolysis, the solid products can be easily recovered by a high vacuum micro-materials colatorium. The IL left under the colatorium could be reused without further treatment. This method provides a new and clean route to convert CO_2 into some carbon nanomaterial products. The possibility to produce carbon nanomaterials by this method with content up to 88.75wt% in the cathode product has been shown; the carbon nanotubes have curved form and most of agglomerate into bundles. The study on the electrochemical mechanism of carbon dioxide electroreduction showed carbon dioxide in ionic liquids was electroreduced as metals oxides were electroreduced in molten salts.

References

[1] K. Hara, A. Tsuneto, A. Kudo, T. Sakata. Electrochemical Reduction of CO_2 on a Cu Electrode under High Pressure. J. Electrochem. Soc. 141 (1994) 2097.

[2] Satoshi Kaneco, Hideyuki Katsumata, Tohru Suzuki, Kiyohisa Ohta. Electrochemical Reduction of Carbon Dioxide to Ethylene at a Copper Electrode in Methanol Using Potassium Hydroxide and Rubidium Hydroxide Supporting Electrolytes. Electrochimica Acta, 51 (2006) 3316.

[3] Howarth J. Oxidation of Aromatic Aldehydes in the Ionic Liquid [Bmim][PF_6]. Tetrahedron Lett., 41 (2000) 6627.

[4] Wolfson A, Wuyts S, de Vos D E et al. Aerobic Oxidation of Alcohols with Ruthenium Catalysis in Ionic Liquids. Tetrehedron Lett. 43 (2002) 8107.

[5] Xue, X., Liu, C., Lu, T., Xing W. Synthesis and Characterization of Pt/C Nanocatalysts Using Room Temperature Ionic Liquids for Fuel Cell Applications. Fuel Cell . 6 (2006) 347.

[6] McFarlane J, Ridenour W B, Luo H et al. Room Temperature Ionic Liquids for Separating Organics from Produced Water. Separation Science and Technology, 40 (2005) 1245.

[7] Howlett P C, MacFarlane D R, Hollenkamp A F. High Lithium Metal Cycling Efficiency in a Room-Temperature Ionic Liquid. Electrochem, Solid-State Lett., 7 (2004) A97.

[8] Wang P, Zakeeruddin S M, Gratzel M, Kantlehner W, Mezger J, Stoyanov E V, Scherr O. Novel Room Temperature Ionic Liquids of Hexaalkyl Substituted Guanidinium Salts for Dye-Sensitized Solar Cells. Appl. Phys., A. 79 (2004) 73.

[9] De Souza RF, Padilha JC, Goncalves RS, Dupont J. Room Temperature Dialkylimidazolium Ionic Liquid-Based Fuel Cells. Electrochemistry Communications, 5 (2003) 728.

[10] Huimin Lu, Yongheng Wang. Refine Aluminum Process in Ionic Liquids. Light Metals 2007, Edited by Morten Sorlie, TMS(The Minerals, metals & Materials Society), 2007, 583-588.

[11] H.Z. Yang, Y.L. Gu, Y.Q. Deng, F. Shi. Electrochemical Activation of Carbon Dioxide in Ionic Liquid: Synthesis of Cyclic Carbonates at Mild Reaction Conditions. Chem. Commun., (2002) 274.

[12] Guoying Zhao, Tao Jiang, Buxing Han, Zhonghao Li, Jianmin Zhang, Zhimin Liu, Jun He, Weize Wu. Electrochemical Reduction of Supercritical Carbon Dioxide in Ionic Liquid 1-n-Butyl - 3-Methylimidazolium Hexafluorophosphate. J. of Supercritical Fuilds, 32 (2004) 287.

[13] I.A. Novoselova, N.F. Oliinyk, S.V. Volkov, A.A. Konchits, I.B. Yanchuk,V.S. Yefanov, S.P. Kolesnik, M.V. Karpets. Electrolytic Synthesis of Carbon Nanotubes from Carbon Dioxide in Molten Salts and their Characterization. Physica E, 40(2008) 2231.

[14] Physicochemistry Teaching and Researching Office in Tianjin University. Physicochemistry. Beijing: High Education Press, 2003,311-314.

[15] Kehe Su, Xiaoling Hu. Physicochemistry. Xian: Northwestern University of Technology Press, 2004.

[16] Suarez P, selbach V, Dullius J, et al. Enlarged Electrochemical Window in Dialkyl – Imidazolium Cation based Room-Temperature Air and Water – Stable Molten Salts. Electrochim. Acta., 1997, 42, 2533 – 2535

Energy Technology Perspectives
Edited by: Neale R. Neelameggham, Ramana G. Reddy, Cynthia K. Belt, and Edgar E. Vidal
TMS (The Minerals, Metals & Materials Society), 2009

SILICON DIOXIDE AS A SOLID STORE FOR CO_2 GAS

Victor Zavodinsky and Sergey Rogov

Institute for Materials Science, 153Tikhookeanskaya, 680042, Khabarovsk, Russia

Keywords: Carbon dioxide, Silicon dioxide, Solid state, Pseudo-potential calculations

Abstract

Pseudo-potential fully relaxed total energy calculations are used to predict a hypothetical $Si_{1-x}C_xO_2$ (x<0.5) compound formed from SiO_2 β-cristobalite by substitution of some SiO_2 complexes by CO_2 molecules. The simulation shows that the $Si_{1-x}C_xO_2$ compound can be quasi stable if the CO_2 content is less than fifty per cent. It is assumed that six-molecule $Si_{6-n}C_nO_{12}$ ($n\leq3$) rings can play a role of nucleuses for formation of the $Si_{1-x}C_xO_2$ compound from SiO_2 and CO_2 molecules. Thus, silicon dioxide can be considered as a possible solid store for gaseous CO_2.

Introduction

The problem of increasing of CO_2 in the Earth atmosphere is one of the main modern problems of mankind. One of conceivable ways to decrease the harmful value of CO_2 is to convert a part of it to solid state. Iota *et al.* [1] reported a polymeric (quartz-like) phase of CO_2 solid synthesized at high pressure (~40 GPa) and at high temperature (~1800 K). Shortly after this experimental work, Serra *et al.* [2] reported a total energy calculation of several possible phases of CO_2 using plane wave pseudo-potential density functional theory. They concluded that there should be a transition from a molecular phase to a phase isostructural to SiO_2 α-quartz in the range of 35 to 60 GPa. The latest calculations by Dong *et al.* [3] have shown that the more favored solid CO_2 phase is β-cristobalite. Recently, Kume *et al.* [4] have demonstrated the polymerization of CO_2 even at room temperature at high pressures up to 80 GPa. Transformations of gaseous CO_2 to different solid phases were studied recently experimentally and theoretically [5-8]. The main obstacle for using synthesized CO_2 solid as a solution for the CO_2 problem is its pressure instability. However the SiO_2 β-cristobalite with the same structure is stable.
The question is whether a silicon-oxide system in which a part of atoms of silicon is replaced by atoms of carbon will be stable? If the answer is positive, this system can be considered, in our opinion, as a possible store for CO_2.

Details of calculations

To study stability of silicon-carbon dioxide we used the generalized gradient approximation and the pseudo-potential plane wave method in the frame of the FHI96md computer code [9]. A β-cristobalite cubic supercell, contained 8 Si and 16 O atoms, was used as a model of silicon

dioxide. The two special *k*-points (1/4, 1/4, 1/4 and 1/4, 1/4, 3/4) and the plane wave cutoff of 40 Ry were applied.
The equilibrium lattice constant for the SiO_2 β-cristobalite supercell was calculated equal to 7.35 Å close to experimental value of 7.16 Å. When some Si atoms were replaced by C (or, in other words, some SiO_2 units were replaced by CO_2 units) all atoms of the modeled $Si_{1-x}C_xO_2$ β-cristobalite supercell were relaxed and a new equilibrium lattice constant was found. In all cases the lattice constant was found less than that for SiO_2 β-cristobalite.

Results and discussions

We analyzed the lattice stability of the $Si_{1-x}C_xO_2$ system (SiO_2 β-cristobalite with some Si atoms replaced by carbon). It has been found that the system kept its stability till content of CO_2 was more than fifty per cent. At larger CO_2 concentrations additional CO_2 units apart from the SiO_2 matrix forming CO_2 molecules and removing from the lattice.
To analyze the energetics of the $Si_{1-x}C_xO_2$ system we have calculated the average binding energies E_{bind} per dioxide molecule (SiO_2 or CO_2), the embedding energy E_{embed} per CO_2 molecule, and the energy barrier $E_{barrier}$ to remove a CO_2 molecule from the β-cristobalite lattice:

$$E_{bind} = \frac{E(Si_{8-n}C_nO_{16}) - (8-n)\cdot E(SiO_2^{mol}) - n\cdot E(CO_2^{mol})}{8};$$

$$E_{embed} = \frac{E(Si_{8-n}C_nO_{16}) - E(Si_8O_{16}) + n\cdot E(SiO_2^{solid}) - n\cdot E(CO_2^{mol})}{n};$$

$$E_{barrier} = E(Si_{8-n}C_nO_{16}) - E(Si_{8-n}C_{n-1}O_{14}) - E(CO_2^{mol});$$

where n is the number of C atoms in the supercell ($n<8$); $E(SiO_2^{mol})$ and $E(CO_2^{mol})$ are energies of free SiO_2 and CO_2 molecules; and $E(SiO_2^{solid}) = \frac{E(Si_8O_{16})}{8}$.
The energy scheme of the CO_2 transfer from the $Si_{1-x}C_xO_2$ system to free molecular state is shown in Figure 1.

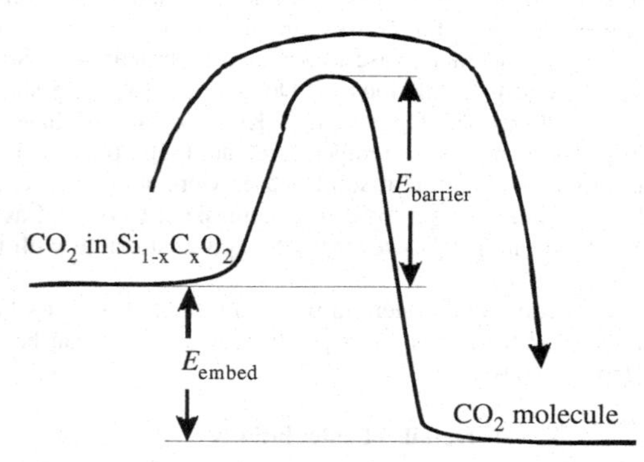

Figure 1. Energetics of CO_2 transfers from the solid silicon-carbon dioxide to the molecular state.

Results of calculations are plotted in Figure 2.

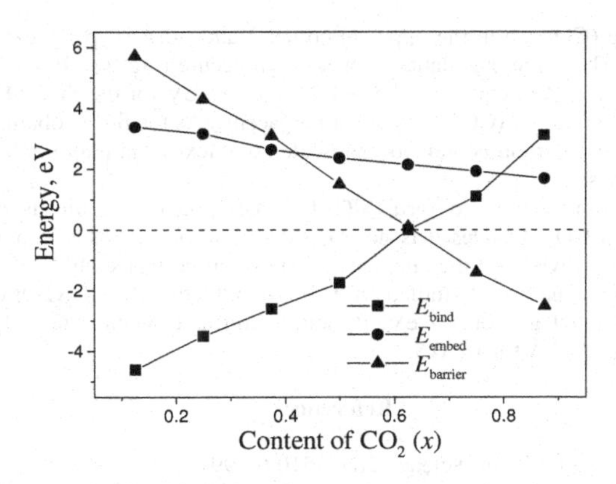

Figure 2. Energy characteristics versus content of CO_2 in the $Si_{1-x}C_xO_2$ solid compound.

One can see that the $Si_{1-x}C_xO_2$ binding energy E_{bind} is negative up to x=0.5. In the other words half of silicon atoms in SiO_2 β-cristobalite can be replaced by carbon atoms without instability of atomic structure. On another hand, the CO_2 embedding energy E_{embed} is positive for all x. It means that CO_2 molecules have a tendency to go out from the β-cristobalite lattice if they can be replaced by SiO_2 molecules. However, the energy barrier $E_{barrier}$ for such transition is positive for $x \leq 0.5$ and is rather large: its value decreases from 5.7 to 1.5 eV when x increases from 0.125 to 0.5. The $Si_{1-x}C_xO_2$ system will be a quasi stable if $E_{barrier}$ is larger than the energy gain (equal to E_{embed}) from transition of CO_2 molecule from solid to gaseous state. This condition is fulfilled for $x \leq 0.375$ when three carbon atoms fall to five atoms of silicon. Thus, the $Si_{1-x}C_xO_2$ lattice can consist approximately on one third of CO_2 solid.
But how can we create this quasi stable compound? Perhaps, directly from molecules. Or from some intermediate nanoparticles.

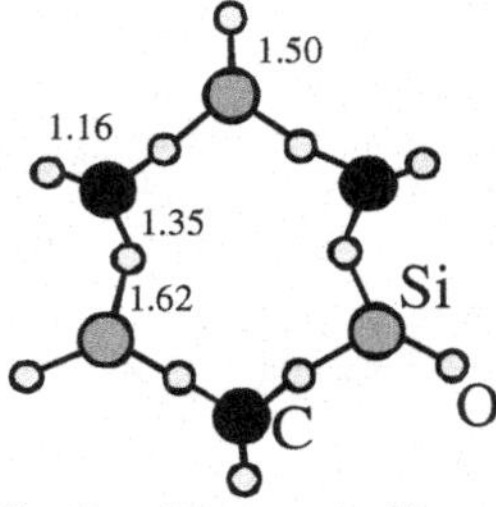

Figure 3. Scheme of $Si_3C_3O_{12}$ ring. Interatomic distances are shown in angstroms.

Our calculations show that CO_2 molecules repel each other, but two SiO_2 molecules attract one to another with the binding energy of –2.5 eV. When two SiO_2 and two CO_2 molecules form a four-molecule ring, such SiO_2-CO_2 system is stable with the average binding energy E_{bind} of

−0.8 eV per molecule (CO_2 are in the opposite corners) and −0.7 eV per molecule (CO_2 are in the nearest corners). The same calculations for a six-molecule ring (see Figure 3) yield similar results. The values of E_{bind} are equal to −1.8, −1.2, and −0.5 eV for the $Si_5C_1O_{12}$, $Si_4C_2O_{12}$, and $Si_3C_3O_{12}$ rings correspondingly (CO_2 units are not adjacent.) As the β-cristobalite lattice contains six-molecule rings, these carbon contained nanoscale complexes can play a role of nucleuses of the $Si_{1-x}C_xO_2$ solid phase.
Certainly, we need some energy to form SiO_2-CO_2 compound, but this is our payment for disposal from gaseous CO_2. Besides it is not excluded, that properties of a new solid material will appear useful, and it will find the application. Those properties should be studied specially. Here we shall only note that in all studied cases the hypothetical $Si_{1-x}C_xO_2$ β-cristobalite is an insulator with a calculated gap of 4-7 eV. Its lattice constant depends linearly on the carbon content and is equal to 6.67 Å for $x = 0.5$.

References

1. V. Iota, C.-S. Yoo, and H. Cynn, Science **283**, 1510 (1999).
2. S. Serra, C. Cavazzoni, G. L. Chiarotti, S. Scandollo, and E. Tosatti, Science **284**, 788 (1999).
3. J. Dong, J. K. Tomfohr, and O. F. Sankey, Science **287**, 11 (2000).
4. T. Kume, Y. Ohya, M. Nagata, S. Sasaki, and H. Shimizu, J. Appl. Phys. **102**, 053501 (2007).
5. M. Santoro, F. A. Gorelli, R. Bini, G. Ruocco, S. Scandollo, and W. A. Crichton, Nature **441**, 857 (2006).
6. V. Iota, C.-S. Yoo, J.-H. Klepeis, Z. Jenei, W. Evans, and H. Cynn, Nature Materials **6**, 34 (2007).
7. A. Togo, F. Oba, and I. Tanaka, Phys. Rev. B **77**, 184101 (2008).
8. J. A. Montoya, R. Rousseau, M. Santoro, F. Gorelli, and S. Scandollo, Phys. Rev. Lett. **100**, 163002 (2008).
9. M. Beckstedte, A. Kley, J. Neugebauer, and M. Scheffler, Comp. Phys. Commun. **107**, 187 (1997).

Energy Technology Perspectives
Edited by: Neale R. Neelameggham, Ramana G. Reddy, Cynthia K. Belt, and Edgar E. Vidal
TMS (The Minerals, Metals & Materials Society), 2009

RECENT DEVELOPMENTS IN CARBON DIOXIDE CAPTURE MATERIALS AND PROCESSES FOR ENERGY INDUSTRY

Malti Goel

Former Advisor and Senior Scientist, Ministry of Science & Technology, Govt. of India and
Adjunct Professor, Jamia Hamdard University, New Delhi 110062, India
Email: malti_g@yahoo.com

Keywords: Carbon capture, Materials, R&D

Abstract

The cost-effective capture of carbon dioxide (CO_2) from the point sources for its reduction the atmosphere offers many challenges in materials science. Novel CO_2 capturing approaches using chemical, physical and biological methods are in the research stage and are aimed to minimize the cost. Appropriate materials development , which can withstand required temperature or pressure as the case may be for CO_2 emanating from coal gas or industrial waste gases form the minimum condition. Other requirements are reclyclability of material and cost of separation. Nano-material composites can be more effective in selective capture of CO_2 and can offer solutions for large-scale separation process. Nano-porous material catalysis can enhance the reaction rate of CO_2 with other chemicals and thus help in capture of CO_2 and convert it into value added products. This paper reviews recent industrial scale developments. In the Indian context, R&D priority areas in CO_2 capture process development with a focus on energy industry are presented.

Introduction

Science of Climate Change is growing at an enormous pace. With the ultimate goal of stabilization of greenhouse gases in the atmosphere, new scientific approaches are being researched worldwide. A shift in policies from low carbon and no carbon energy sources to fossil fuel sources with capture of carbon and its permanent storage is taking place. The process of carbon capture is both energy and materials intensive and the cost of power generation may increase up to 40% when carbon capture and storage (CCS) technology is implemented. Development of cost effective R&D technology therefore remains the greatest challenge before the scientific community.

At the current rate of growth of fossil fuel use, the global CO_2 concentration is expected to double in the next 30 years. Global CO_2 emissions during 2000-2005 have increased 13.4%. from 23.2 Gt in 2000 to 26.3 Gt in 2005 [1]. Electric power and manufacturing industry together have almost 57% share of total CO_2 emissions (Fig. 1). Approximately 8000 point sources have been identified emitting more than 0.1MtCO2 per annum [2]. Levels of carbon dioxide in the Earth's atmosphere are seen to increase at an average annual rate 1 to 2 ppm per year. The average CO_2 concentration in the atmosphere was measured at 379 ppm in 2005 (30% higher than in pre-industrial era). Fourth Assessment Report of UN Intergovernmental Panel on Climate

Change (IPCC) states that "warming of climate system is univocal" and reported an observed rise in temperature by about 0.74 °C in the last 100 years (1906-2005).

Looking at the current energy situation in India, coal is dominant fuel in power generation as well as for industrial production. In the total electricity installed capacity of 138 Giga Watts (GW) in 2007, coal accounted for 62% share. Indian economy is currently growing at a rate of 8 to 9% per annum. India is facing formidable challenges in meeting its energy demand, the

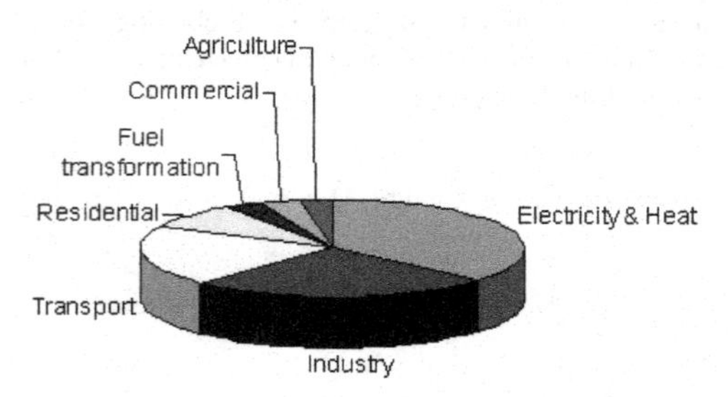

Fig. 1: Global CO2 emissions by sector in 2005 (source- IEA Greenhouse Gas R&D)

process of economic development essentially requires increasing production and consumption of energy. It is projected that the installed power generation capacity would increase by six times by 2031-32 approaching 800 GW and coal requirement will grow at least five times and therefore CO_2 emissions are bound to increase. The total CO_2 emission per annum according to the national greenhouse gas Inventory 1990, prepared by National Physical Laboratory, is 508 million tones from the fuel & industry sectors. In 2003, carbon dioxide emissions from fuel combustion were estimated to be 1049.57 million tones. The energy intensive industrial processes like iron & steel production, oil refineries, cement and others also emit large amount of CO_2. Yet, per capita emissions on average from India are about one twentieth of many advanced countries and one fourth of global average (Fig. 2).

India has adopted United Nation Framework Climate Change Convention (UNFCCC) and has signed Kyoto Protocol in August 2002, which has been ratified in 2005. India's compliance on CDM has led to capturing one-third share in total number of projects registered. India has also made a beginning in Carbon Capture and Storage research. My brief technical association from the Department of Science and Technology, Govt. of India with the carbon sequestration leadership forum, a programme of Department of Energy, USA, has thrown open the challenges of sustaining the coal use for our energy security through application of science & technology [3]. Important challenges ahead of us in research and development (R&D) and need to examine the technical issues in the national perspective was highlighted. In this paper, a review on challenges in materials science in carbon capture research is presented.

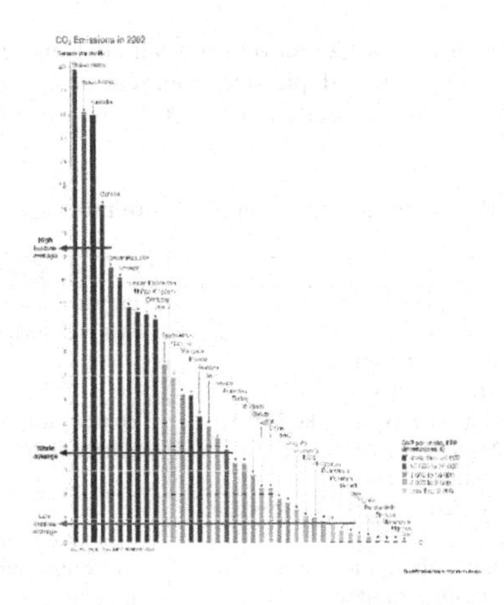

Fig. 2 Per capita CO_2 emissions of Nations of the world

The Science of Carbon Capture

Carbon dioxide (CO_2), is a naturally occurring colorless and odorless gas in the atmosphere. It is one of the major greenhouse gas responsible for maintaining earth's temperature its increasing concentration is contributing to global warming. It has molecule weight of 44 and is a critical molecule for both living and non-living matter on the earth. Carbon monoxide (CO), which is a toxic gas, gets converted into CO_2 upon oxidation. A STM picture on birth of CO_2 molecule from carbon monoxide (CO) and oxygen (O) reaction is shown in Fig. 3.

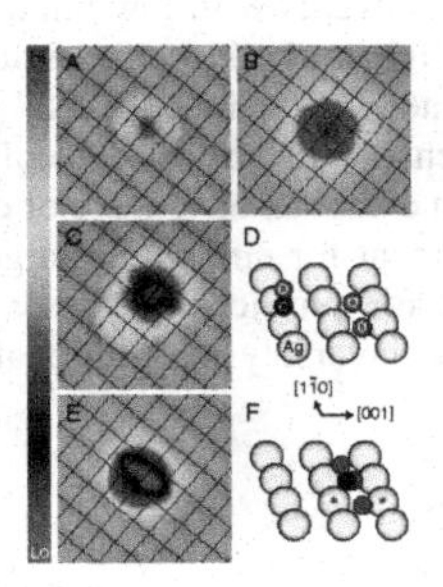

Fig.3: STM of CO2 molecule - Panel (A) shows an isolated carbon monoxide atom adsorbed on a silver surface. Panel (B) shows a pair of oxygen atoms on the surface. (C) shows CO and two O atoms separated and (E) an intermediate O-CO-O complex. The O-CO-O complex ultimately leads to the formation of a CO2 atom and an oxygen atom. Panels (D) and (F) are depicting the arrangement of oxygen and CO for panels (C) and (E) respectively. Source- [4]

This single-molecule experiments has provided important chemical insights into better CO_2 emission control from automobiles, in air purification as well as in chemical sensing and will provide new insights into CO_2 capture. The process of carbon capture at source is not new, CO_2 as gas has been separated from natural gas and in other industrial processes using well developed techniques. However, when these methods are applied to power plant flue gas (containing 6-12% CO_2 in presence of other impurities) the cost of power generation increases immensely. In search for high adsorption capacity adsorbents for CO_2 at power plant environment, a series of physical sorbents, chemical solvents, membranes and physico-chemical sorbents/ solvents have been investigated for CO_2 capture [5]. Desirable process parameters and materials under research are listed in Table I.

Table 1: Post-combustion Carbon capture processes

Process	Solvent Materials	Desirable Parameters
Physical Absorption	Methyl based solvents Selexol, perfluorinated compounds Poymeric sorbents	Temperature >75°C High pressure>3M Pa
Chemical Solvents	Amine based solvents like MEA, DEA, MDEA, TEA Biomimetic solvents Amino functionalized polystyrene	Temperature >40°C Intermediate Pressure
Solid Adsorbants	Zeolites Activated carbon, molecular sieves Metal organic frameworks Composite materials, hydrotalcites Lithium Silicates Nano porous materials	Ambient conditions to different range of temperatures
Mineral Carbonization	Metal oxides like CaO, MgO, Natural Silicates Alkaline and alkaline earth oxides Industrial waste	High temperature High pressure
Biological Capture	Photosynthesis Microalgae, Micro Organisms	Ambient temperature Ambient pressure

Physical processes

Physical capture processes are based on cryogenic cooling or solvents with low binding energy. A cryogenic cooling result in high purity CO_2 stream through multiple compressions and is highly energy consuming process. It has not been demonstrated on 'technical scale' for power plant flue gas [6]. The liquid physical absorption solvents such as polyethylene glycol di-methyl ether (Selexol), fluorinated based and propylene carbonate can create an advantage of continuous process but energy penalty and additional equipment requirement for circulating large volumes of absorbents add significantly to cost. Material selection for physical processes relies on pressure induced recovery and the cost of separation is inversely proportional to CO2 sequestration.

Chemical solvents

A large number of chemical solvents have been investigated for carbon capture. High molecular weight amines are extensively studied and can be loaded with carbon dioxide capture. Monoethanolamine (MEA), diethanolamine (DEA), mixed amines and tertiary amines have been used. As chemical processes rely on heat induced recovery and strong bonding make them less efficient compared to physical processes, but their efficiencies are constantly improving. Polymer system amines like amine functional polystyrene perform better and can be regenerated at 60°C temperature [7]. However, amidine polymer binding processes are not yet fully understood.

Solid Adsorbents

Solid adsorbents are promising as they reduce regeneration and recirculation costs and increase binding capacity for CO_2. Materials such as zeolites, activated carbon, carbon molecular sieves (CMS) are being studies extensively. Novel approaches, materials, and molecules for the abatement of carbon dioxide (CO_2) at the pre-combustion stage of gasification-based power generation point sources include membranes that consist of CO_2-philic ionic liquids encapsulated into a polymeric substrate for permeability and selectivity [8]. Binding energy estimates and thermo gravimetric analysis are used as screening tools to identify possible options. New class of nano-materials, which attracts CO_2 molecules in large number at 220°C, i.e. nearer the temperature at which water gas shift reaction occurs are being developed. Materials for both hydrogen selective membrane and CO_2 selective membrane are targeted.

Mineral Carbonation

Mineral carbonation process is an interesting concept, which involves permanent storage of CO_2 in metal oxides or natural silicate minerals and hazardous solid wastes. Chemical reactions produce carbonate minerals, which are stable and can be deposited on earth. The advantages are that fixation is permanent and the potential is large as the material, either manmade or natural, is freely available. The carbonation process is however energy intensive and need more field studies on the ways to assess its technical feasibility.

Biological Capture

Application of flue-gas treatment using algae bioprocesses for the absorption of CO_2 is a growing field of research. Microalgae acts as a biocatalyst for the photosynthetic conversion of flue gas CO_2 to hydrocarbons that are able to convert a significant fraction of the CO_2 outputs from a power plant into biofuels. Genome analysis of green algae has uncovered hundreds of genes associated with CO_2 capture [9]. New strategies for biology based solar energy capture; carbonic anhydrase enzymes, carbon assimilation and detoxification of soils by employing algae to remove heavy metal contaminants are being discovered. Suitable algal composition for hydrogen production and optimum design of bioreactor using computational fluid dynamics are key area of research. The area is vast and enormous possibilities exist.

Carbon Capture Challenges in Materials research

New processes in carbon capture are based on combination of more than one process significantly reducing energy penalties associated with carbon capture. Novel materials and processes are under development.

Air Capture of CO2

Nanomaterials based metal-polymer frameworks such as chromium terephathalates have shown promise in capturing CO_2 directly from air or flue gas. A nanomaterial structure measuring 2.9 to 3.3 nm with a surface area of 6000m2 is capable of absorbing about 2 mcum of gas [10]. Such carbon capture devices mounted on towers would allow reduction to take place irrespective of where carbon emissions occur, thereby make active management of global CO2.

Chilled Ammonia Process

Energy major ALSTOM has proposed capture of CO_2 using chilled ammonia under ambient conditions from a coal based power plant [11]. It isolates CO_2 in a highly concentrated, high-pressure form. The cooled flue gas flows upwards in counter current to the slurry containing a mix of dissolved and suspended ammonium carbonate and ammonium bicarbonate. More than 90% of the CO_2 from the flue gas is captured in the absorber. The process is energy efficient and can be used for low CO_2 concentrations in the flue gas.

New Material Synthesis

Metal-Organic Frameworks (MOFs) are versatile metal ion clusters having porous structures and are used for gas separation. Using technique of high throughput synthesis used in drug delivery systems and screening of drug molecules, Zeolite Imidazolate Framework (ZIF) as porous metal structures are shown to exhibit unusual selectivity of CO_2 capture from CO_2/ CO mixtures [11]. A no. of metal organic framework compositions are examined in different topologies and a comparative study of gas separation properties have shown that ZIF-69 and ZIF-70. have significantly higher porosity and selectivity as much as five times compared to current commercial materials. It is estimated that 1 lt. of molecular sponge ZIF-69 can hold 83 lt. of CO_2 under ambient pressure and $0^{o}C$ temperature.

Membrane Separation

Fitting membranes are possible to develop for both pre combustion and post combustion capture. There are overlapping areas with solid adsorbents. Polymeric and ceramic membranes are under development for low temperature and high temperature applications. Denaturation card type polyimide membranes, asymmetric hollow yarn membranes, dendrimer membrane with CO2 molecular gate function from pressurized gas are some of the recent developments. In another approach a low cost reusable material hyperbranched polyamidoamine polymer-ultra fine silica hybrid composite is suggested for capturing CO2 from stack gases waste of a coal fired plant [12]. When such a material is heated to 100-120 ^{o}C, the CO_2 can be extracted and absorber could be used again.

Application on Industrial Scale

The material requirements in some energy related carbon capture choices are provided in Table II. Examples from power generation and manufacturing sector are included.

Power Generation

Clean coal technologies in power production are categorized in pre-combustion, oxy fuel and post combustion. These technologies improve efficiency of generation and reduce greenhouse gas emissions per unit of electricity generated. To suit flue gas characteristics, which are heavily dependent on technology, fuel quality and other local environment different materials are required. Pre-combustion options are decarbonization of fuel before combustion, or coal conversion such as coal gasification. Product of coal gasification is synthesis gas comprising of CO and H_2 at high pressure. When reacted with steam products are CO_2 and H_2, the CO_2 is separated and sequestered, while H_2 is sent to gas turbine. Membrane separation and nanoporous solid adsorbents are potential capture processes. DYNAMIS an integrated programme of research in CO_2 capture aims at decarbonised power production using H_2 separation membranes, energy production using novel power cycles and safe storage of CO_2.

Combustion based methods of reduction of CO_2 emissions in the atmosphere are oxy-fuel technology, chemical looping, In-situ gasification and super-critical pulverized coal firing [13,14]. Besides CO_2 capture material, each combustion process has different material needs for igniters and boilers depending on the process parameters. Oxyfuel is the most innovative approach to capture CO_2 from a power plant. It envisages coal combustion with oxygen, in place of air, the flue gas has no nitrogen and is almost pure CO_2, which can be liquified and then separated using membranes. The construction of the 30 MW Vattenfall thermal pilot plant at Schwarze Pumpe in Germany is an important milestone in carbon capture in O2/ CO_2 cycle. It has an oxyfuel boiler and plans to capture 0.1Mt of CO_2 per year.

Full scale demonstration of post combustion CO_2 capture on industrial size power plant has been made under CASTOR project of European Union [15], where flue gases are directed to an absorber, where they are mixed with a solvent having more affinity with CO_2 than those with other components of flue gas, the solvent is then fed to a regenerator. Device is heated to 120°C to break bonds between CO_2 and solvent and CO_2 is then isolated, with solvent re-injected into the absorber.

Table II : Industry Specific Carbon Capture Material needs

Capture process	Energy Cycle	Material needs
H2/CO2 separation	Gasification, IGCC H2/CO2 separation	Silicates, Metal compounds Hydrotalcite porous membranes Material for refractory, Gas turbine
O2/CO2 Cycle	Oxyfuel combustion	Membrane for air separation, Polymers, Ceramic membranes, Metal- organic frameworks
Solid adsorption	Chemical Looping	Calcium Oxides and Carbonates
Chemisorption	Iron & Steel works	Steel slag
Residue treatment	Alumina production	Bauxite waste

Manufacturing industry

Manufacturing industry accounted for about 40% of total energy use in India. Globally direct and indirect CO_2 emissions from industry in 2005 are estimated at 345, about 70 % of these are from manufacturing processes and remaining from fuel combustion. Besides improvement in energy efficiency [16] and use of available waste heat, use of oxy fuel combustion in boilers with the application of post combustion carbon capture process can cut emissions up to 85%. In fertilizers industry carbon dioxide capture is usually adopted for production of urea. First industrial size Carbon Di-Oxide Recovery (CDR) unit in India has come up at Aonla and Phulpur plants of Indian Farmers Fertiliser Cooperative Ltd. (IFFCO). CO_2 captured from the steam reformer flue gases could be captured and utilized to meet the CO_2 requirement for producing urea [17]. The project has 450Mtpd capacity of CO_2 with a turn down capacity of 40% and employs structure packing in the flue gas water cooler and CO_2 absorber and is based on Mistubishi Industries CO_2 recovery technology, which employs KS-1 solvent.

In steel production, a large no. of solid residue are present in the steel slag generated during production through blast furnace, basic oxygen furnace or electric are furnace route [18]. Acidic nature and basicity of metal oxides present show high strength for chemisorption of CO_2 leading to carbonate formation in the following grading CaO, MgO, AzO3, Cr2O3, TiO2, MnO2, iron oxides > SiO2. In aluminum industry, ALCOA has proposed innovative residue treatment for carbon dioxide capture that involves mixing bauxite residue from alumina production containing alkaline liquor. Mixing with CO_2 reduces pH and the resulting product can be beneficially used in applications such as for road base to improve soil strength.

Need for Research Efforts in India

In India, coal is dominant energy fuel. India has adopted sustainable development policy and is taking steps to maximum use of renewable sources and improve energy efficiency for sustainable energy future. National Action Plan on Climate Change, which has identified measures to be taken in eight core areas to maintain high economic growth with the commitment that at no point of time India's per capita Greenhouse gas emission would exceed that of developed countries at present. Several new initiates in Power sector are in place to enhance efficiency of future power plants as well as existing power plants. In addition, research in carbon capture has also begun on novel amines based, multi-phased absorbents and innovative adsorptive materials and processes in different R&D laboratories [19]. R&D efforts are directed to development of cost effective regenerative adsorbents and membrane materials. To develop highly efficient carbon based composite materials as solid adsorbents for CO_2 capture at low temperature (70-200°C) studies and on carbon dioxide adsorption over Ca/Al hydrotalcite and Mg/Al hydrotalcite are being reported in the temperature range 40 to 700 °C, (at different pressures and mol ratio) [20]. Taking the lead Department of Science & Technology, Govt. of India has organized a number of national and two international workshops since 2004 on R&D related issues in CCS. Other Government Departments/ public sector industry has also supported research in the R&D centers across the country. The success of such a programme in the long term would depend on networking and integration of data with user's perspective.

Conclusions

It is noteworthy that carbon dioxide capture research worldwide has been advancing rapidly attracting prompt response from materials scientists, and the interest in field trials on industrial scale demonstration plants has been growing. Climate change is the driving force. Coal is dominant energy fuel and while carbon capture helps in stabilizing of concentrations, it induces high energy penalty to a coal based power generation. Cost-effective carbon capture is still a challenge, worldwide. Regenerability and recyclability of the suitable materials are other issues. Novel material synthesis and processing techniques are developing. New materials and processes in microporous, nano-porous structures, polymers, composites and MOFs are under development.

In India, energy technology is being up-graded for achieving sustainable economic growth and a strategy is required for research aimed at low carbon future. Carbon capture research has just begun. In this context, a survey of CO_2 capture technology is needed to assess technologies capacities. Issues of CO_2 sequestration in some direct and indirect comparisons with other clean technology options are also envisaged. A widespread network research program to manage CO_2 from energy industry will require adoption of techniques to separate and capture CO_2 from the point sources and its utilization. Many R&D challenges exit in Materials science. These challenges could be met through strengthening of research network and taking capacity building measures through an Institutional mechanism.

References:

1. IEA GHG (2007), CO2 Capture Ready Plants, Technical Study 2007/4, May 2007, U.K.
2. B. Metz, et al., Eds., 2005, IPCC Special Report, Working Group III, Carbon Capture and Storage.
3. M. Goel, Carbon Capture and Storage, Energy Future and Sustainable Development: Indian Perspective, in CARBON CAPTURE AND STORAGE R&D Technologies for Sustainable Energy Future, Eds. Malti Goel, B. Kumar & S. N. Charan, 2008, p3-14.
4. http://www.aip.org/png/2001/139.htm
5. M. Gupta, I. Coyl and K. Thambmuthu, 2003, Capture technology and opportunities in Canada, Paper presented at the First of the Canadian Workshop on Co2 capture and storage Technology Research, 18-19 September, Calgary, AB Canada.
6. SA Gattuso, Ph.D thesis, University of Pittsburgh 2007, CO2 Capture by tertiary amidine functional absorbent.
7. A Chakma, *CO2 Capture process opportunities for improved energy efficiency Energy conserves,* Mgmt, 38, 51-56 (1997)
8. H.W Pennline., et al, "Progress in carbon dioxide capture and separation research for gasification-based power generation point sources" *Fuel Processing Technology* 89 (9), 2008 pp. 897-907.
9. P. Lata, Coupled Solar Bio Process for Industrial Effluent Treatment in the Proceedings of the International Conference on Cleaner Technologies and Environmental Management, Pondicherry Engineering College, Pondicherry, India, P.308, No: 57, Allied Publishers Pvt.Ltd. 4th – 6th January 2007.
10. Smith A.R and J. Kbsek(2001), A review of Air Separation Technologies and their integration with Energy Conversion Process, Fuel Processing Technology, Vol. 70, No.2, p115-134.
11. R.Banerjee et al, High-Throughput Synthesis of Zeolitic Imidazolate Frameworks and Application to CO2 Capture.
12. United States Patent 67842, Polymers containing hyperbranched monomers.
13. D Singh et al, *Technoeconomic study of CO2 capture format existing coal fired plant,* Energy Conv. And Management ;44 3073-3097 (2003).
14. M. Goel and Alok Kumar, Eds., 2007, Carbon Capture and Storage Technology: R& D Initiatives in India, , Joint Publication of Department of Science & Technology and Ministry of Power, Government of India, pp 167.
15. *www.co2castor.com/QuickPlace/castor/Main.nsf/h_Index/*
16. M. Goel, Recent Developments In Technology Management For Reduction Of CO2 Emissions In Metal Industry In India, CO2 Reduction Metallurgy, Edited: Neale R. Neelameggham & Ramana G. Reddy, The Minerals, metals & materials Society Publication, USA, p71-82.
17. N. K.Verma, 2008, (Personal communication)
18. D. Bonenfant et al, Molecular analysis of carbon dioxide adsorption processes on steel slag oxides, (Science Direct , Elsevier Ltd. 2008).

19. M. Goel, 2006, Short Term Opportunities and Challenges for CCS in fossil fuel sector: Current Status of CCS In India, In-session workshop on carbon dioxide capture and storage, UNFCC 24th Session of SBSTA, 20th May 2006, Bonn, Germany.
20. K. N. Shinde and A. G. Gaikwad, Studies on carbon dioxide adsorption over Ca/Al hydrotalcite, to be presented at the symposium, CATSYMP-19 (Jan. 2009).

Energy Technology Perspectives:

Conservation, Carbon Dioxide Reduction and Production from Alternative Sources

Ferrous and Titanium Metallurgy

Session Chairs

Jean-Pierre Birat
Malti Goel

Energy Technology Perspectives
Edited by: Neale R. Neelameggham, Ramana G. Reddy, Cynthia K. Belt, and Edgar E. Vidal
TMS (The Minerals, Metals & Materials Society), 2009

REDUCTION OF CO_2 EMISSIONS IN STEEL INDUSTRY BASED ON LCA METHODOLOGY

Ana-Maria Iosif[1], Jean-Pierre Birat[1], Olivier Mirgaux[3], Denis Ablitzer[3]

[1]Arcelor Research, Voie Romaine, BP30320, Maizieres-les-Metz, 57283, France,
[3]LSG2M, Ecole des Mines de Nancy, Parc de Saurupt, F-54042 Nancy cedex, France,

Keywords: CO_2 emissions, LCA, steelmaking, Aspen Plus™ software

Abstract

Integrating environmental considerations into the product traditional process design is now the major challenge for steel industry. Life Cycle Assessment (LCA) is nowadays considered as an appropriate method for assessing environmental impact and selecting new technologies to reduce CO_2 emissions for steel industry.

In this paper we propose a new methodological concept which combines LCA thinking with process simulation software in order to carry out the life cycle inventory of classical steelmaking process.
Using Aspen Plus™ software, a physicochemical model has been developed for the integrated steelmaking route. This model gives the possibility to carry out life cycle inventories for different operational practices in order to optimise the use of energy, to calculate CO_2 and other emissions and to control the mass and the heat balances of processes.

It is also shown that such approach can be used to design and assess new technologies for steelmaking without large industrial application.

Introduction

During recent years, it was recognised that CO_2 emissions are the major factor of the global warming effect. In order to meet Kyoto requirements, the steel industry is embracing the challenge of sustainable development by improving its competitiveness and economic success while reducing its environmental impacts. Integrating environmental considerations into the traditional product design process, for powerful eco-efficiency, is now one of the major priorities for steelmakers.
The European Steel Industry has measured up to this challenge by creating ULCOS (Ultra Low CO_2 Steelmaking) consortium that has taken up the mission to develop breakthrough steelmaking process with the potential to reduce CO_2 emissions (Eurofer, 2005).
In order to develop technologies to reduce emissions, it is necessary to assess the environmental impact of the classical steelmaking route (coke plant, sinter plant, blast furnace, basic oxygen furnace, continuous casting and hot rolling).
Life Cycle Assessment (LCA) method has been undertaken in ULCOS as the most holistic approach of assessing environmental impact and selecting new technologies. Previous to the impact assessment, it is essential to carry out the Life Cycle Inventory (LCI) of the process which is the core part of LCA methodology. According to the current LCA standards (ISO 14040, 1997), the quality of the data used for carrying out this inventory is one of the most important limiting factors. The LCI analysis involves data collection, calculation and procedures

to quantify the relevant inputs and outputs of the product system. The data collection should correspond to some conditions: precision, completeness, representativeness, consistency and reproducibility (ISO 14040, 1997). It is obvious that such conditions are not easy to respect. The environmental performance of complex steelmaking processes such as iron ore sintering are strongly dependent on operational conditions and can change significantly when different types of fuels and recycled wastes are used in the process.

Moreover, data used for inventory calculation should respect also exigencies such are the age, the geographical and technological coverage. In many cases, the environmental data related to the steelmaking industry cannot be used due to the lack of measurement of certain pollutants or because the information is much too summarized. It is clear that LCA practitioners are confronted with serious difficulties in respecting these conditions when the LCI is based only on data coming from literature and/or industrial practice. Also, it is generally recognized that the classical approach of assessing LCI takes time, and usually it cannot guarantee the mass and energy balances of flow rates which are considered in the system boundaries.

That's why it is important to improve the way of assessing the LCI of the steelmaking industry in order to guarantee the quality of the data and to predict the change of the environmental performances with respect to the operational conditions.

New methodological framework proposal

The objective of the current study was to develop a new way of carrying out the LCI of the classical steelmaking route so that the quality of data and the mass and energy balances are guaranteed. The classical route of steel production is based on the production of hot metal from iron ore, and its conversion to steel in a converter.
The proposed methodological framework is based on the interconnection between the environmental tool (LCA) and the process simulation software Aspen Plus™ (Advanced System for Process Engineering) as shown in figure 1.

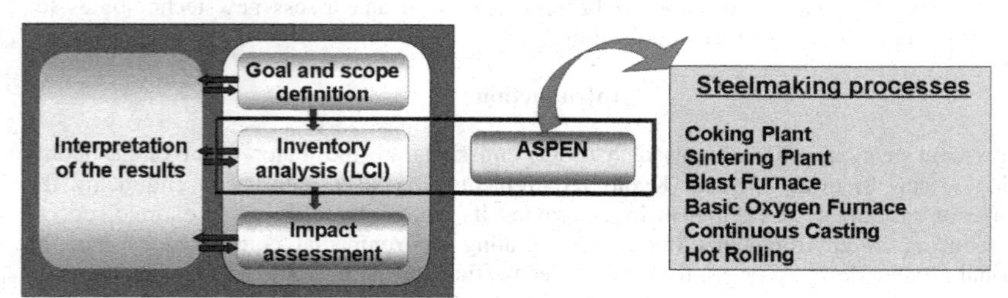

Figure 1. New methodological framework for LCI analysis.

The concept of life cycle assessment originated in the late 1960's when it became clear that the only sensible way to examine industrial systems was to examine their performance, starting with the extraction of raw materials from the earth and tracing all operations until the final disposal of these materials as wastes back into the earth (cradle to grave). The LCA concentrated on energy, raw materials, air emissions, water emissions and solid waste calculations has four main stages as shown in figure 1.
Aspen Plus™ is a process engineering software package that is used to simulate processes based on the thermodynamic models, properties of materials and several ready-made unit operation models.

Based on the proposed approach, the LCI has been carried out for the classical route of steel production. Hence, simplified physicochemical model has been developed for each process of the integrated steelmaking plant as defined by the system boundaries.
In terms of system boundaries, the study covers the foreground processes: coking plant, sintering plant, blast furnace, converter (or basic oxygen furnace) and hot rolling plant. The interconnections of these processes is given in figure 2.

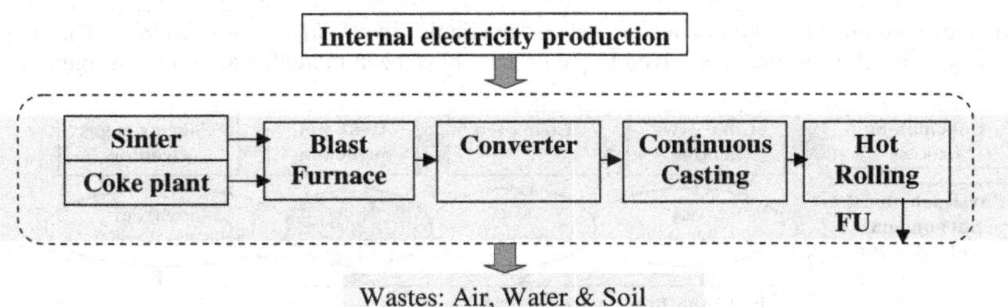

Figure 2: Steelmaking system boundaries.

The continuous casting was excluded from our study because this system is not adapted to Aspen simulation. Actually, the continuous casting process is more characterized by thermal phenomena than by physicochemical mechanisms.
The electricity required to operate the process was also considered in the system boundaries. It was assumed that the electricity requested by the integrated process is supplied by an internal power generation plant which uses steelworks gases (namely blast furnace gas, coke oven gas and converter gas). For the first stage of our study, the system considered does not include the extraction of raw materials, their transportation to the plant and the waste storage.
In order to ensure the comparability of LCA results, the selected functional unit (FU) for the current study is one ton of hot rolled coil produced in a classical integrated steelmaking plant.

Modelling of integrated steelmaking plant

The five main processes characterizing the production of steel have been modeled using Aspen Plus^{TM} commercial simulation software equipped with a thermodynamic database, which is used to design process flow sheets and to calculate mass and energy balances, emissions, and the chemical compositions of products and by-products simultaneously. Finally, the association of the five separately developed modules builds the complete flow sheet of the integrated steelmaking plant. For clarity reasons, the steelmaking process modelling is briefly described in this paper.

Modelling of cokemaking plant

Metallurgical coke is one of the raw materials used in blast furnaces for pig iron production. The main process which characterises coke production is the thermal decomposition of coal in a closed reactor (heating in the absence of air). Coke manufacturing includes the following stages: coal grinding and blending; heating of the coke oven; carbonisation of coal; coke quenching; cleaning the coke oven gas and finally wastewater treatment. During the pyrolysis process, the primary gas leaving the coke oven contains a great amount of tarry matter, considerable moisture and various hydrocarbons compounds.

In the case of cokemaking plant, the facilities considered in the model are: coke oven heating facility, coke oven batteries, coke quenching and coke gas cleaning units.
For simplicity sake and because of their low environmental burden, some facilities were not considered in the frame of the model.

Full presentation of cokemaking model is available in (Iosif et al., 2008[a]). The main physicochemical phenomena involved in the process have been modelled as shown in figure 4.

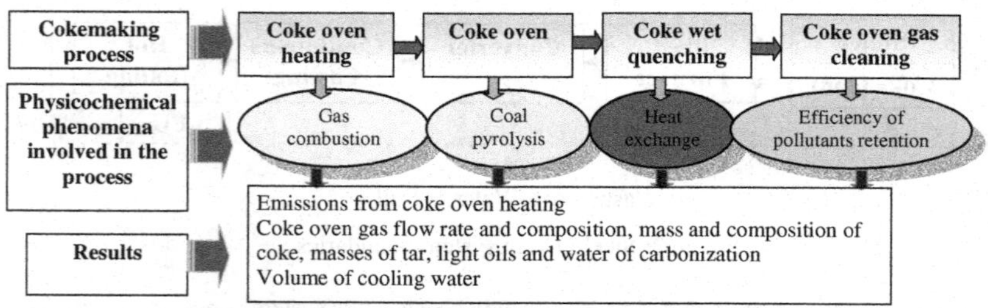

Figure 4. Schematic representation of coke plant modelling.

The input data requested are: the energy supplied for the coke oven heating, the coal mass flow and its elementary composition. Some other parameters are imposed such are the tar average composition, the partition of input of sulfur between the gas phase and solid phase the final temperature of products, etc.
The model provides with good precision the volume and the final composition of the coke oven gas (CO, CO_2, CH_4, C_2H_4, C_2H_6, H_2, N_2, NH_3, H_2S), the mass and the elementary composition of coke (C, H, S, N) and the mass of by-products (tar, light oils and ammonium sulphate). Also the model allows the calculation of CO_2, CO and SO_2 emissions realised during the operation of coke oven heating by the combustion of steelmaking gases (coke oven and blast furnace gases).
Thanks to available industrial data, the model was validated through simulations carried out for various mixtures of coals with different characteristics in terms of chemical composition. Comparison of some results with industrial data are given in figure 5.
Good results have been also obtained concerning the coke oven gas composition. Finally, it was proved that the model is mature enough to be undertaken in a complete Aspen flowsheet developed for the integrated steelmaking plant.

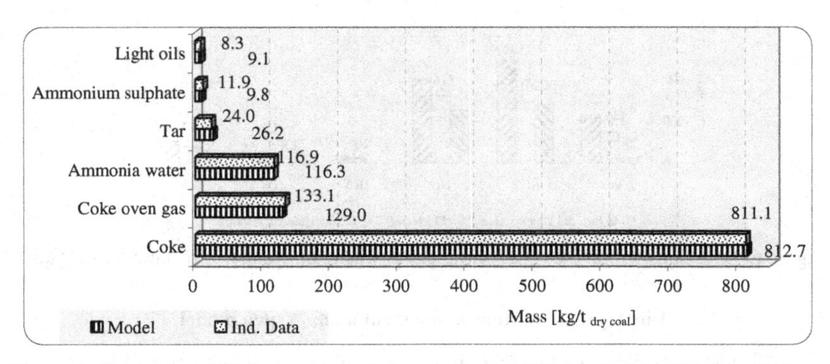

Figure 5. Comparison between coking products calculated by the model and industrial data.

Modelling of sinter plant

The iron ore sintering is the process which allows the preparation of iron ores into the sinter before charging into the blast furnace. Before the agglomeration, the iron ores are first mixed with various substances such as solid fuels, dolomite and lime. The iron ores are agglomerated on conveyor sinter installations thanks to the combustion of a solid fuel.
The sinter plant is considered the most polluting unit among all the steel works and the modelling of this facility is the most complex in comparison to the others. Details about the modelling of this process are available in (Iosif et al., 2008[b]). The processes considered into the module boundary are: the ignition process, the agglomeration of iron ores on the sinter strand, the exhaust gas cleaning and the sinter cooling.
The model has been built as a mathematical matrix, based on chemical reactions and correlations between emissions and different parameters of the process in order to allow an accurate simulation of various mechanisms.
The main pollutants evolved by the system, namely CO, CO_2, NOx, SOx, VOC ($CH_{4\ equ.}$), HCl, heavy metals and dust are the most important outcomes of the model. Also, the mass and the chemical composition of the sinter have been calculated and the energy balance has been attentively checked.
After the model building, simulations of different European industrial configurations have been carried out and the results were successfully compared with industrial data. Some of emissions calculated by the mode are plotted out in figure 6. The other constituents of sintering fumes calculated by the model are: 7.1% CO_2, 1.3% CO, 10.8% H_2O, 77% N_2 and 14.6% O_2.
Using a similar approach, the other processes defined for the integrated steelmaking plant: the blast furnace, the basic oxygen furnace and hot rolling have been modelled with Aspen software. Each module has been validated with industrial data and finally connected together in order to build the integrated steelmaking model. Full details on modelling these processes are given elsewhere (Iosif, 2006).

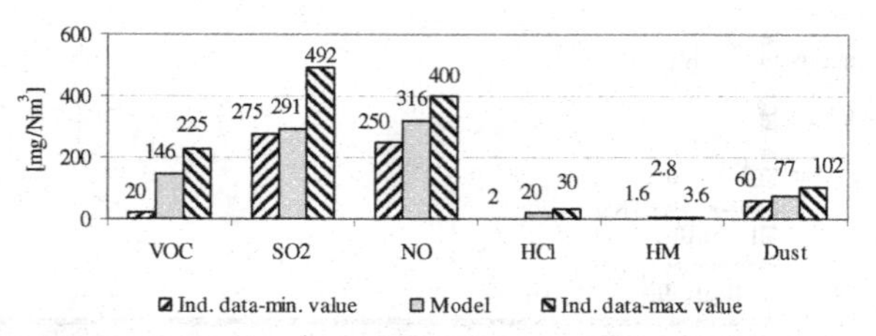

Figure 6: Emissions within the sinter waste gas downstream from electrostatic precipitator.

Life Cycle Inventory assessment using Aspen model

Using the global model and industrial data, the LCI of an existing European integrated steelmaking plant has been successfully calculated. The inventory calculated by the model has been inserted into GaBi software in order to assess the environmental impacts for the analysed case. GaBi allows us to quantify emissions which are not calculated by the model (i.e. uncontrolled emissions). As illustration, a part of the LCI calculated with the model for the given European integrated plant is summarized in table 1.

Table 1. Summary of the integrated steelmaking plant inventory calculated by the model.

Flux	Identification	Unit [a]	Quantity
Materials inputs	Iron ore	kg/FU	1321
	Coal for cokemaking	kg/FU	430
	BF injection coal	kg/FU	154
	Scraps	kg/FU	127
	Pellets	kg/FU	139
	Lime	kg/FU	40
Energy inputs	Internal electricity	MJe/FU	884
Intermediates products	Sinter	kg/FU	1403
	Coke	kg/FU	336
	Hot metal	kg/FU	1020
	Liquid steel	kg/FU	1077
	Slabs/ Blooms	kg/FU	1027
	Coke oven gas	Nm^3/FU	132
	Blast furnace gas	Nm^3/FU	1478
	Converter gas	Nm^3/FU	82
Material outputs (by-products)	Slag	kg/FU	423
	Tar	kg/FU	10
	Ammonium sulphate	kg/FU	4
Product	Hot rolled coil	kg/FU	1000
Emissions to air	CO_2 [b]	kg/FU	1587
Emissions to land	Sintering dust	kg/FU	1

[a] FU: one ton of hot rolled coil ; [b] "gate to gate" system boundaries according to figure 2.

Validation of the new approach for LCI assessment

In order to check the maturity of the developed approach, two Aspen simulations have been carried out using:

a) industrial data supplied by a European integrated steelmaking plant,

b) "benchmark" data supplied by the ULCOS project.

In the frame of the ULCOS project, the "benchmark" data defined an integrated steelmaking plant characterised by the best available technologies.

The inventories calculated by the model for both cases have been compared between them and to one another, considered as a "reference LCI". This reference LCI has been developed by the International Iron and Steel Institute (IISI) to quantify the use of resources, energy and environmental emissions associated with the processing of fourteen worldwide steel plants (IISI, 1998). For all three cases, LCI are reported at the same functional unit derived via the blast furnace/basic oxygen furnace route. The comparison between these three "gate to gate" LCI has been possible thanks to GaBi software. The objective of this comparison was to demonstrate the maturity of the model for given reliable inventories of real cases such as the integrated European plant but also of "virtual" cases such as the ULCOS benchmark. As illustration, in the current paper is showing only the comparison of CO_2 emissions between the three scenarios. This comparison is outlined in figure 7.

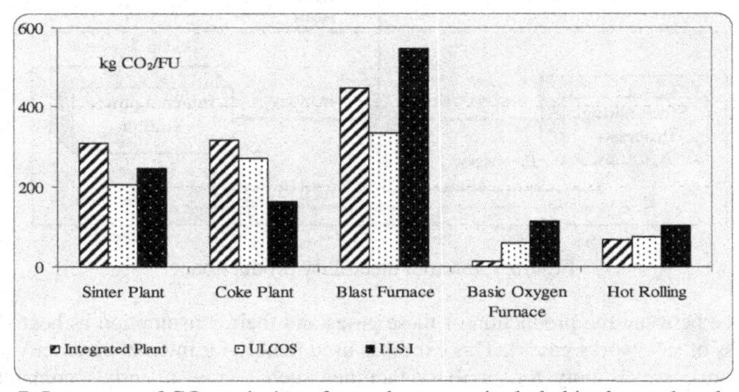

Figure 7. Inventory of CO_2 emissions for each process included in the analysed system.

According to this figure, the total CO_2 emission released by the defined processes is: 1147 kg/FU (European integrated plant); 937 kg/FU (ULCOS benchmark) and 1165 kg/FU (IISI reference case). The CO_2 emissions are calculated for each process without taking into consideration the consumption of electricity. CO_2 variance between the analysed cases is basically linked to the difference masses of raw and intermediate materials (coke, sinter, hot metal…) used in the process. Moreover, the use of various types of fuels for heat supply: blast furnace, coke oven, converter and natural gases, can also contribute to the variation of emission due to the chemical composition and the heat capacity of these gases. As mentioned, the system electricity is supplied by internal production using steelworks gases, namely blast furnace gas, coke oven gas and converter gas, which are considered as by-products of coke, hot metal and steel plants. It is important to stress that the entire amount of steelworks gases produced in the system is not consumed only for electricity generation. In reality, prior to electricity production, some of the gases are used in the system as heat supply for coke oven heating, hot rolling stoves, and blast stoves as indicated in figure 8.

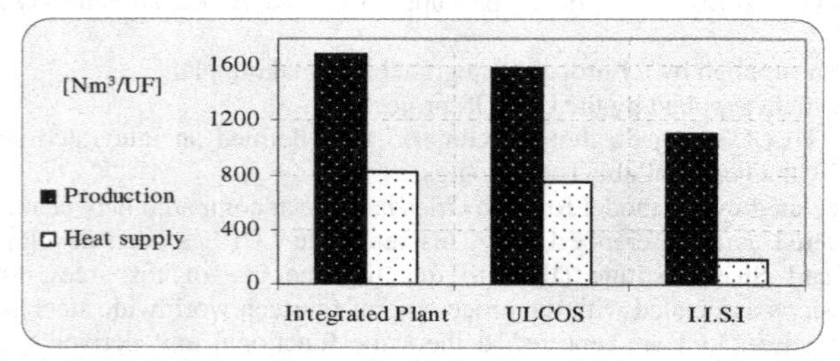

Figure 8. Steelworks gases production and internal consumption as heat supply.

A schematic representation of the steelworks gases used in an integrated plant is given in figure 9.

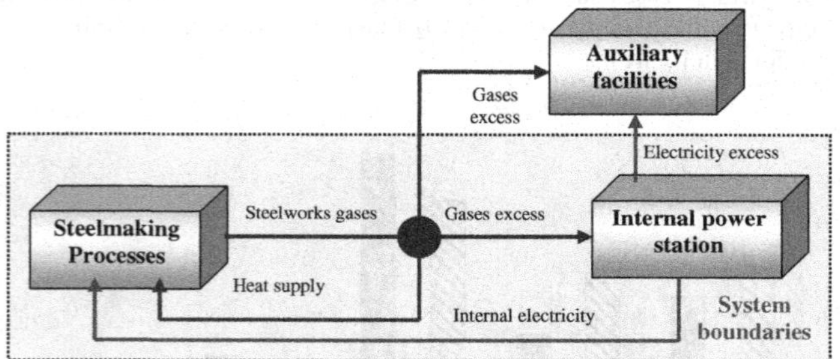

Figure 9. Internal electricity production.

The difference between the production of these gases and their consumption as heat supply was called "excess of steelworks gases". This excess is used mainly for internal electricity production but also as an energy supply for auxiliary facilities such as raw materials preparation, lime production and steam production. For the current LCA study these facilities are not considered in the system boundaries. The production of electricity using steelworks gases was also simulated with a simplified Aspen module developed for an internal power station. The CO_2 emissions calculated with this module for the production of electricity required by the system are given in table 2.

Table 2. CO_2 emissions involved in the integrated steelmaking plant.

CO_2 emission [kg/FU]	**Integrated Plant**	**ULCOS**	**IISI**
Steelmaking Processes	1147	937	1165
Electricity production	440	510	499
System total emissions	**1587**	**1447**	**1664**
Auxiliary facilities	362	293	344
Total	**1949**	**1740**	**2007**

Table 2 also sums the amount of CO_2 released by the production of one ton of hot rolled coil as defined by the system boundaries and all the CO_2 emissions are summarized in figure 10. CO_2 emissions for the ULCOS benchmark are significantly lower than the other scenarios. This can be explained by the fact that ULCOS scenario is based on the consumption of a large amount of pellets in the blast furnace. The environmental burden of pellet production is lower than the burden of sinter production. In addition, in the ULCOS scenario, the coal injection in the blast furnace reaches the maximum rate and consequently the consumption of coke is reduced. Indeed, the decrease of coke demand leads to lower environmental burden for the system. The CO_2 emissions calculated by the model for the European steelmaking plant were successfully compared to the average value of fourteen plants derived from the IISI inventory. This result emphasizes the maturity of the model and fortifies the reliability of the proposed approach.

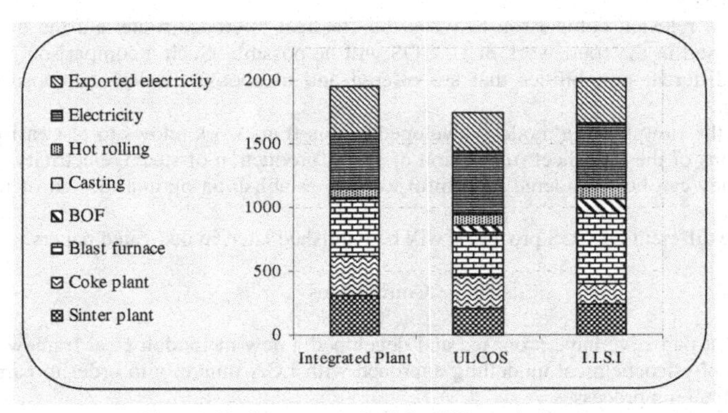

Figure 10. Summary of all the CO_2 emissions.

According to the steel people (Iosif, 2006), the industrial experience shows that there is no excess of energy when all the auxiliary facilities (raw materials preparation, lime production, steam production) are part of integrated plant. If the system boundaries are extended towards auxiliary facilities, the environmental burden involved in the total consumption of steelworks gases in the frame of the system should be integrated into the LCA study (see table 2 for CO_2 emissions). Thanks to the current approach, it has been shown that by using the best available techniques and optimizing the use of resources, the environmental burden can be notably reduced. To exemplify, the CO_2 emissions can be reduced by 200 kg of CO_2/FU for the ULCOS benchmark case. Finally, it is important to stress that the agreement between results obtained for the ULCOS benchmark and the European plant cases validates the current approach.

Modelling of new steelmaking processes

The integrated steelmaking plant constitutes the actual reference for the classic steelmaking route. The modules developed to simulate the different processes of this integrated steelmaking plant were all validated on the basis of industrial data and proved their ability to generate accurate life cycle inventories. Once this method has proved its maturity it can be extended to other systems, which have the ability to be modelled with Aspen Plus^{TM}. In this framework, models for the Direct Reduction route are currently elaborated, on the basis of MIDREX^{TM} process, in order to draw a comparison with the blast furnace route, in terms of CO_2 emissions.

But the main feature of this methodological coupling between flowsheeting and Life Cycle Analysis, lays in the ability to study processes that do not exist yet.
Indeed for processes that do not exist yet, the establishment of the life cycle inventory with the classical data collection is not possible. The most relevant way to produce an accurate life cycle inventory with such processes is to go through a physicochemical modelling step that ensure mass and energy balance, and verify fundamentals laws of physics and chemistry.
In this way, flowsheeting has to be considered as a powerful way to generate LCI and thus allows relevant prospective LCA.
In the framework of ULCOS, several breakthrough processes were imagined and designed to produce iron with limited CO_2 emissions. Those processes are currently being modelled via simplified physicochemical models and integrated into complete ASPEN flowsheets to generate LCI of those new routes.
Once done, a relevant comparison between the classical reference route and the breakthrough routes proposed in the framework of ULCOS will be possible. Such a comparison may help to assess the different possibilities that are offered and to choose the less burdensome for the environment.
In addition the flowsheeting models developed in this framework allows to test easily different configurations of the flowsheet (repartition of gases, production of steam, electricity, ...) and in this way, they can be considered as helpful tools to establish an optimal design of the process flowsheet.
Study of the different ULCOS processes will be published later, in dedicated papers.

Conclusions

In the present paper we have proposed and developed a new methodological framework which combines a physicochemical modelling approach with LCA thinking, in order to carry out the LCI of steelmaking processes.
The integrated classical steelmaking route (via blast furnace/converter) has been modelled with Aspen software and the results were successfully compared with industrial data. It was shown that the validated model is a powerful tool in order to provide a rigorous "gate to gate" inventory. Moreover, the model allows the calculation of the chemical compositions of products and by-products such as the steelworks gases. This information is very important because the steelworks gases are used as fuels by all of the steelmaking processes. Hence, the contribution of these gases to the total environmental burden of the system can be easily estimated. Also, the developed model helps different companies rapidly assess their environmental impacts with respect to their own industrial configuration.
The main advantages offered by this new methodology framework for LCI assessment are the ability to predict emissions for different flowsheets, and to control the mass and the heat balances of the analysed process. Consequently, the quality of data used in LCI is improved remarkably. Moreover, the environmental burden for special conditions such as gas and waste recycling can be rapidly calculated, and the best scenario for each processing route can be easily selected. These attributes give a strong credibility to the calculated inventory and allow LCI analysis to proceed more quickly.
In addition this methodology can be extended to processes that do not exist yet. This allows carrying out the prospective LCA of those processes, in order to be able to choose the less burdensome for the environment. Such a methodology is currently applied for the breakthrough processes which are investigated under the ULCOS umbrella.

References

A-M Iosif (2006): Modélisation physico-chimique de la filière classique de production d'acier pour l'analyse de l'Inventaire de Cycle de Vie. PhD, INPL, 68-95.
A-M Iosif, F. Hanrot, D. Ablitzer (2008[a]): Application de l'inventaire du cycle de vie en sidérurgie. Techniques de l'Ingénieur, volume M7 160, p. 1-10.
A-M Iosif, F. Hanrot, D. Ablitzer (2008[b]): Process integrated modelling for steelmaking Life Cycle Inventory analysis. Environmental Impact Assessment Review, volume 28, issue 7, p. 429-438.
Eurofer. Steel strategic research agenda to 2030, EU steel technological platforms. Brussels: European Commission; 2005. Eurofer.
IISI.Worldwide LCI database for steel industry products. International Iron and Steel Institute; 1998.
Standard ISO 14040 (1997): Environmental management. Life Cycle Assessment. Principles and framework, ISO 14040.

Energy Technology Perspectives
Edited by: Neale R. Neelameggham, Ramana G. Reddy, Cynthia K. Belt, and Edgar E. Vidal
TMS (The Minerals, Metals & Materials Society), 2009

Electrolytic reduction of ferric oxide to yield iron and oxygen

A Cox[1] and D.J. Fray[1]

[1]Department of Materials Science and Metallurgy,
University of Cambridge, Cambridge CB2 3QZ
United Kingdom

Keywords: Ferric Oxide, Reduction, Electro-deoxidation, Sodium Hydroxide

Abstract

Hematite pellets were electro-deoxidized to iron in molten sodium hydroxide to produce iron containing about 10-wt % oxygen. The anodic reaction was the evolution of oxygen on an inert nickel anode. Before the electrochemical reaction, there is a reaction between sodium hydroxide and ferric oxide to form sodium ferrite and, in order to compensate for the reduction in sodium oxide content, 1 wt % sodium oxide is added to the melt. In a laboratory cell, the cell voltage was 1.7 V with a current density of 5000 A m^{-2}, a current efficiency of 90% and an energy consumption of 2.8 kWh kg^{-1}.

Introduction

The iron and steel industry contributes about 2.3 bn tonnes/annum of the carbon dioxide to the atmosphere, which is about 10% of the total, and some believe that, in order to prevent a catastrophic rise in the earth's temperature, carbon dioxide emissions must be reduced by 50% come 2050. However, by 2050, it is likely that the iron and steel industry would have grown significantly so that the production may have risen to 2 or 3 bn tonnes/year so in order to meet the 50 % reduction on today's figure, the carbon dioxide may have to be reduced to 25% or less, which is a very demanding requirement. There are probably three ways of reducing the carbon dioxide output:

1. By capture and sequestration, but this may be technically very demanding.

2. Reduction using hydrogen in place of carbon.

$$3H_2 + Fe_2O_3 = 3H_2O + 2Fe \quad (1)$$

However, the reaction does not go to completion and most of the world's hydrogen is derived from methane with carbon monoxide and carbon dioxide as by-products[1]. Hydrogen could be produced by the electrolysis of water but this approach is not commercial at the present time and would require two processes, the electrolysis of water followed by the reduction of iron oxide with the hydrogen.

3. The use of electrolysis to directly reduce iron oxide to iron and oxygen using electricity from renewable sources. Obviously, this is not possible at the present time but it has been estimated that by 2050, a significant contribution to electricity generation will be made by renewable energy.

There have been two approaches to the electrowinning of iron, one is aqueous electrowinning [2] and the other is the dissolution of the iron oxide in molten salt and its subsequent electrochemical reduction [3]. There are several challenges with these approaches in that the iron can exist in different valence states which means that the reduction of ferric ions to ferrous can occur at the cathode. The ferrous ions then diffuse to the anode where they are oxidized back to the ferric state with the net result that electricity is consumed but no iron is produced. This problem can be overcome by surrounding the anode with a diaphragm and although this is feasible in aqueous solutions it is particularly demanding in molten salts that have been selected to dissolve oxides. There are also problems on the cathode side as in order to get a planar type deposits a low current density is required (250 A m^{-2}) in both aqueous and molten salt electrolytes. Lastly, in aqueous solutions, it is relatively easy to find a material that can act as an inert anode to evolve oxygen but, in molten oxide melts, this problem is far more demanding. The challenges to be overcome are to avoid the effects of variable valency, to run the cell at a significant current density to create a reasonable production rate and lastly to find a suitable inert anode.

Several years ago, the concept of electro-deoxidation was developed [4], in which instead of dissolving an oxide in an electrolyte, the oxide is made the cathode and on the application of a voltage, the oxygen in the oxide ionizes, dissolves in the electrolyte and diffuses to the anode where it is discharged. Since in this process ferrous/ferric ions are not present in the electrolyte, the valency of the cation in the oxide is irrelevant and, in fact, is an advantage as most oxides containing mixed valance cations have high electronic conduction. It may be possible to apply this concept to an aqueous electrolyte but the selection of the electrolyte is important as although ferric oxide is not a particularly stable oxide, its stability is similar to that of water and, therefore, hydrogen is likely to be evolved at the cathode as well as the reduction of the iron oxide, thereby, decreasing the current efficiency. The original work on electro-deoxidation used molten calcium chloride as the electrolyte but it has been difficult to find an inert anode that is stable in chloride melts.

A process that operated very successfully in the early part of the last century was the Castner process for the production of sodium from sodium hydroxide using a nickel inert anode [1] This process was highly reliable and was only discontinued due to the fact that hydrogen was co-deposited at the cathode making the maximum current efficiency for sodium only 50 %. Given the proven fact that nickel is an excellent anode in this melt, it is worth giving consideration to molten sodium hydroxide as the electrolyte for the electro-deoxidation of ferric oxide. Sodium hydroxide can dissociate into sodium oxide and water:

$$2NaOH = Na_2O + H_2O \tag{2}$$

$$K = \frac{(a_{Na_2O}).(a_{H_2O})}{(a_{NaOH})^2} \tag{3}$$

and either of these species can be electrochemically decomposed depending on their activities and the potential on the cathode.

Sodium hydroxide will react with ferric oxide to form sodium ferrite which is stable in sodium hydroxide:

$$Fe_2O_3 + 2NaOH = 2NaFeO_2 + H_2O \quad (4)$$

The effect of this is to increase the activity of water which obviously affects the decomposition potential of the melt. Table I shows the decomposition potentials for H_2O, Na_2O and $NaFeO_2$ at various activities at 530°C.

Table I: Decomposition voltages for H_2O, Na_2O and $NaFeO_2$ at various activities at 530°C.

	a(H_2O)	**V°(H_2O)**	**a(Na_2O)**	**V°(Na_2O)**	**V°($NaFeO_2$)** a($NaFeO_2$) = 1
Unit activity species	1	0.862	1	1.294	1.368
[1]**Neutral NaOH**	5.00×10^{-6}	1.284	5.00×10^{-6}	1.716	1.227
NaOH + 1wt% Na_2O	1.74×10^{-10}	1.640	1.52×10^{-1}	1.359	1.335

As can be seen, at unit activities water will decompose first, followed by sodium oxide and then sodium ferrite. In neutral NaOH, the sequence should be sodium ferrite, water and then sodium oxide but when a small amount of sodium oxide is added, the sequence is sodium ferrite, sodium oxide and then water. Given the fact that sodium ferrite will always form, and remove sodium oxide from the sodium hydroxide, it is important to have a slight excess of sodium oxide to prevent the activity of the water rising and allowing the decomposition of water to take place. This work reports preliminary measurements on this system.

Experimental Procedure

High purity ferric oxide (Aldrich) with a particle size of 1-5 μm was pressed at 20 tonnes in 20 mm die. The pellets were sintered in the range 900 to 1300°C for 30 minutes to form semi dense pellets. A single pellet (porosity in the range 15-20% except for experiment 6) was held in a small nickel basket for each experiment which was placed in a nickel crucible containing about 600g of sodium hydroxide plus 6g Na_2O. The crucible was placed in an Inconel 600™ reactor through which argon was passed to sweep out the evolved hydrogen (Figure 1).

A two-electrode arrangement was used with a Thurlby Thander PL154 power supply and the current was determined by measuring the voltage drop across a precision resistor. A constant voltage was applied.

After reduction, the nickel basket containing the pellet was raised to the cooler part of the reactor and left to cool. It was then placed in water or a mixture of ethanol and water to dissolve the residual sodium hydroxide which was checked by measuring the pH of the wash liquid.

The effect of temperature, time, applied voltage and sodium oxide additions were investigated.

[1] Based on Lux-Flood Scale of pO^{2-}

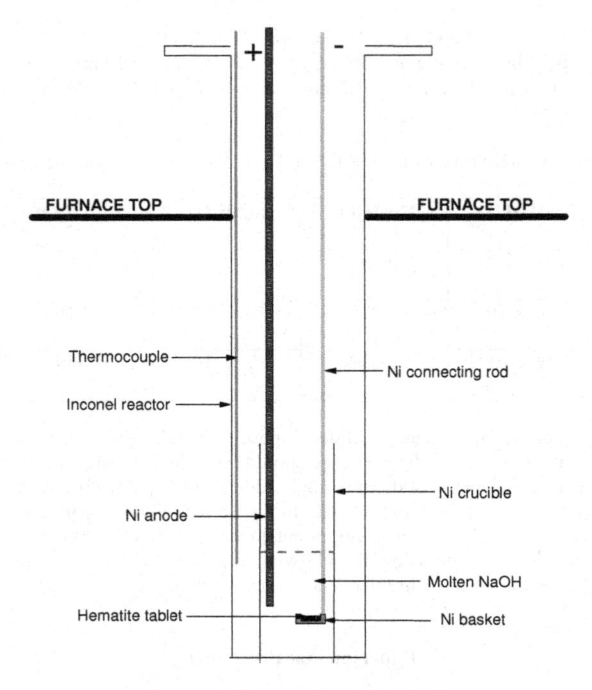

Figure 1: Electro-deoxidation cell

The precursors and the reduced pellets were characterized by scanning electron microscopy (SEM, JOEL 5800), electron dispersion analysis by X-rays (EDAX Oxford Instruments INCA x-sight 6587) and X-ray diffraction analysis (XRD Phillips X'Pert PW3020) and Inductively Coupled Plasma Spectrophotometry (ICPAES, Varian axial Liberty Series II).

Results

Sintering of pellets

Pellets sintered below 1050° C were too fragile to use whereas those sintered above this temperature were less porous but remained intact during the reduction. When sintering at 1300° C, the shrinkage was 43-vol % which can be compared with an 8.5-vol % at 950° C.

Reduction Experiments

The conditions and results of the reduction experiments are shown in Table II and comments summarized in Table III. The microstructure of the iron oxide and the reduced iron are shown in Figure 2 and the EDS spectrum shown in Figure 3 confirms that the product is iron.

Figure 2a: Sintered oxide

Figure 2b: Iron powder

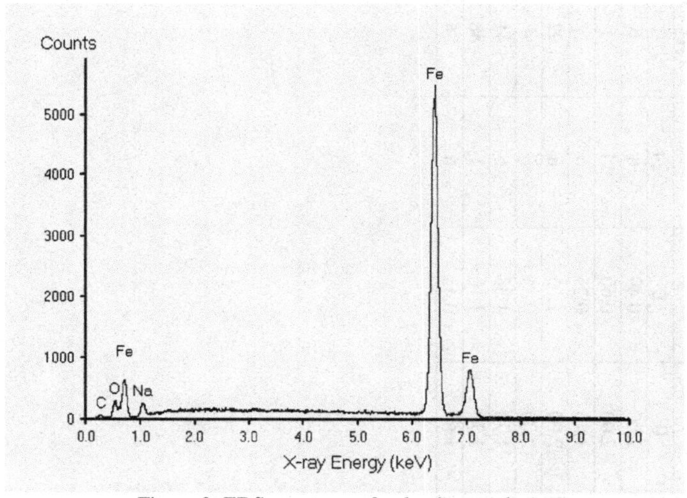

Figure 3: EDS spectrum of reduction product

The results show that although sodium ferrite readily forms at around 400°C, there is little electrochemical reduction indicating that the kinetics are slow which is in accordance with the work of McKewan who found that the kinetics for the reduction of hematite using hydrogen as a reductant was very slow below 500°C [5]. On increasing the temperature to 530°C but applying a lower voltage showed that little reaction occurred which was due to IR and polarization losses reducing the actual voltages on the electrodes below the decomposition potential of the sodium ferrite, sodium oxide or water. On increasing the potential to 1.7 V or greater, the ferrite was reduced but, at the same time, significant hydrogen was evolved which lowered the current efficiency and increased the energy consumption. As shown in Table 1, the addition of sodium oxide increases the activity of the sodium oxide and decreases the activity of water so that the decomposition voltage of water is not exceeded so that the reduction of sodium ferrite is the favored reaction. As can be seen, the addition of sodium oxide is highly beneficial in terms of current efficiency and yield.

Table II: Experimental conditions and results

Experiment Number	Electrolyte Composition	Temperature (°C)	Applied Voltage (V)	Steady Current (A)	Time (Hr)	Iron Content (Wt%)	Current Efficiency (%)	Energy Consumption (kWh kg^{-1})
1	NaOH	450	0	0	24	0	-	-
2	NaOH	400	2.00	0.49	6	6	-	-
3	NaOH	420	2.00	0.60	21	1	-	-
4	NaOH	530	1.40	0.25	6	5	-	-
5	NaOH	530	2.00	1.65	6	92	19	15.3
6[2]	NaOH	530	2.00	0.55	6	4	-	-
7	NaOH	530	1.70	1.33	6	41	32	-
8	NaOH + Na_2O	530	1.70	1.10	1	88	89	2.8
9	NaOH + Na_2O	530	1.60	0.45	6	86	78	3.0

[2] Porosity of pellet was only 4-vol%

Table III: XRD analysis and summarizing comments

Experiment Number	Non-metallic Phases Present	Comments
1	$NaFeO_2$, Fe_2O_3, Na_4FeO_3, NaOH	Major Phase: $NaFeO_2$ and Minor Phase: Na_4FeO_3. Formed spontaneously.
2	Fe_2O_3, $NaFeO_2$	$NaFeO_2$ reduction kinetics too slow. H_2 formation from H_2O decomposition occurred.
3	Fe_2O_3, $NaFeO_2$, Fe_3O_4	$NaFeO_2$ reduction kinetics too slow. H_2 formation from H_2O decomposition occurred.
4	Fe_2O_3, Na_2FeO_4, $NaFeO_2$, NaO_2, $Fe_{0.925}O$, Fe_3O_4	Voltage not sufficient to decompose $NaFeO_2$.
5	$NaFeO_2$, Fe_3O_4, Fe	Good yield but competition from H_2 formation leads to poor current efficiency.
6	Fe_2O_3, $NaFeO_2$	Very dense pellet, reduction restricted to surface, unreduced in bulk of pellet.
7	FeOOH, Fe_2O_3, $NaFeO_2$, Na_2FeO_3, Fe	Current efficiency improved but poor yield and H_2 formation still evident.
8	$NaFeO_2$, Fe_3O_4, Fe	Excellent current efficiency and yield. H_2 formation suppressed due to very low H_2O activity.
9	$NaFeO_2$, Fe_3O_4, Fe	Possesses attributes of Exp.8 but slower rate of reduction due to approach of minimum voltage to reduce $NaFeO_2$.

Experiment 6 shows the importance of having a pellet with significant open porosity so that the electro-deoxidation reaction can take place through out the pellet rather than just on the surface which would be the case of a pellet with only 4-vol % porosity.

Discussion

From the phases found during the experiments, it is concluded that the initial step is a chemical reaction between the sodium oxide in the sodium hydroxide to form sodium ferrite.

$$Fe_2O_3 + Na_2O_{(l)} = 2NaFeO_2 \qquad (5)$$

This phase then undergoes reduction to form several sodium iron oxide compounds before being finally reduced to iron.

$$2NaFeO_2 + Na_2O_{(l)} + 2e = Na_{2+2x}FeO_{2+x} + O^{2-} + FeO \qquad (6)$$

$$Na_{2+2x}FeO_{2+x} + FeO + 4e = 2Fe + 2O^{2-} + (1 + x)\ Na_2O_{(l)} \qquad (7)$$

The overall cathodic reaction is:

$$Fe_2O_3 + 6e = 2Fe + 3O^{2-} \qquad (8)$$

In the melt the oxide ions react with the water in the melt to form hydroxyl ions

$$3O^{2-} + 3H_2O = 6OH^- \qquad (9)$$

which are then oxidized at the anode to liberate water, which stays in the melt, and oxygen:

$$6OH^- - 6e^- = 3H_2O + 1\tfrac{1}{2}\ O_{2(g)} \qquad (10)$$

The overall cell reaction is:

$$Fe_2O_3 = 2Fe + 1\tfrac{1}{2}O_2 \qquad (11)$$

The nickel anode proved to be particularly successful in that no contamination of the melt by the anode after many hours of use was detected. The anode was unaffected except for a very thin layer of black oxide. This is accordance with the results from the Castner Cell [1].

From the thermodynamics it would appear that with the addition of 1 wt% sodium oxide that reduction of sodium ferrite becomes the predominant reaction on the cathode. This inference is verified by the fact that neither hydrogen nor sodium readily reduce sodium ferrite, that the product at the cathode is predominantly iron and that the current efficiency is very high (~90%). The composition of the melt changed when the hematite pellet was introduced to the melt, as the hematite reacted with some of the sodium oxide in the melt. During the electro-deoxidation process further sodium oxide was taken up by the pellet but in the final reduction step, the sodium oxide was returned to the melt.

From the results, an applied voltage of about 1.7 V is sufficient to reduce sodium ferrite, the current density can be up to 5000 A m^{-2} with a current efficiency close to 90%. This gives an energy consumption of 2.8 kWh kg^{-1}. The rate of reduction works out at approximately 3 $mg/cm^2/min$ which is close to the rate of reduction at the same temperature of hematite by hydrogen [5]. Obviously, these are preliminary results but it does give an indication of the possibilities of this approach when operating at higher temperatures where the rates of electro-reduction are likely to be higher.

The product from the process is very similar to directly reduced iron with about 10 wt % oxygen remaining in the product but, unlike iron from the blast furnace, is totally free of carbon.
The prospects for scale-up are good as the reduction was performed on a significantly sized pellet, rather than an individual grain. It can be envisaged that a cell would consist of a cathode basket containing a packed bed of pellet with the reduction starting at the periphery of the basket and as the products are conducting, the electro-deoxidation reaction would gradually move through the bed.

Conclusion

Hematite was successfully reduced to iron in molten sodium hydroxide at 530°C by the method of electro-deoxidation. The reduction was performed below the decomposition voltage of the electrolyte provided the activity of sodium oxide was enhanced in order to decrease the activity of water in the melt and prevent its electro decomposition. For this simple cell, the operating voltage 1.7 V which resulted in a current efficiency of 90% and an energy consumption of 2.8 kWh kg^{-1}.

Acknowledgements

The authors are grateful to the EPSRC for financial support.

References

1. F. Habashi, ed., *Handbook of Extractive Metallurgy* (Chichester, Wiley, 1997)

2. F. A. Lowenheim, ed., *Modern Electroplating* (New York, Wiley, 1974)

3. G. M. Haarberg, private communication, University of Trondheim, 2005

4. G.Z. Chen, D.J. Fray and T.W. Farthing, "Direct electrochemical reduction of titanium dioxide to titanium in molten calcium chloride" *Nature* 407 (2000), 361-364

5. W. M. McKewan. "Kinetics of Reduction of Iron Ores" *Iron and Steelmaking* Chipman Conference, eds. J. F. Elliot and T. R. Meadowcroft (Cambridge, MIT Press, 1962), 141-155

Energy Technology Perspectives
Edited by: Neale R. Neelameggham, Ramana G. Reddy, Cynthia K. Belt, and Edgar E. Vidal
TMS (The Minerals, Metals & Materials Society), 2009

Enhanced Energy Efficiency and Emission Reduction through Oxy-Fuel Technology in the metals industry

Jupiter Oxygen Corporation, Dietrich Gross – CEO, Mark Schoenfield – COO, Thomas Weber – VP Marketing, Brian Patrick – Director of Development, Norman Bell[1] – Chief Project Engineer

TMS (The Minerals, Metals & Materials Society);
184 Thorn Hill Rd.; Warrendale, PA 15086-7514, USA

Jupiter Oxygen Corporation, 4825 N. Scott St., Schiller Park, IL, 60176

Keywords: Oxy-fuel, Carbon Dioxide Reduction, Sequestration, Energy Savings

Abstract

The reduction of carbon dioxide emissions and the resultant greenhouse gas affect is hampered by the fact that all fossil fuels contain copious quantities of carbon. Process tweaking can keep the amount of carbon dioxide to a minimum but can not significantly reduce the amount of carbon dioxide generated. The very nature of utilizing fuels requires that they be burned to release the heat required for the various processes. Reduction of metal oxides to manufacture or modify pure metals, generation of power, transportation and many other forms of fuel usage result in much of the generation of man made carbon dioxide. Oxy-fuel is a unique contributor to carbon dioxide emissions. While creating higher concentrations of carbon dioxide the total volume of carbon dioxide generated is reduced. This paper will examine the unique characteristics and potential of oxy-fuel that make the generation of lower volumes of carbon dioxide possible.

Nearly ten years ago a commercial application of oxy-fuel was installed at Jupiter Aluminum Company in Hammond, Indiana. This process was funded and developed by Dietrich Gross CEO. This aluminum remelt facility now runs completely on oxy-fuel and has continuously for over ten years. This has reduced the melt rates to as low as 750 BTU per pound in continuous melting periods. The best alternate processes can achieve in 2008 is around 1350 BTU per pound with significant capital and maintenance requirements not required by

[1] Presenting at TMS 2008; E-mail; normanb@jupiteroxygen.com

oxy-fuel burners or ancillary equipment. Several factors influence why oxy-fuel is able to reduce the volume of carbon dioxide generated:

1. **Furnace Retention Time** – Since oxygen and fuel are the only inputs to the furnace the volume of flue gas has been reduced by approximately 80% through the elimination of nitrogen in the combustion air. This translates to longer furnace retention time for the flue gas and better heat transfer resulting in lower stack temperatures for the flue gas exiting the furnace.
2. **More Uniform Heat Transfer** – The green house gas effect of carbon dioxide is a very valuable asset within the furnace. Due to superior radiation ability compared to nitrogen a carbon dioxide rich atmosphere is an excellent atmosphere to transfer heat to metals. Temperature uniformity is excellent with very little variation anywhere within the heating area.
3. **Higher Flame Temperature** – Oxy-fuel flame temperatures are among the hottest available. Properly utilized and placed they allow the radiant effect of carbon dioxide molecules to disperse the heat of combustion in a very efficient manner.
4. **Heating Efficiency** – As a result the efficiency of the furnace is substantially increased resulting in a lower input required to do the same amount of work.

Table one indicates the advantages from installing oxy-fuel on the aluminum melt and holding furnaces at Jupiter Aluminum Company.

Year	Estimated Casted Aluminum Pounds	Production FCE Natural Gas MM Btu	Waste Oil MM Btu	Waste Oil Gallons	Percent Oxy-fired	CO2 - Oil TPY	CO2 - NG TPY	CO2 TPY
1997	134,797,919	605,370.38	**0**	**0**	**0.00%**			
1998	164,041,703	570,063.62	0	0	**26.90%**	0	32,001	32,001
1999	181,984,940	441,750.48	0	0	**46.00%**	0	24,798	24,798
2000	184,773,584	400,809.52	0	0	**53.50%**	0	22,499	22,499
2001	161,286,676	125,018.00	84,516	612,742	**100.00%**	5,930	7,018	12,948

Table 1: OXY-FUEL TECHNOLOGY IMPLEMENTATION AT JAC 1998 TO 2001

Note that the tonnage in column one increased by approximately 12.5% over the time period. In addition the over all BTU reduction per year firing both natural gas or oil dropped by nearly 80% and that carbon dioxide emissions dropped to approximately 40% of the amount before conversion to oxy-fuel. There are very few methods through which a company can maintain current production levels while reducing carbon dioxide emissions by 60%. Currently all emissions substantially exceed AP-42 standards by a wide margin.

NATURAL GAS OXY-FUEL			
	AP-42	**JUPITER**	**% REDUCTION**
Nox	0.098039	0.000494	**99.50%**
Sox	0.000588	0.000326	**44.56%**
CO	0.082353	0.023507	**71.46%**
VOC	0.005392	0.000598	**88.91%**

Table 2: RESULTS FROM TEST FURNACE OPERATING ON NATURAL GAS

The result for the oxy-fuel process user is a much more efficient process combined with carbon credits that can be banked for future expansion or sold on a carbon exchange.

As a result of this work Jupiter Oxygen Corporation was formed to market this patented process to many industries. In addition to aluminum remelt, current research in conjunction with the Department of Energy – NETL is centered on the usage of oxy-fuel for boiler applications. Jupiter Oxygen Corp. has very successfully fired a 15 MW boiler at our Hammond research facility utilizing natural gas and oxy-fuel. Currently testing is being done utilizing Illinois #6 coal with oxy-fuel. This is showing very promising results. According to DOE-NETL personnel the Sox generated is not the very aggressive SO3 form of Sox which is very important as the Sox generated is much easier to contain and remove from the flue gas. A slip stream DOE-NETL process titled "Integrated Pollutant Removal System" is also being evaluated with the boiler. This process lends itself very well to a process that only generates carbon dioxide and water as the components in the flue gas. This system partially compresses the gas stream causing the water vapor to condense stripping most of the Sox as dilute acid. The remaining component is a high percentage carbon dioxide stream that through further compression can be liquefied. The total volume of flue gas generated is less than 20% of a normally aspirated boiler making sequestration much more affordable. In addition through the use of the proper equipment a food grade purity of carbon dioxide can be produced. In many areas the carbon dioxide stream can become a revenue generator rather than a liability.

This process is not limited to the aluminum remelt producer. Current projects under consideration on a world wide basis include a side fired walking beam billet furnace in a major steel mill and other steel mill furnace applications as well as commercial and industrial power plants.

When Jupiter Aluminum Corporation undertook this project a cryogenic oxygen plant was purchased and is operated by Jupiter Aluminum Corporation. This remains the most economical method of obtaining oxygen. Recent events have allowed Jupiter Oxygen Corporation to partner with a major supplier of air plants. This agreement provides the end user with very low cost oxygen in return for a long term contract of ten years or longer. The installer of oxy-fuel now has the benefit of the carbon dioxide reduction without the need to become versed in the operation and maintenance of an air plant or provide personnel. The result is a much lower capitalization entry point to install oxy-fuel on existing or new furnaces. In one case, an aluminum remelt company had an ROI of less than 8 months. Most applications will have a payback approaching one year. Oxy-fuel can now be installed by an end user for less money than any other process while significantly reducing carbon dioxide emissions. As a side benefit generation of NOx is limited to minor air infiltration into the process and fuel borne nitrogen. The overall advantages of oxy-fuel as a proven method of reducing carbon dioxide emissions can not be disputed. The barriers to consideration have been removed. The metals industries must consider this potential asset in their operating plan immediately if the lofty goals of significant carbon dioxide reduction are to be achieved. They are achievable today.

Energy Technology Perspectives
Edited by: Neale R. Neelameggham, Ramana G. Reddy, Cynthia K. Belt, and Edgar E. Vidal
TMS (The Minerals, Metals & Materials Society), 2009

SUSPENSION IRONMAKING TECHNOLOGY WITH GREATLY REDUCED CO_2 EMISSION AND ENERGY REQUIREMENT

H. Y. Sohn, Moo Eob Choi, Y. Zhang and Joshua E. Ramos
Department of Metallurgical Engineering,
University of Utah,
Salt Lake City, Utah 84112

Keywords: carbon dioxide emission, hydrogen reduction, iron oxide concentrate, rate analysis

Abstract

A new technology for ironmaking based on direct gaseous reduction of iron ore concentrate is under development. This technology would drastically lower CO_2 emission and reduce energy consumption by nearly 38% of the blast furnace requirements. Experiments were performed using iron oxide concentrate at 1150°C in a bench-scale facility, which was the highest temperature that could be reached in the facility. The reduction extent was determined at residence times of 3.5 - 5.5 seconds in which the reduction extent approached 43% with 0% excess H_2 and 95% with 860% excess H_2. Separate kinetics measurements showed that the rate is much faster at 1200 - 1400 °C. Experiments were also carried out using syngas (mixtures of H_2 + CO). About 90% reduction in 3.5 seconds with 860% excess hydrogen contained in the syngas demonstrated sufficiently fast reduction for a suspension process and the feasibility of using syngas instead of pure hydrogen.

Introduction

Despite the improvements in the modern blast furnace (BF) process, it suffers from many drawbacks. The process requires the feed in the form of sinters or pellets and high-grade coking coal. Thus, BF iron production is projected to decrease gradually [1]. The main factors for this are the environmental regulations [2] and the high capital investment cost.

A large number of new ironmaking technologies have been developed or are under development [3]. Most of these processes, however, are not sufficiently intensive to replace the blast furnace because they cannot be operated at high temperatures due to the sticking and fusion of particles. Therefore, a new technology is under development for producing iron directly from fine concentrates by a gas-solid suspension technology, which would reduce energy consumption by about 38% of the amount required by the blast furnace and drastically lower environmental pollution, especially CO_2 emission, from the steel industry.

Another important factor in the development of the suspension reduction technology is the large quantities of fine iron oxide concentrates produced in the U.S. and elsewhere that are well suited for suspension reduction. Up to 60% of the U.S. iron production is based on taconite concentrates (50 - 55 million tons/year) [4] with particle sizes ranging from –400 mesh to –500 mesh. In the new process, the concentrate is reduced in suspension in a hot reducing gas generated by the partial combustion of coal, natural gas, heavy oil, hydrogen, other materials like waste plastics, or a combination thereof. Hydrogen is the best material as a reductant and/or fuel from the viewpoint of environmental concerns and reduction kinetics. Although too costly at present, hydrogen is expected to become much less expensive with the development of hydrogen economy, for which the U.S. government is devoting much effort and resources [5].

Material Balance and CO_2 Reduction

The input and output quantities for three possible fuel types used with the new technology are given in Table I, together with those for the conventional blast furnace operation. Of particular interest is the greatly reduced generation of CO_2 for the new technology, even with the use of coal, compared with that from the blast furnace. The proposed technology will have other environmental benefits from the elimination of cokemaking/burning and the pelletization and sintering steps.

Table I. Material balance comparison between the proposed technology and the blast furnace (BF) (for 1 metric ton of molten hot metal).

	BF [kg/tonHM]	Prop'd(H2) [kg/tonHM]	Prop'd(CH4) [kg/tonHM]	Prop'd(Coal) [kg/tonHM]
Input				
Fe_2O_3	1430	1430	1430	1430
SiO_2	151	100	100	125
$CaCO_3$	235	162	162	235
$O_2(g)$	705	227	411	463
$N_2(g)$	2321	749	1354	1525
C(coke)	428			
$H_2(g)$		83		
$CH_4(g)$			211	
Coal ($C_{1.4}H$)				301
Sub-Total	5270	2751	3668	4079
Output				
Fe	1000	1000	1000	1000
$CaSiO_3$	273	191	191	257
Si	5			
$CO_2(g)$*	1671	71	650	1145
H_2O		740	473	152
$N_2(g)$	2321	749	1354	1525
Sub-Total	5270	2751	3668	4079

*The value for the preparation process for BF was not included into this calculation (7%).

Energy Benefits

Comprehensive overall energy balance calculations performed in this work followed the procedure established in a U.S. DOE Report [9]. Table II lists the energy requirements per metric ton of molten hot metal for the blast furnace (BF) and the proposed technology using different types of fuel used, i.e. hydrogen, methane (natural gas), and bituminous coal (considered as $C_{1.4}H$). It is seen that even with the assumption of full combustion of CO to CO_2 in the blast furnace, equivalent to giving full credit for the heating value of the BF off-gas, the proposed technology using any of the three possible reductants and fuels requires a much smaller amount of energy with ~ 38% reduction of consumption. It is also seen from Note 1 below Table II that the proposed technology compares favorably with other commercial alternate ironmaking processes.

Table II. Energy requirement comparison between the proposed technology and the blast furnace (BF) (for 1 metric ton of molten hot metal).

	BF [GJ/tonHM]*	Prop'd (H_2) [GJ/tonHM]	Prop'd (CH_4) [GJ/tonHM]	Prop'd (coal) [GJ/tonHM]
Energy required (Feed at 25°C to products)				
1) Enthalpy of iron-oxide reduction (25°C)	2.09	- 0.31	-0.61	1.73
2) Sensible heat of molten Fe (1600°C)	1.36	1.36	1.36	1.36
3) Slag making	- 0.21	- 0.15	- 0.15	- 0.21
4) Sensible heat of slag (1600°C)	0.46	0.32	0.32	0.46
5) SiO2 reduction	0.09			
6) Limestone ($CaCO_3$) decomposition	0.42	0.29	0.29	0.42
7) Carbon in pig iron**	1.65			
9) Heat loss and unaccounted-for amounts (assumed the same for all processes)	2.60	2.60	2.60	2.60
10) Sensible heat of offgas (90°C)	0.25	0.25	0.21	0.20
Sub-Total	8.70	4.36	4.02	6.56
Energy value for reductant				
Heating value of feed used as Reductant	5.37	7.70	8.00	5.67
Total for Iron oxide reduction	**14.07**	**12.06**	**12.02*****	**12.23**
Preparation				
1) Pelletizing	2.87			
2) Sintering	0.62			
3) Cokemaking	1.93			
Sub-Total for preparation	**5.42**	**0**	**0**	**0**
Total for molten hot metal making	**19.49**	**12.06**	**12.02**	**12.23**

* Gigajoules per metric ton of hot metal [BF numbers compiled based on data from: "Energy and Environmental Profile of the U.S. Iron and Steel Industry," DOE Report No. EE-0229, August 2000; DOE Report: Theoretical Minimum Energies to Produce Steel for Selected Conditions by Fruehan, Fortini, Paxton and Brindles, March 2000; DOE Report: Energy Use in the U.S. Steel Industry: An Historic Perspective and Future Opportunities by J. Stubbles, Table 2, page 8.]

** Carbon in pig represents the heating value of dissolved C. We recognized the fact that its heating value is used in subsequent converting, but decided to leave it in as an energy item because the carbon removal is an added required step that requires other energy and costs, and we are not doing the overall energy balance for entire integrated steelmaking. Even if this item is removed from BF numbers, the proposed process has much lower energy requirement than BF.

***The difference between this number and that for BF agrees closely with the value of 2.3 GJ/metric ton calculated by the use of the Steel Energy Savings Estimator Tool [Energetics web page provided by AISI].

Note 1: COREX 16.9 – 20.2 GJ/TonHM, MIDREX 13.3 GJ/TonHM, and HYLIII 12.3 GJ/TonHM. Heat to melt solid products has been added. (source: L. Price et al., "Energy use and carbon dioxide emissions from steel production in China," Energy, 27, 429-446 (2002)., and G. Wingrove et al., "Developments in ironmaking and opportunities for power generation," Gasification Technologies Conference, San Francisco, CA, Oct. 17-20, 1999.)

Note 2: C and H_2 were assumed to be converted to CO_2 and H_2O, equivalent to crediting their heating values.

Note 3: The amounts of silica in the proposed (H2), (CH4) and (coal) were assumed to be, respectively, 70%, 70% and 100% of that for BF. Thus, the same ratios were used for 'sensible heat of slag', 'slagmaking' and 'limestone' values.

Feasibility Tests

A high temperature drop-tube reactor system was designed and fabricated for accurate determination of the reduction rate of individual iron oxide concentrate particles (Minorca concentrate, screened to 25 - 32 μm). In this apparatus, over 90% reduction was accomplished in 2.8 seconds and the particles were almost completely reduced in 4.0 seconds at 1200°C, as shown in Figure 1-(a). Complete reduction of the concentrate was achieved at a residence time shorter than 2 seconds at 1300°C and 1400°C in pure hydrogen, as shown in Figure 1-(b). The results indicate that the rate is sufficiently fast for the reduction of currently available concentrate to be carried out in the suspension process. It should also be noted that there was a considerable leap in the hydrogen reduction rate of iron oxide concentrate when the reduction temperature increased from 1100°C to 1200°C, which indicates the possibility of much faster reaction at higher temperatures in pure hydrogen.

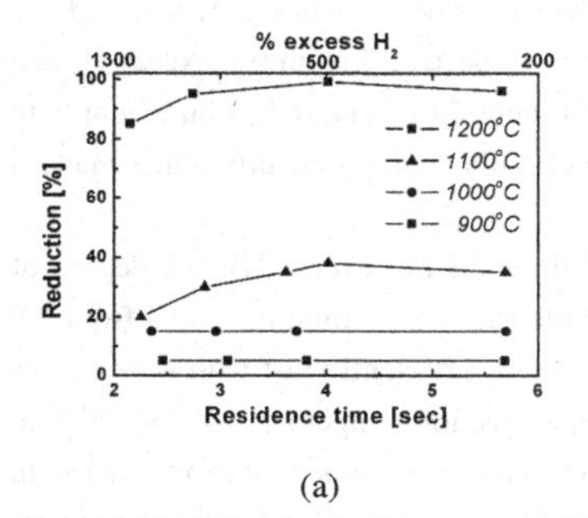

(a)

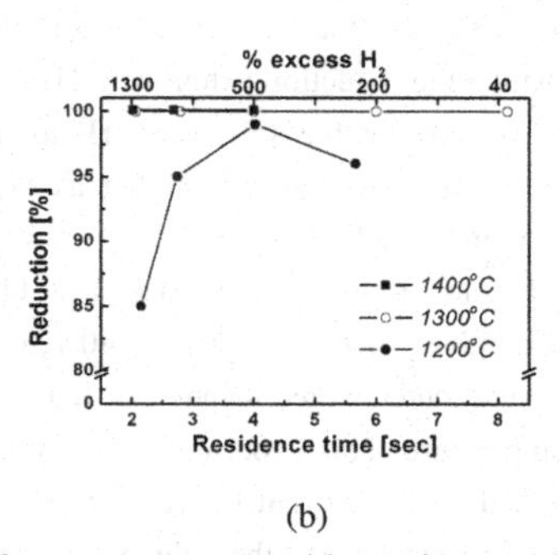

(b)

Figure 1. Hydrogen reduction extent of iron oxide concentrate vs. residence time and % excess H_2 at (a) 900 - 1200°C and (b) 1200 - 1400°C.

Bench-scale Tests

Based on the feasibility tests, preliminary scale-up tests were carried out using a bench-scale test facility as shown in Figure 2. The apparatus consists of five subsystems; a vertical furnace system, an electric power control system, a gas delivery system, a preheating system, and a powder feeding system. A tubular steel reactor (20.3 cm ID, 244 cm long) was electrically heated by six SiC heating elements, which were grouped into two and managed by two SCR controllers. The maximum temperature obtained with the set-up was 1150°C in 76 cm of isothermal zone.

The input gas mixture was preheated to about 500°C before entering the main reactor. The unscreened iron oxide concentrate particles from Ternium were fed into the reactor by a pneumatic powder feeding system at about 1.5 g/min. The residence time of the particles in the isothermal zone ranged from 3.5 to 5.5 seconds and % excess H_2 was varied from 0 to 860 %.

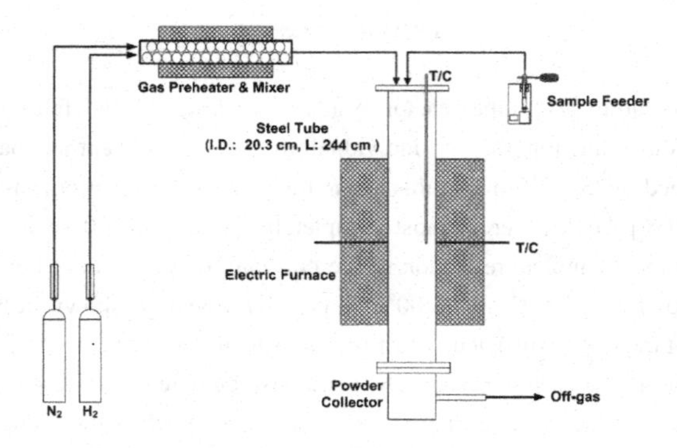

Figure 2. Bench-scale flash furnace for testing suspension reduction of fine iron oxide particles.

The reduction extent was determined at three different residence times; 3.5, 4.0, and 4.5 seconds in which the reduction extent was 21, 29 and 43%, respectively, with 0% excess H_2 and approached over 90% with 860% excess H_2, as shown in Figure 3-(a). The reduction extent with longer residence time was always higher as expected and only 0.5 second difference made a notable change in the reduction rate.

It is also noted in Figure 3-(b) that at 1150°C with about 2% excess H_2, the degree of reduction is relatively low from 20% to 40%, even when the residence time increases from 3.5 seconds to 4.5 seconds, which means that reduction rate is not sufficiently fast with little excess H_2 at the temperature. This is because of the relatively low reaction temperature of 1150°C that can be provided by the current bench-scale facility. According to the kinetic experiments with the drop-tube furnace system, the reduction rate should be faster at higher reaction temperatures. Furthermore, in an industrial operation, the size of furnace will be much larger than the current one, and thus the residence time will be much longer. Then, higher degrees of reduction are expected especially with a higher reaction temperature than 1150°C.

To further verify the kinetic feasibility and better simulate the process in an industrial operation, more experiments were conducted with still moderate but higher % excess H_2 from 50 to 100% and longer residence times from 5.0 to 5.5 seconds at 1150°C. As shown in Figure 3-(c), 71% and 64% reduction degrees were reached in 5.5 and 5.0 seconds, respectively, with 100% excess H_2.

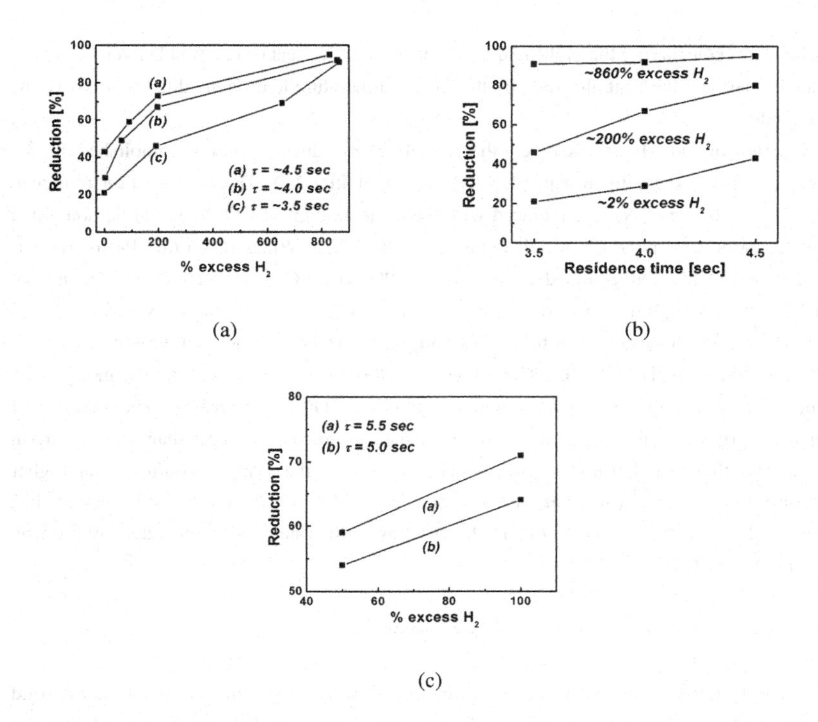

(a) (b)

(c)

Figure 3. Reduction extent of iron oxide concentrate vs. (a) % excess H_2 in 3.5 - 4.5 seconds (b) residence time with 2 - 860% excess hydrogen and (c) % excess H_2 in 5.0 and 5.5 seconds.

Because about 90% reduction was accomplished only with large excess hydrogen, it is obvious that the operating temperature of the facility needed to be increased by 100 - 200°C to obtain a sufficiently high reduction rate with smaller excess H_2 (50 - 100%). To overcome the technical issue supplying additional heat, the design and fabrication of a new bench-scale facility with an internal burner is in progress.

Although hydrogen is best as a reductant and/or fuel from the environmental and reduction kinetics viewpoint, it is currently expensive. Instead, the syngas, which is mainly composed of H_2 and CO from the reforming of natural gas or the coal gasification, has been used as a reducing gas for the majority of direct reduction processes. Thus, the use of syngas was examined in the bench-scale facility.

The syngas was simulated with a mixture of H_2, CO, and N_2 while keeping the composition ratio the same as that obtained by mixing hydrogen with a combustion gas from an internal burner. The input amount of each component was calculated based on material and energy balances considering the preheated gas temperature, the heat of reaction between CH_4 and O_2 in the burner, and the heat absorbed by the room-temperature hydrogen added to improve the reduction rate.

A series of experiments showed that about 90% reduction was accomplished in 3.5 seconds at 860 % excess hydrogen (32.7 kPa H_2, 53.4 kPa N_2). The % reduction remained similar when 11.6 kPa of N_2 was replaced with the same amount of CO. It should be noted that the average barometric pressure of Salt Lake City is 86.1 kPa. When 10% (7.6 kPa) of H_2 was replaced with CO the conversion decreased from 90% to 80% even at about 4.5 seconds of residence time. Although there was a decrease in reduction rate by introducing syngas instead of H_2, it is still of interest especially at higher operating temperatures. As temperature increases, the equilibrium ratio of H_2/H_2O and CO/CO_2 changes in the opposite way, i.e. the thermodynamic reducing power of H_2/H_2O increases whereas that of CO/CO_2 decreases. The amount of hydrogen in a typical syngas used in the direct reduction process is larger than that of carbon monoxide and thus the thermodynamic reducing power of the syngas would rise at higher temperature. In other words, more H_2 can be replaced with CO. Furthermore, the water-gas shift reaction would convert some of CO to H_2 by reacting with water vapor generated by the iron oxide reduction by H_2.

Conclusions

A new technology for producing iron directly from iron ore concentrates by a gas-solid suspension reduction has been proposed. The material and energy balance showed that the process would reduce energy consumption by about 38% of the amount required by the blast furnace and drastically lower environmental pollution, especially CO_2 emission, from the steel industry. As a result of the feasibility tests with iron ore concentrates, the reduction rate of fine concentrate particles by hydrogen-containing gases was determined to be sufficiently fast for a suspension reduction. From the bench-scale tests in addition to the kinetics measurements, it was found that the reduction temperature should be higher than 1200°C to obtain a sufficiently high reduction rate at moderate excess H_2 and with increased fine concentrate feeding rates. The use of syngas from the reforming of natural gas or coal gasification as part of the reducing gas mixture was considered. Tests showed that syngas was adequate as a reducing gas, especially at higher temperatures.

Acknowledgments

This work was supported by American Iron and Steel Institute under Project TRP 0403C for services with respect to its Technology Roadmap Program under Cooperative Agreement No. DE-FC36-97ID13554 with US Department of Energy. Special thanks are extended to Mr. H.R. Kokal and Dr. P.C. Chaubal of ArcelorMittal and Ing. R. Viramontes B. of Ternium for providing the iron oxide concentrates used in this work and technical discussion.

References

1. R. J. Fruehan, "Fruehan: Blast-furnace output will continue to fall," *New Steel*, 14 (5) (1998), 34-36.

2. A. Ritt, "Reaching for Maximum Flexibility," *New Steel*, 16 (1) (2000), 20-26.

3. Lockwood Greene Technologies, "Ironmaking Process Alternatives Screening Study" (Report LG Job No. 010529.01, U.S. Department of Energy, 2000)

4. USGS, "2005 Minerals Yearbook - Iron Ore" (Report U.S. Department of the Interior, U.S. Geological Survey, 2007)

5. U.S. Department of Energy, "Hydrogen Posture Plan - An Integrated Research, Development and Demonstration Plan" (Report U.S. Department of Energy, 2006)

6. A.D. Brent, P.L.J. Mayfield, and T.A. Honeyands, "The Port Hedland FINMET Project - Fluid Bed Production of High Quality Virgin Iron for the 21st Century," *International Conference on Alternative Routes of Iron and Steelmaking (ICARISM '99)*, eds. V. N. Misra and R. J. Holmes (Victoria, Australia: The Australasian Institute of Mining and Metallurgy, 1999), 111-114.

7. R. Husain, S. Sneyd, and P. Weber, "Circored and Circofer Two New Fine Ore Reduction Processes," *International Conference on Alternative Routes of Iron and Steelmaking (ICARISM '99)*, eds. V. N. Misra and R. J. Holmes (Victoria, Australia: The Australasian Institute of Mining and Metallurgy, 1999), 123-132.

8. D. Macauley, "Options increase for non-BF ironmaking," *Steel Times International*, 21 (1) (1997), 20.

9. Energetics Inc., U.S. Department of Energy, "Energy and Environmental Profile of the U.S. Iron and Steel Industry" (Report DOE/EE-0229, U.S. Department of Energy, 2000)

Energy Technology Perspectives
Edited by: Neale R. Neelameggham, Ramana G. Reddy, Cynthia K. Belt, and Edgar E. Vidal
TMS (The Minerals, Metals & Materials Society), 2009

INVESTIGATION OF CARBONIC ANHYDRASE ASSISTED CARBON DIOXIDE SEQUESTRATION USING STEELMAKING SLAG

C. Hank Rawlins[1], Simon Lekakh[2], Kent Peaslee[2], Von Richards[2]

[1]The Center for Advanced Mineral and Metallurgical Processing
Montana Tech of the University of Montana; 1300 West Park Street; Butte, MT 59701, USA
[2]Missouri University of Science and Technology, Department of Materials Science and Engineering; 223 McNutt Hall, 1400 N. Bishop; Rolla, MO 65409-0330, USA

Keywords: Steelmaking slag, calcium leaching, carbon dioxide sequestration, carbonic anhydrase enzyme

Abstract

Batch aqueous leaching and carbonation tests were conducted using industrial steelmaking slags to determine the effect of carbonic anhydrase enzyme as a catalyst. Calcium leaching is a strong function of particle surface area, and the extent can be expressed as a function of time and particle size. Carbonic anhydrase did not affect the calcium-leaching rate, however, it did catalyze calcium carbonate formation to achieve a neutralization time near the theoretical rate. Additionally, carbonic anhydrase modified the precipitate morphology due to accelerated particle nucleation. Time controlled tests in which the pH dropped to ~6 decreased the amount of carbonate produced, and this effect was exaggerated by carbonic anhydrase, while pH controlled tests (>8.5) exhibited the highest rate of carbonation. Because the leaching rate was ~50% faster than the carbonation rate, a further increase in the amount of carbonation may be realized by using carbonic anhydrase however pH must be >10.3.

Introduction

Steelmaking slag contains high fractions of alkaline earth oxide based phases (i.e., CaO and MgO) that exothermically form carbonates. Thus, it is being considered as a means of permanent carbon dioxide sequestration. At current U.S. steel production rates, approximately 13-17 million tons/year of slag are generated from basic oxygen furnace (BOF), electric arc furnace (EAF), and ladle metallurgy furnace (LMF) processes [1]. BOF and EAF slag is used as high quality mineral aggregate or cement clinker, while LMF slag, produced at ~15% the rate of BOF and EAF slags, has limited use as blast furnace flux or acid mine neutralization with the majority going to landfill [2]. Based on slag's chemical nature and immediate availability as a co-product from steel production, the steel industry initiated a project to investigate hydrous carbonate formation in steelmaking slag as a method of sequestering carbon dioxide emitted in steelmaking offgas [3]. The goal of this project is to determine the reactor parameters suitable for the design of an industrial system for treating steelmaking offgas with raw or minimally processed slag.

While, thermodynamically, slag can capture 6-11% of the carbon dioxide emitted by integrated mills and 35-45% emitted by mini-mills, the actual reaction kinetics are very slow and full equilibrium may not be reached at atmospheric conditions for 3-6 months [4]. Retardation of the carbonate formation reactions is caused by binding the alkaline earth oxides into complex oxide phases, which reduces their activities, encapsulation of these complex oxides by inert phases, and formation of a dense product layer upon initial carbonation that inhibits diffusion of the reacting species [1]. Therefore, the primary focus of this investigation is to improve the process kinetics of carbonate formation from the alkaline earth oxide phases.

Both physical and chemical methods of improving the reaction kinetics for large-scale mineral deposit based sequestration systems have been investigated. These efforts have primarily focused on geologic features containing serpentine, olivine, and wollastonite, but the results are directly applicable to slag based sequestration [5-7]. Physical methods include increasing temperature and/or carbon dioxide partial pressure to enhance diffusion rates, pre-treating particles through grinding or acid leaching to increase surface area, and exfoliation of the product layer by high shear/abrasive reactors. Only grinding pre-treatment (with associated metal recovery) was deemed an appropriate physical method for improving the kinetic reaction rate [2]. Slag leaching and carbonation studies showed that particle size is a controlling factor for the conversion of calcium oxide in slag [1]. Physical methods alone cannot achieve industrial scale rates; therefore, improvements to the process chemistry are required. Chemical enhancements have primarily focused on aqueous based processing, which has a much higher inherent reaction rate than gas-solid reactions. Additives to the aqueous slurry include NaCl, $NaHCO_3$, citric acid, or EDTA [5-8]. These additives have moderately improved the carbonation rates, but only when used in high pressure high temperature reactors.

The current work investigates a catalytic additive for calcium carbonate precipitation in aqueous systems operating under atmospheric conditions. The use of purified carbonic anhydrase in a biomimetic-based industrial scale carbon dioxide sequestration process has been proposed for this use [8]. Carbonic anhydrase (CA) is an enzyme found in animals, plants, algae, and bacteria that catalyzes a wide range of metabolic functions including respiration, photosynthesis, and calcium carbonate formation (in both avian shell and mollusk nacre). Carbonic anhydrase is of interest for sequestration as it catalyzes the reversible hydration of carbon dioxide at or near the diffusion-controlled limit. One molecule of the CA II (an isozyme of carbonic anhydrase) can hydrate 1.4×10^6 molecules of carbon dioxide per second, which is equivalent to processing more than 2 million m^3/min of carbon dioxide (at atmospheric conditions) with one mole of CA II.

Carbonic anhydrase assisted slag-based carbon dioxide sequestration must be considered in both the rate of cation leaching from slag and the subsequent precipitation of calcium carbonate. Previous aqueous kinetic studies of slag showed that leaching occurs faster than carbonation, but both processes are inhibited by the calcium carbonate product layer [1]. The leaching rate is governed by the rate of calcium (complex) oxide dissolution and the tortuous porous surface layer of the slag particle. The dissolution rate may be increased with temperature and the porous layer effect may be reduced in proportion to particle size. Carbonic anhydrase has been shown to increase the release rate of Ca^{2+} from limestone by a factor of 11.7, so it may assist in cation release from slag particles [9]. The carbonation rate is governed by product layer diffusion, and any mechanism that can increase the diffusivity of the reacting species through this layer, or remove this layer to continually expose fresh surface, will improve the carbonation kinetics. Carbonic anhydrase greatly improves the precipitation rate of calcium carbonate [8,10] and thus, may affect the growth rate or morphology of the product layer on a slag particle during carbonation. This project investigates the role of carbonic anhydrase in assisting cation leaching and carbonate formation in an aqueous sequestration system using steelmaking slag.

Material Characterization and Experimental Procedure

Three industrial slag samples were selected for use in leaching and carbonation tests. Raw slag samples were collected from BOF, EAF, and LMF steelmaking facilities and stored in a sealed container with a desiccant until each test commenced. Physical and chemical characterization on each slag sample was undertaken in an effort to understand the nature of the starting material.

Each slag sample was analyzed by x-ray fluorescence (XRF) for chemical composition, x-ray diffraction (XRD) for phase identification, and helium pycnometry for true density. The grindability of each slag was measured as part of a separate study to determine the work index values for each slag sample [2]. Table I lists the results from the chemical and physical characterization analyses.

Table I Chemical and Physical Properties of Steelmaking Slags Studied

	BOF	EAF	LMF
XRF Comp. (wt.%)			
CaO	40.9	35.9	50.0
SiO_2	12.9	9.9	4.3
FeO	21.7	28.0	6.3
MgO	12.0	10.1	4.5
Al_2O_3	5.2	9.2	32.3
MnO	4.7	4.3	0.9
Density (kg/m^3)	3614	3822	3069
Work Index (kWh/st)	21.4	19.9	13.8
Spec. Surf. Area (m^2/g)	1.09	0.29	0.12

Each slag contained a high fraction of calcium. Approximately one-half of the volume of the LMF slag was composed of lime containing phases. BOF and EAF slags contained a much higher fraction of iron oxide, which reflects their processing history. The turbulence of BOF and EAF processes results in more entrained iron oxide (and metallic iron) compared to LMF slag. This particular LMF slag had a very high alumina content, which resulted from the deoxidation process used at the mill from which it was collected. The higher iron oxide and manganese oxide levels in BOF and EAF slags lead to their greater density than LMF slag. Additionally, BOF and EAF slags require 40-50% more energy (kWh/st) for grinding than LMF slag. This greater energy requirement can also be correlated to the iron and manganese oxide contents.

Leaching studies were conducted using individual batch reactors. Each (ground) slag sample was sieved to obtain 45-53 μm, 90-106 μm, 150-212 μm, 500-600 μm, and 800-1180 μm size fractions. Seventy-five milligrams from each size fraction of each slag was added to a sterilized 50 ml polypropylene tube, which served as the leaching reactor. Fifty milliliters of distilled-deionized (DD) water (18.0 MΩ resistivity) was added to each reactor based on gravimetric analysis. For the initial LMF slag leaching tests (water only) 2 wt.% TES (2-[(2-hydroxy-1,1-bis(hydroxy-185-methyl)ethyl)amino]ethanesulfonic acid) pH buffer was added to the reactor. For LMF leaching tests with BCA and the BOF and EAF tests, the amount of TES buffer was reduced to 1 wt.%. Each reactor was sealed, placed in a wrist-action shaker, and shaken for a prescribe period of time (1-1440 minutes) to prevent boundary layer concentration build-up of the leaching species. At the end of each test, a 10 ml aliquot was drawn off through a 0.45 μm filter using a syringe. The aliquot was placed into a separate 15 ml polypropylene sample tube to which 1 vol.% nitric acid was added to prevent metal precipitation. Each sample was analyzed using inductively coupled plasma optical emission spectroscopy (ICP-OES) to determine the calcium, magnesium, and iron concentrations, however only calcium leaching is discussed at current. Using the concentration of calcium in the original slag samples, the percent leached could be determined. For the leaching tests using BCA, a 10x stock solution of bovine erythrocyte carbonic anhydrase (Sigma-Aldrich Co.) was prepared using Tris-EDTA solution (2.5 mg BCA in 25 ml of TE). This solution was kept at 4°C until required for the leaching test. One milliliter of the stock solution was added to the 50 ml reactor, which produced a 66.6 nM concentration of BCA, which matches the concentration used in previous studies [8]. Leached samples were analyzed for specific surface area using BET analysis and for morphology using scanning electron microscopy (SEM) analysis.

Carbonation tests were conducted in a 500 ml spherical glass 3-hole flask placed on a mixing plate. For each test, 400 ml of water was added, into which argon or carbon dioxide gas was bubbled through a glass frit sparger introduced into the top hole. Separate pH and temperature probes were introduced through another entry hole, leaving the third-hole open for adding the powdered sample. Argon gas was bubbled through the water while a magnetic bar stirred the solution for ten minutes prior to the start of each test. Slag carbonation was conducted by adding 3000 mg of slag to the reactor. The slurry was mixed for five minutes with argon bubbling and the pH was measured continually. After five minutes of leaching, carbon dioxide was introduced through the sparger for a prescribed period of time (5-360 minutes). The slurry was mixed by the stir bar throughout the entire process. Upon completion of the reaction time, the sparger, pH probe, and temperature probe were removed. The mixture was (vacuum) filtered through 11 μm paper and oven dried. The dried samples were placed into plastic bags and stored in a desiccant cabinet. Tests using 66.6 nM concentration of BCA were formed by adding 8 ml of 10x stock solution. A photograph of the test setup is shown in Figure 1.

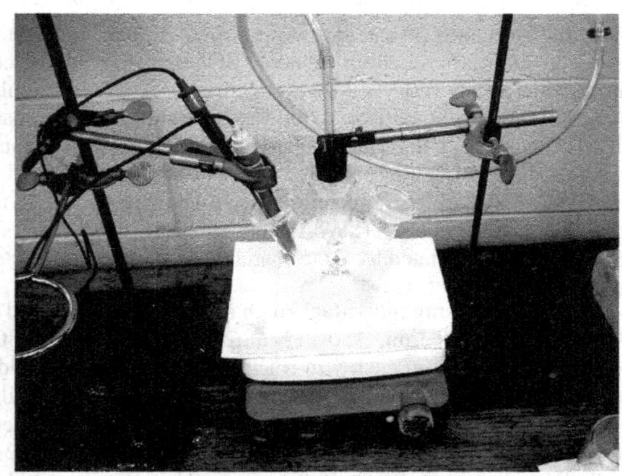

Figure 1. Carbonation test apparatus using 3-hole flask; pH and temperature probes at left and glass frit sparger for gas introduced at the top.

Two sets of carbonation tests were conducted. The first set was time based and used no buffer in the solution. Tests ran for a prescribed time, regardless of the solution pH. As carbonate speciation is a function of pH, a second set of tests used a buffer to maintain an alkaline pH. The buffer used was CAPS (3-Cyclohexylamino-1-propanesulfonic acid) titrated with sodium hydroxide to an initial pH of 10.5. Instead of a prescribed test time, each test was conducted until the pH dropped to 8.5. Increasing the buffer concentration (0.0-0.200 M) allowed for longer test duration. Carbonated samples were analyzed for surface morphology and composition using SEM and energy dispersive x-ray spectroscopy (EDS). The hydroxide and carbonate content of each carbonated sample was measured by thermogravimetric analysis (TGA) in argon to 920°C. Weight loss at <600°C indicated hydroxide content, while weight loss >600°C indicated carbonate content.

Leaching Results and Discussion

The three slag types all exhibited calcium leaching directly related to time and indirectly related to particle size. Figure 2 shows the calcium leaching curves for LMF slag at five particle sizes. The leaching of calcium provided the best straight-line proportionality on a log-normal scale.

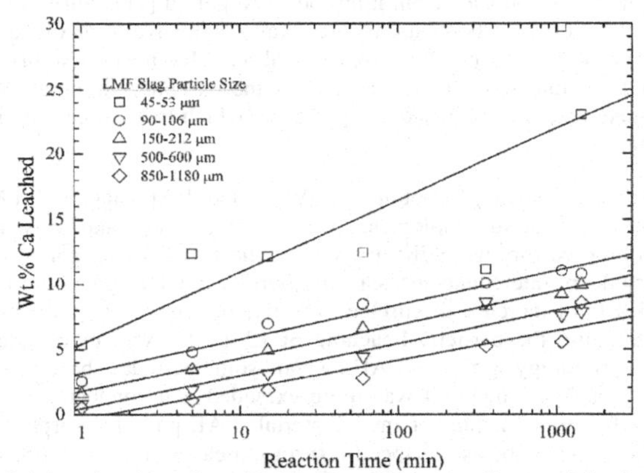

Figure 2. Calcium leaching from LMF slag with time at five particle size fractions.

Curve fits of all particle size fractions >90 μm showed parallel proportionality with time. The overall amount leached with the >90 μm particles was low with a maximum of ~11% leached at 24 hours. The 45-53 μm size fraction exhibited a different slope of the proportionality and the interim points showed no change in leaching, but the final point was much higher at 23-30% at 18-24 hours. Observation of the leached material showed that all particles >90 μm remained free flowing in the reactor, but the 45-53 μm particles had agglomerated together as a result of a chemical reaction. Because they had to be scraped from the reactor tube, these particles were not turbulently mixed to prevent boundary layer formation. Phase maps produced from EDS analysis of raw LMF slag particles (850-1180 μm fraction) showed equal distribution of 20-80 μm Fe and Ca rich regions. Magnesium was associated with the Fe while Si was associated with the Ca. Aluminum was located between the Fe and Ca rich regions. EDS analysis of the leached particles showed increased Al and Si concentrations in their respective regions. The same type of analysis on particles from the 45-53 μm fraction showed that the surface of the leached agglomerate contained mostly Al and Si with very little Fe, Ca, or Mg present.

Calcium leaching curves for BOF and EAF slags showed similar magnitude and proportionality of the amount of calcium leached by particle size, however, the overall amount of calcium leached is significantly higher than that exhibited by the LMF slag. These two slags showed five times the amount of calcium leached at 18 hours for the 90-106 μm size fraction than in the LMF slag. While the BOF slag showed a greater amount of calcium leached in proportion to the amount of CaO present, than did the EAF slag (Table I), both slags greatly surpassed the amount leached from LMF slag, which had the largest amount of calcium oxide. EDS analysis of leached BOF slag particles (90-106 μm fraction) showed alternating regions rich in Si and Mg and a high amount of Fe present throughout.

BCA added to the leaching solution had no affect on the amount of calcium leached from LMF slag (BCA was not tested with BOF or EAF slag). Calcium leaching curves for two particle size fractions (90-106 μm and 500-600 μm) of LMF slag with and without BCA in the leaching solution show the amount of calcium leached under both conditions is nearly identical and no significant difference could be measured. Because BCA is a carbon dioxide active catalyst and leaching took place in the absence of this component, it had no effect on the leaching rate. The effect of BCA on the release of Ca^{2+} from limestone as reported [9] is attributed to the carbonate present in the parent mineral. Carbonic anhydrase catalyzes the hydration of carbon dioxide. This reaction reverses when gaseous carbon dioxide is absent. By means of this reverse reaction, carbon dioxide and calcium ions are released from the carbonate mineral. Because raw slag contains no carbonate material, BCA had no species on which to act rendering it inert in regards to calcium leaching.

The difference between leaching amounts for BOF and EAF slag and LMF slag can be understood from their surface morphologies, specific surface areas, and phase analyses. Surface morphology was observed through SEM analysis of the leached slag particles. The BOF slag particles show very deep, interconnected leaching pores extending up to 25% into the particle. EAF slag showed a high increase in surface porosity, but its pores were less connected and extended up to 10% into the particle. Leaching of LMF slag was more selective with some particles showing morphology similar to EAF slag and some particles showing no leaching at all. The distribution of leachable material was more extended and connected in BOF slag, while LMF slag contained a high fraction of inert material. EAF particle morphology fell between these two extremes. Table II shows the specific surface area for the raw and leached particles of the three slag samples at two size fractions. The increased porosity in the raw and leached particles is shown by specific surface area analysis. Raw BOF slag has approximately 9-10 times the specific surface area of LMF slag and up to four times the specific surface area after leaching. EAF slag specific surface area values lie closer to LMF slag before leaching and closer to BOF slag after leaching. The increased porosity of BOF and EAF slag leads to higher specific surface areas, which support a higher leaching rate than in LMF slag.

Table II. Effect of Leaching on Specific Surface Area

Specific Surface Area (m^2/g)	BOF	EAF	LMF
90-106 μm fraction			
Raw (post-grind)	1.09	0.292	0.125
Post-Leach	16.76	13.37	4.87
500-600 μm fraction			
Raw (post-grind)	0.997	0.145	0.086
Post-Leach	10.86	7.65	6.18

The leaching curves for calcium all follow similar trends based on particle size and time, and can be approximated by a common function for each slag. The general approximation is shown in Equation 1 and the equation constants are given in Table III. The weight percent of calcium leached (m_{Ca}) is a function of the particle diameter (d_p in μm) and time (t in minutes). The values A-D are equation constants for each slag type.

$$m_{Ca} = A(d_p)^B (t)^{C(d_p)^D} \tag{1}$$

Table III. Calcium Leaching Constants for Equation 1

	A	B	C	D
BOF	1110.4	-0.997	0.0568	0.3228
EAF	642.27	-0.918	0.1029	0.2103
LMF	68.306	-0.66	0.046	0.302

Carbonation Results and Discussion

Several sets of carbonation tests were conducted in which the time or presence of BCA was a variable. The results of these tests are listed in Table IV. Thermogravimetric analysis was used to determine the amount of calcium hydroxide and calcium carbonate produced during the carbonation tests.

Table IV. Carbonation Results with 90-106 μm Size Fraction Slag Particles

Slag Type	Reaction Time (min)	Water Only		W/ BCA	
		Wt.% Hydr.	Wt.% Carb.	Wt.% Hydr.	Wt.% Carb.
EAF	0 (raw)	0.71%	0.00%	0.71%	0.00%
EAF	5	0.70%	0.00%	-	-
LMF	0 (raw)	0.00%	0.04%	0.00%	0.04%
LMF	5	1.98%	0.24%	-	-
BOF	0 (raw)	3.20%	0.09%	3.20%	0.09%
BOF	5	1.02%	3.98%	-	-
BOF	60	1.75%	2.31%	1.94%	1.31%
BOF	360	2.90%	1.74%	4.78%	0.77%

The raw slag material contained varying amounts of hydroxide and carbonate. BOF slag started with the most hydroxide at 3.2 wt.%, while EAF slag had 0.71 wt.% and LMF slag had none. All of the slags had <0.1 wt.% carbonate in the raw state. Of the three slags, BOF slag had the most product material from handling and storage, showing it was the most reactive slag (corresponding to the leaching results). The EAF and LMF slags showed different results after the five-minute carbonation test. The hydroxide and carbonate amounts of EAF slag did not change, while both products increased for LMF slag. The results of the BOF test are the most intriguing. In the raw state the hydroxide amount was 3.2 wt.%, while the carbonate amount was near zero. After five minutes of the carbonation test, the hydroxide amount decreased to one-third the level, while the carbonate amount increased to nearly 4 wt.%. This result was more than twice the amount of carbonation achieved in previous carbonation tests with LMF and EAF slag at five minutes [1]. However, as the carbonation test proceeded to longer times the amount of hydroxide increased and the amount of carbonate decreased, contrary to the desired effect. These results were exaggerated with the use of BCA at each time. The most hydroxide and the least carbonate were achieved with BCA at six hours carbonation.

These results show the need for pH control when precipitating calcium carbonate. A significant portion of the five minute carbonation tests with the three slags took place at pH>7 (43% for BOF, 47% for EAF, and 74% for LMF). However, for the one and six hour tests, the slag (w/precipitate) was exposed to low pH (pH=5.7 for water only and pH=6.1 for BCA tests) for most of the test period (>97% of the time for the one hour test and >99.9% of the time for the six hour test). The carbonate ion is most abundant at pH>10.33 and is almost non-existent at ph<8.4, as shown in Figure 3, which illustrates the speciation of carbonate as a function of pH based on thermodynamic analysis.

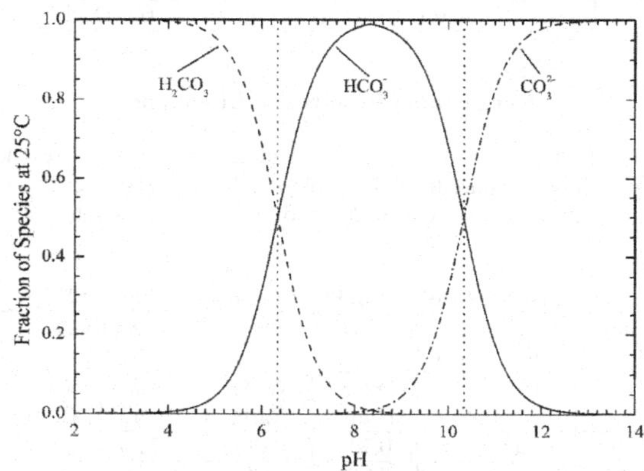

Figure 3. Predominance of carbonate species as a function of pH.

As the carbonate ion is required to react with Ca^{2+} for calcium carbonate precipitation, allowing the pH to drop to the 5.7-6.1 level reduced the activity of this specie in the system to extremely low levels. At the pH range of 5.7-6.1, the most abundant carbonate specie is aqueous CO_2. Calcium carbonate is generated in the first few minutes of each carbonate test when the pH is high. As the pH drops, the carbonate species equilibrium shifts such that calcium carbonate species are robbed of carbon dioxide to maintain a high concentration of aqueous CO_2 in the system. Removing carbon dioxide from calcium carbonate frees up the calcium, which then precipitates as calcium hydroxide. This effect is magnified with the use of BCA, which catalyzes the reversible carbon dioxide hydration reaction, thus increasing the rate at which carbon dioxide is robbed from calcium carbonate. To produce a large amount of calcium carbonate, the solution pH should be maintained at a value of at least 8.5.

After one-hour carbonation without BCA, the slag particle was evenly coated with evenly sized granular precipitates that covered the particle with a regular spacing, as shown in Figure 4a. The precipitates were semi-blocky with straight regular sides, and overlapping plates or ledges can be seen on some particles. The majority of the particles are 2-3 µm in diameter with some 0.5 µm particles scattered throughout. The morphology of the BCA-assisted precipitate coating is less regular with granular particles that are heterogeneously distributed with several large pores indicating less even precipitate growth, as shown in Figure 4b. At high magnification the particles to be more dendritic in construction and less evenly distributed. While some 2 μm particles can be seen, most of the precipitates are <0.5 μm in diameter. BCA increased the rate of nucleation, resulting in a higher population of smaller nucleates. More nucleation events occurred on the existing precipitates, resulting in an irregular agglomeration of small nucleates as opposed to growth of the initial nucleating particles. The result should be a less dense precipitate layer containing interconnected pores. This morphology should allow continued carbonation as the layer thickens because the decreased structure density has a higher diffusion rate than does a fully dense product layer.

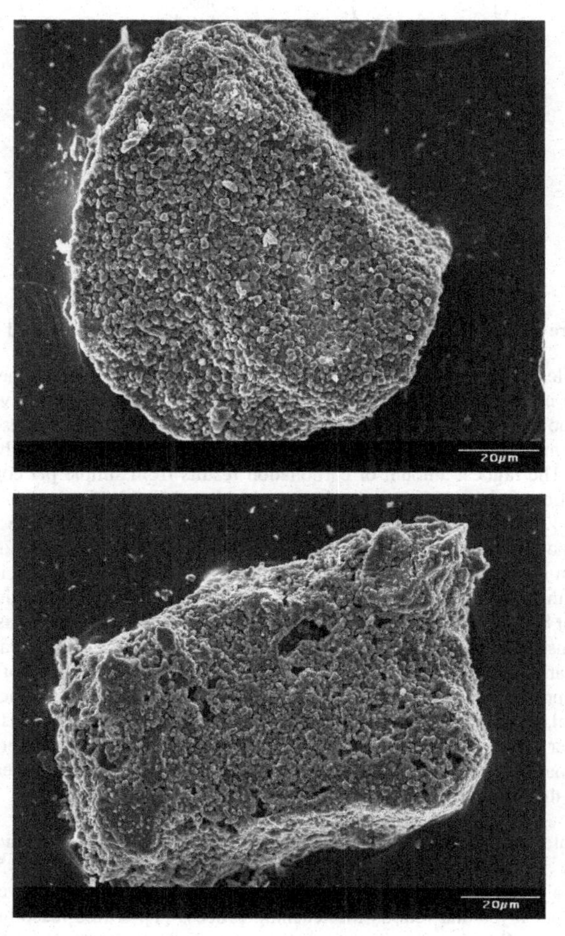

Figure 4. BOF slag particles (90-106 μm fraction) after one hour carbonation (a) without BCA and (b) with BCA added to the aqueous mixture.

The second set of carbonation tests was conducted to maintain a pH of above 8.5. A CAPS buffer was used to allow increased test length at higher pH. Several tests were conducted with and without BCA added. Figure 5 compares the carbonation results for BOF slag in water only, buffer, and buffer with BCA condition to EAF and LMF carbonation results from previous test work.

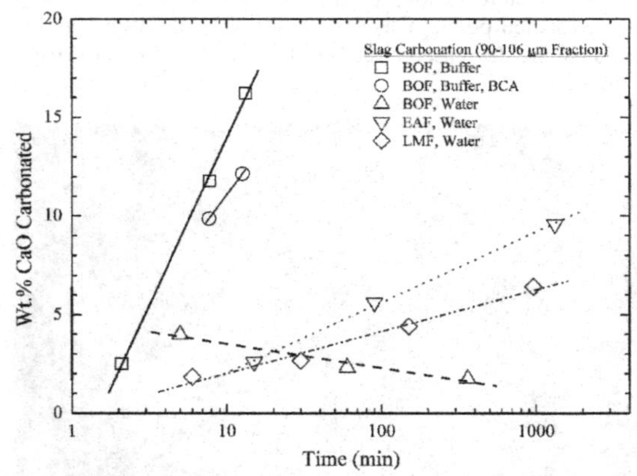

Figure 5. BOF slag carbonation results from time and pH controlled tests.

The water-only leaching results for BOF slag ("BOF, Water") from the time controlled tests show a decrease in carbonation due to low pH. The amounts of carbonation when the pH was kept >8.5 are shown as "BOF, Buffer" and "BOF, Buffer, BCA". Both sets of results show increasing carbonation with time and at a much higher magnitude than those of the time controlled tests. The highest amount of carbonation results from simple pH control without the BCA enzyme. Greater than 16% carbonation was achieved in 13.25 minutes. Extrapolation of this rate yields 47% carbonation at 24 hours. The addition of BCA to the solution slightly decreased the amount of carbonation, but increased the amount of hydroxide formed. At 8.5 pH the carbonate ion is still present, but the bicarbonate ion is the dominant species. The effect of BCA is so rapid that it pulls the carbonate species towards equilibrium and begins to rob calcium carbonate even at 8.5 pH. To realize the full potential of the BCA enzyme in this system, a pH of at least 10.33 must be maintained. Comparing of the buffered test results to data from previous EAF and LMF carbonation tests [1] shows that a significantly higher amount of carbonation was achieved. These previous EAF and LMF tests were conducted at prescribed times and did not use pH buffer control, but a higher slag/water ratio (30 g slag/250 ml water) was used. This ratio produced a higher concentration of Ca^{2+} in solution, thus maintaining a higher pH for a longer time. For batch-based carbonation, the pH controlled method showed an increased carbonation rate of 5-6 times that of the previous tests (BOF slag versus EAF or LMF slag).

The leaching rate of calcium from BOF slag is calculated using Equation 1, while the carbonation rate is shown in Figure 4. Leaching occurs approximately 50% faster than carbonation (on a molar conversion basis) at a particle size of 100 μm. However, carbonation results in a product layer that retards the leaching process [1], so the leaching rate in a batch carbonation system will be less than in a two-stage system. According to the carbonation results, as long as the system pH stays above 8.5, the slag system can process 151 kg carbon dioxide per metric ton of slag in a 24-hour period, which is 47% of the slag theoretical calcium carbonate capacity. The amount of carbon dioxide sequestered is 0.4% of that emitted per ton of steel produced for BOF production. As a first-order approximation, assuming the carbonation rate is proportional to leaching activity, the pH-controlled process should sequester 1.8% of the carbon dioxide emitted from EAF steelmaking. Under the same conditions, LMF slag can add another 0.1% and 0.6% to the sequestration amounts for BOF and EAF mills, respectively. In regards to slag stabilization, BOF and EAF slags will achieve approximately 50% stabilization in 24 hours, while LMF slag should reach the same amount in five days.

Conclusions

Tests conducted on steelmaking slags showed the rates and mechanisms underlying both aqueous leaching and carbonation processes, and the effect of carbonic anhydrase enzyme as a catalyst. The amount of calcium leached is a strong function of particle surface area, which is a more important factor than calcium oxide content. BOF slag exhibited a calcium-leaching rate of approximately five times the leaching rate of LMF slag even though it had 80% of the calcium content. This is because BOF slag had 8.7 times the specific surface area in its raw state. Like LMF slag, EAF showed an increase in calcium leached. Carbonic anhydrase did not affect the calcium leaching rate because it only catalyzes reactions involving carbon dioxide. The specific surface area increased from 15-40 times due to leaching, with BOF particles exhibiting the most extensive leaching depth. All leaching curves exhibited similar trends, and the extent of calcium leaching as a function of time and particle size can be expressed by a mathematical relationship.

Carbonic anhydrase catalyzed the reaction between calcium oxide and carbon dioxide in water to decrease neutralization time by 50%. A 66.6 nM concentration of BCA reduced the neutralization time to near the theoretical rate. Carbonic anhydrase modified the structure of the precipitating layer on the slag particles from an overlapping block structure to a dendritic morphology with smaller particles because carbonic anhydrase increased the nucleation rate of the precipitating particles at the expense of the particle growth. The rate of carbonation is a strong function of pH. Time controlled tests in which the pH dropped to ~six resulted in decreased carbonate production for a given time. This decrease was accelerated by carbonic anhydrase, which acts on the reversible hydration of carbon dioxide so it will pull the system rapidly towards equilibrium. Thus, if the pH drops to a level at which carbonate dissolution would be necessary to reach equilibrium, the enzyme will accelerate that dissolution. Tests in which the pH was maintained >8.5 exhibited the highest rate of carbonation. At the rate exhibited, 47 wt.% carbonation would be achieved by BOF slag at 100 μm particle size at 24 hours in a batch reactor. Because the leaching rate was found to be ~50% faster than the carbonation rate, a further increase in carbonation may be realized by using carbonic anhydrase (but the pH must be maintained >10.33).

Acknowledgements

This paper was prepared as an account of work sponsored by the U.S. Department of Energy in cooperation with the American Iron and Steel Institute (AISI) and its participating companies under Agreement DE-FC36-97ID13554. Such support does not constitute an endorsement by DOE or AISI of the views expressed in the article.

References

1. S.N. Lekakh et al., "Aqueous Leaching and Carbonization of Steelmaking Slag for Geological Sequestration of Carbon Dioxide," *Metallurgical and Materials Transactions B*, 39B (2008), 122-134.

2. C.H. Rawlins, "Grindability Study of Steelmaking Slag for Size-by-Size Recovery of Free Metal," (Paper presented at the 2008 SME Annual Meeting and Exhibit, Salt Lake City, UT, 24-27 February 2008).

3. C.H. Rawlins et al., "Sequestration of CO_2 from Steelmaking Offgas by Carbonate Formation with Slag," *AISTech 2006 Proceedings, Vol. II*, (Warrendale, PA: Association for Iron and Steel Technology, 2006), 1133-1144.

4. C.H. Rawlins et al., "Steelmaking Slag as a Permanent Sequestration Sink for Carbon Dioxide," *Steel Times International*, 30.7 (2006), 25-28.

5. W.K. O'Connor et al., "Continuing Studies on Direct Aqueous Mineral Carbonation for CO_2 Sequestration," *Proceedings of the 27th International Technical Conference on Coal Utilization & Fuel Systems*, ed. B.A. Sakkestad (Gaithersburg, MD, 2002), 819-830.

6. W.K. O'Connor, "Carbon Dioxide Sequestration by Direct Aqueous Mineral Carbonation," *Proceedings of the 26th International Technical Conference on Coal Utilization & Fuel Systems*, ed. B.A. Sakkestad, (Gaithersburg, MD, 2001), 765-776.

7. S. Teir, S. Eloneva, and R. Zevenhoven, "Production of Precipitated Calcium Carbonate from Calcium Silicates and Carbon Dioxide," *Energy Conversion and Management*, 46.18-19 (2005), 2954-2979.

8. G.M. Bond et al., "CO_2 Sequestration Via a Biomimetic Approach," *Proceedings of the Sessions and Symposia sponsored by the 1999 Extraction and Processing Division Congress*, ed. B. Mishra (Warrendale, PA: The Minerals, Metals, & Materials Society, 1999), 763-781.

9. W. Li et al., "Enhancement of Ca^{2+} Release from Limestone by Microbial Extracellular Carbonic Anhydrase," *Bioresource Technology*, 98 (2007), 950-953.

10. P. Mirjafari, K. Asghari, and N. Mahinpey, "Investigating the Application of Enzyme Carbonic Anhydrase for CO_2 Sequestration Purposes," *Industrial & Engineering Chemistry Research*, 46 (2007), 921-926.

Energy Technology Perspectives
Edited by: Neale R. Neelameggham, Ramana G. Reddy, Cynthia K. Belt, and Edgar E. Vidal
TMS (The Minerals, Metals & Materials Society), 2009

Novel Alkali Roasting of titaniferous minerals and leaching for the production of synthetic rutile

Abhishek Lahiri and Animesh Jha
Institute for materials research, Houldsworth Building,
University of Leeds, Leeds LS2 9JT, UK

Keywords: Roasting, rare earths, cations

Abstract

A new technique for the selective separation of rare earths and impurities from lower grade titaniferous ores has been developed which will reduce the overall reduction in the energy costs, and therefore the CO_2 generation by at least an order of magnitude during the chlorination and waste handling stages of TiO_2 to $TiCl_4$ reaction. The process includes roasting the ilmenite ore with alkali followed by water washing step, which removes nearly 80% of oxides of lanthanide and actinide elements present in lower grades of titaniferous minerals. In the context of preferential separation of rare-earth and actinide oxides, we have examined the roles of Al^{3+} and K^+ ions on the lattice structures of ilmenite and complex titanate as the product layer forms during the roasting in air in the presence of potash. The composition of the mixtures of oxides derived from the separation process shows that the mixture of rare-earth oxides may be a high-value co-product of the overall beneficiation process.

Introduction

Worldwide the production of synthetic rutile is carried out either by the sulphate or the slag-making followed by fluidised bed chlorination of TiO_2 enriched slag granules. The chlorinated gas is then re-oxidized to make white pigment grade TiO_2. The waste generated by the sulphate process is predominantly iron sulphate, which has limited use in the water treatment plant. The remaining wastes from the sulphate process undergo further treatment using filtration followed by evaporation and thermal decomposition after which it is disposed into deep sea. On the other hand the chloride waste is extremely hazardous and requires special case by neutralisation, followed by impoundment in a managed pond, where it is monitored for long-term for potential contamination to the soil and ground water. Rare-earth and prelanthanide wastes often end up in the pond where they are diluted and irretrievably lost for further use or recycle. Certain parts of the world the chloride process wastes are pumped via deep boreholes into porous geological strata. The present techniques of disposal are not only expensive but also pose continued threat groundwater and soil and marine life in case of sulphate waste [1, 2].

Prior to chemical beneficiation via chlorination or sulphate process, the separation of heavy minerals from the ore is achieved via a combination of physical methods, namely the density differences, paramagnetic susceptibility and magnetisation and surface ionization potential [3]. Following the physical beneficiation, the ore is further beneficiated via slag making [4, 5] or acid leaching [6-8] processes to remove iron. However neither of the above mentioned processes selectively removes the rare earths and other impurities. Since a vast majority of rare-earth oxides depict vivid colours, as do the transition metals oxides, the lack of an effective methodology for either physical or chemical technique potentially risks whiteness of the pigment grade TiO_2 [9, 10].

Recently we reported the importance of alkali roasting and leaching of ilmenite as a beneficiation step for making synthetic rutile [11]. In this method, the ore is roasted with an alkali, e.g. the sodium carbonate, in air above 800°C for few hours, followed by washing and leaching.

In this investigation we report on the enhanced tendency for separation of rare-earth oxides when the roasting was carried out especially in the presence of alumina. When a small proportion of alumina (<5 wt%) is incorporated, a sodium-iron-aluminate complex also forms together with sodium titanate and sodium ferrite. Since the rare-earth oxides have strong chemical affinity for alumina by forming complexes such as the garnet type structure (e.g. in yttrium garnet, $YAlO_4$ [12]), the propensity for complex formation between the rare-earth oxides and alumina increases during the roasting reaction. For enhancing the selective separation of rare-earth oxides, we will also examine the role of potash as an alkali in the presence of alumina. The roasted product with and without alumina, when washed with de-ionised water, has been found to preferentially separate into layers, which we have analysed chemically. The complexes formed as a result of potash roasting are also identified using the analytical scanning electron microscopy and X-ray fluorescence spectroscopy. Based on the experimental results and thermodynamic analysis, the mechanism of physical separation is explained.

The impact of the preferential separation of rare-earth species is then examined in terms of the reduction in CO_2 emission in chlorination step and chloride waste generation and treatment. On this basis of the composition of synthetic rutile analysed we also estimate the reduction in the chlorine and coke consumption in the fluidised bed reactor. The preferential separation of rare-earth oxides and the beneficiated ilmenite therefore offer a novel route for upgrading TiO_2 for the production of $TiCl_4$ and Ti metal, either via the chloride route (Kroll versus Hunter-Armstrong routes), or the oxide electro-reduction (FFC process) [13].

Experimental

The roasting and leaching experiments were carried out using the Bomar ilmenite, for which the composition is shown in table 1. The magnetic and non-magnetic fractions were separated magnetically. For the chemical analysis the ore was digested in potassium hydrogen sulphate and dissolved in 15% concentrated sulphuric acid. Various elements and compounds were analysed using the titration technique for Ti^{4+} and the atomic absorption spectroscopy for Fe, Al, Ca and other minor oxides, shown in table 1 . The concentrations of rare earth oxides were estimated using the X-ray florescence (XRF) technique, for which the standard samples were used for calibration.

For the roasting experiments, ilmenite was mixed with sodium carbonate in the ratio of 1:1.5 and then 5 wt% of alumina was incorporated in the mixture for studying its effect on roasting which was carried out at 1173K for 4 hours in air. The overall roasting reaction can be represented by equilibrium in equation 1.

$$3Na_2CO_3 + 2FeTiO_3 + 0.5O_2 = 2Na_2TiO_3 + Na_2Fe_2O_4 + 3CO_2 \quad ..1$$

Experiments were also performed by roasting a mixture of stoichiometric amounts of potassium hydroxide with ilmenite ore at 773K in air. The roasting reaction with KOH does not yield CO_2, but another greenhouse gas which is H_2O. The water cycle on earth at present is not believed to be a threat for climate change and is equally, if not more, important for the survival of planet than is CO_2.

$$6KOH + 2FeTiO_3 + 0.5O_2 = 2K_2TiO_4 + K_2Fe_2O_4 + 3H_2O \quad ..2$$

Ten grams of weighed mixture was transferred inside an alumina crucible, which was then placed in the constant temperature zone of an air circulated muffle furnace, which was set at a prerequisite temperature. The roasted samples after quenching to room temperature were leached with water for 30 minutes at 300 K. The phase composition and microstructures of the samples were analyzed using the analytical scanning electron microscopic techniques, described elsewhere specifically for ilmenite minerals.

Results and discussion

Characterisation of ore

Ilmenite by virtue of being rich in iron often co-exists with small concentrations of Mn_3O_4 spinel. The presence of iron often then leads to the incorporation of Cr_2O_3, MgO, Al_2O_3 and small amounts of CaO [2], essential for forming complex spinels. The presence of these oxide constituents for the magnetic and non-magnetic parts is evident from the chemical analysis of the mineral phases in table 1. The rare earth oxides in the non-magnetic fractions of ore were present as a complex with phosphorus which was identified as a calcium phosphate phase, shown in the backscattered scanning electron image in figure 1a and confirmed by the accompanying energy dispersive X-ray (EDX) spectrum. The EDX shows that thorium, cerium, lanthanum are present as a complex with phosphorous and calcium. The small peaks of aluminium and silicon correspond to the presence of a complex aluminium silicate. It is evident that the concentrations of rare-earth oxides are much larger in the non-magnetic fractions than in the magnetic phase.

Roasting with soda ash and alumina

As indicated in Table 1 that the non-magnetic fractions of Bomar ilmenite are richer in the rare-earth oxides, for the preferential separation of rare-earth oxides, we mixed sodium carbonate with the non-magnetic concentrate of ilmenite in the stoichiometric ratio, shown in reaction 1. In this mixture 5 weight percent (wt %) of alumina was also mixed for enhancing the separation tendency of rare-earth oxides. The mixture was roasted in air for 4 hours at 1173K. After roasting the sample was cooled and was analysed for the phases present. In figure 2, the powder diffraction pattern of soda-ash roasted ilmenite confirms the presence of Na_2TiO_3 (ICDD 50-0110) and a solid-solution of $NaFeO_2$ with $NaAlO_2$, which appear to have a composition of $Na_2O.5[(Fe_{0.95},Al_{0.05})_2O_3]$ (ICDD 40-0024) [14].

The backscattered electron image and the accompanying EDX spectra in figure 3 shows the concentrations of various elements in a soda ash roasted sample. The backscattered image depicts the atomic number contrast which appears brighter than the lighter elements (dark). The presence of rare-earth oxides along the cracks is evident from the EDX analysis which shows dominant actinide (Th) and lanthanide elements (Nd, Ce). The lines of lanthanum overlap with other lanthanide and actinide elements. By contrast the dark grey regions are richer in Ti, Ca, O, Al, Mg, Fe and Si. Since the concentrations of Ca, Mg and Si in table 1 are much less than 2 wt%, and unlike the rare-earth elements, they are much more homogeneously distributed through out the grey colour titanate matrix, it is rather difficult to pinpoint their mineral composition and precise elemental association easily.

Roasting with potassium hydroxide

The non-magnetic fraction of ilmenite was roasted with potassium hydroxide in the ratio of 1:3, in accordance with the stoichiometry of reaction 2, described above. The mixture of KOH with Bomar ilmenite was roasted at 773 K for 4 hours in air.

The micrograph in figure 4 demonstrates the effect of roasting with KOH in air on the morphology of ilmenite. The microstructural and EDX analyses confirm that the product layer around the ilmenite grain in Figure 4 is potassium titanate. The EDX spectrum of inner core is unreacted ilmenite, which is evident from the characteristically very weak peak of potassium element in the spectrum and the strong presence of iron, titanium and oxygen peaks. We believe this weak peak might be due to the cross-contamination during polishing, and not due to the diffusion of K^+ ions in the dense ilmenite structure.

Washing and leaching of roasted minerals

The soda-ash and potash roasted mineral samples were dumped into 100 ml of cold de-ionized water at 300 K and stirred for 5 minutes. After stirring the soda-ash roasted mineral residue settled at the bottom of the beaker. By contrast the potash roasted material produced three different media – one which floated at the top was very rich in Ca, pre-lanthanide (Zr, Y), lanthanide, and actinide elements. The titanate rich residue settled at the bottom of the beaker, leaving majority of iron and aluminium in the leachate in the form of alkali aluminate and ferrite.

The composition of the filtered orange-brown floating layer was analysed using the XRF and is shown in Table 2, from which we find that more than 24 elemental oxides are present in the filter residue. When the dried floating residue was examined using analytical SEM, we also found occasionally large flakes of sodium-aluminium silicate phase, which is shown in figure 5. The EDX confirms the presence of relevant elements in this flaky microstructure. It is for this reason we also observe a relatively higher concentrations of silicates in the floating layer residue, as shown in Table 2.

By comparison, when ilmenite was roasted with 5 wt% alumina at 1173K in air for 4 hours and then treated with water, no such physical separation of rare-earth oxide rich layer was observed. Note that the roasting temperature for sodium carbonate was 400K higher than the KOH, and yet no such a distinct difference in the separation was observed. Amongst the rare-earth and non rare-earth elements shown in Table 2, we have chosen only the CeO_2, La_2O_3, P_2O_5, and Al_2O_3 as the tracker oxides in Table 3, for determining their relative concentrations with respect to the oxides of iron and titanium in the beneficiated rutile. We have adopted this approach because we were unable to analyse the compounds of thorium, neodymium, samarium and uranium using XRF due to unavailability of their standards and significant overlap of the flame absorption spectra, which made quantification process unreliable. A similar problem of overlapping was also observed in the X-ray fluorescence analysis, which is manifested in the EDX spectra of lanthanide and actinide elements in figures 1 and 3.

Following physical separation in the aqueous media, the settled mineral residue was aeration leached with water, NH_4Cl and acetic acid, as described in ref.11. The leaching removed a majority of iron and alkali and yielded more than 95 wt% pure TiO_2.

In figure 6 we have compared the microstructure of fully reacted ilmenite with potash at 773K in air, which was washed and leached at room temperature. We find that the bright regions in the microstructure are rich in rare-earth elements and have fragmented away during washing and leaching, thereby leaving characteristic holes which identify their parental site in the overall mineral microstructure. Note that there are very fine bright colour particulates in this microstructure which support the evidence that the fragmentation of mineral phase as a result of chemical reaction is the root cause of physical separation of lanthanide and actinide element rich particles from the mineral matrix.

Suitability of beneficiated rutile for chlorination

By adopting potash roasting in air, washing and aeration leaching we produced more than 95 wt% of TiO_2 with uniform particle size. The average particle size was in the range of 100 to 250 μm, which we believe is smaller than that required for the chlorination of beneficiated titaniferous minerals, via slag-making process. This characteristic particle size range in the fluidised bed may result in the loss of unreacted particles in the flue gas. The particle size re-engineering is therefore essential for adapting the beneficiated rutile particles for chlorination.

It should however be noted that the reduction in the concentrations of rare-earth oxides, determined by the compositions of residual tracker oxides in Table 4, is more than 80%. This analysis presents a significant step-forward for the beneficiation of lower grade titaniferous minerals, which currently cannot be processed via either slag-making or acid leaching process. The reduction in the concentrations of oxides of lanthanide and actinide together with Ca, Al and phosphorus will reduce the formation of complex chloride liquid within the pores of beneficiated rutile and lead to enhanced chlorination at low chlorine and coke consumption. The reduced consumption of coke and chlorine will also reduce the volume of waste generated and refractory corrosion rate.

Conclusions

The alkali roasted ilmenite is a novel method for reducing the concentrations of undesirable impurities in the mineral. In particular the use of potash as an alkali enables more than 80% removal of lanthanide, actinide, calcium, aluminium, and phosphorus from the mineral phase. The reduction in the concentrations of these elements will also improve the chlorination of beneficiated rutile. However the particle size of beneficiated rutile may be smaller than that currently required for the fluidised bed chlorination.

Acknowledgements

The authors acknowledge the financial support from the EPSRC, DTI, and Millennium Inorganic Chemicals for the project.

References

1. P. Doan, Sustainable development in the TiO2 industry: Challenges and opportunities, TiO_2 2003 intertech conference, Miami, Florida.
2. Heins et al, Titanium handbook of extractive metallurgy vol 2. Habashi, F. (ed). Weinheim, Wiley-VCH, 1997. pp 1140.
3. C.K. Gupta and N. Krishnamurthy, Extractive metallurgy of rare earths, CRC press, London, 2004, pp134.
4. G.E. Viens, R.A. Campdell and R.R. Rogers, Experimental electric smelting of ilmenite to produce high-titania slag and pig iron, Trans. Can. Inst. Mining Met. (1957), 60, 405-10.
5. Ercag, Erol; Apak, Resat., Furnace smelting and extractive metallurgy of red mud: recovery of TiO_2, Al_2O_3 and pig iron., J. Chem. Tech.& Biotech. (1997), 70(3), 241-246.
6. Jeffrey A. Kahn, Non-rutile feedstock for the production of titanium, Journal of Metals (1984), 36(7), 33-8.
7. W.A. Weyl and Tormod Forland, Photochemistry of rutile, J. Indus. Engg. Chem., 1950, vol 42, 2, 257-263.
8. T.A.I. Lasheen, Chemical beneficiation of Rosetta ilmenite by direct reduction leaching, Hydrometallurgy, 2005, 76, 123-129.
9. F.A. Grant, Properties of rutile, Rev. Mod. Phy., 1959, vol 31, 3, 646-673.
10. Jelks Barksdale, Titanium, second ed., Ronald press company, New York, 1966.
11. C. Bale, A.D. Pelton, W.D. Thompson, J Melancon and G. Eriksson, FACTSAGE ver. 5.5, Ecole Polytechnique CRCT, Montreal, Quebec, Canada.
12. David R. Lide, Handbook of chemistry and physics, 72 ed. (CRC, 1992).
13. D. J. Fray, T. W. Farthing, Z. Chen, Nature, 407 (2000) pp. 361-363.
14. International Commission on Diffraction Data

Table 1: Chemical composition of Bomar ilmenite ore

Chemical composition	Magnetic ilmenite	Non magnetic ilmenite
TiO_2	61	75.6
Fe_2O_3	36.8	15.5
Al_2O_3	0.47	1.2
MgO	0.2	0.41
Mn_3O_4	0.17	0.34
CaO	0.14	0.1
SiO_2	1.24	2.22
P_2O_5	0.71	0.89
Cr_2O_3	0.19	0.16
CeO_2	0.19	0.9
La_2O_3	0.0	0.27
LOI at 1100°C	1.05	1.69

Table 2: Chemical composition of the filter residue derived from the aqueous treatment of roasted ilmenite with KOH at 773K in air. 5 wt% of alumina was incorporated in the roasted mass.

Elements	Concentration (%wt)	Elements	Concentration (%wt)
Ca	69.8	Cu	0.44
La	5.72	Mg	0.358
Ce	5.35	Sm	0.13
Si	5.05	Al	0.12
Th	3.82	Ni	950ppm
Nd	2.97	Nb	860ppm
Pr	1.86	Gd	840ppm
P	1.72	Ti	460ppm
Y	0.642	V	420ppm
Zn	0.594	Cr	250ppm
Fe	0.531	Cl	160ppm
Zr	0.468	S	110ppm

Table 3: Concentrations of tracker rare earth elements in the two fractions of ilmenite

	CeO_2	La_2O_3	Al_2O_3	P_2O_5
Magnetic ilmenite.	0.356	0	1.369	0.031
Non magnetic ilmenite	1.969	0.274	1.838	0.418

Table 4: Comparison of concentrations of rare earth oxides in the initial and treated samples of non magnetic fractions of ilmenite ore

Phase type	CeO_2 (wt%)	Al_2O_3 (wt%)	CaO (wt%)	P_2O_5 (wt%)
Non magnetic fraction of ilmenite ore.	1.969	1.917	0.148	0.418
Treated non magnetic fraction of ilmenite ore	0.374	0.010	0.017	0.00

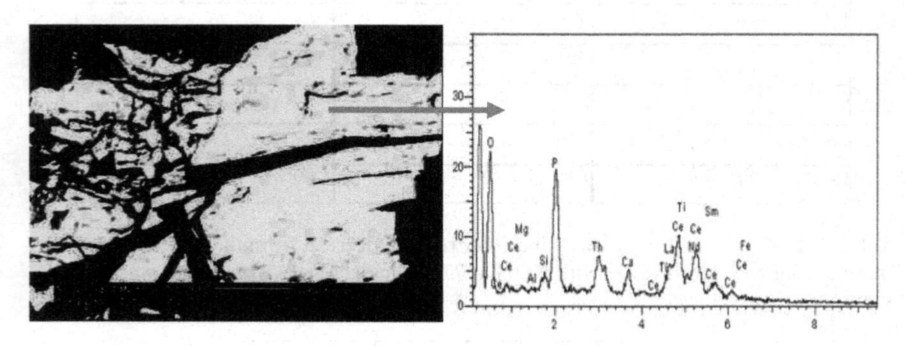

Fig 1: Backscattered image of rare earth oxide rich grains and the corresponding EDX spectrum.

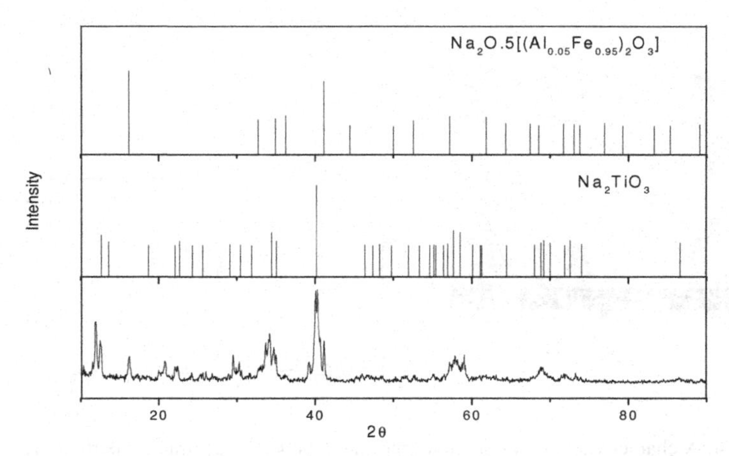

Fig 2: A comparison of X-ray diffraction pattern of roasted Bomar ilmenite with sodium carbonate and 5 wt% alumina in air at 1173K, which is compared with the standard ICDD data for sodium titanate and sodium aluminium ferrite

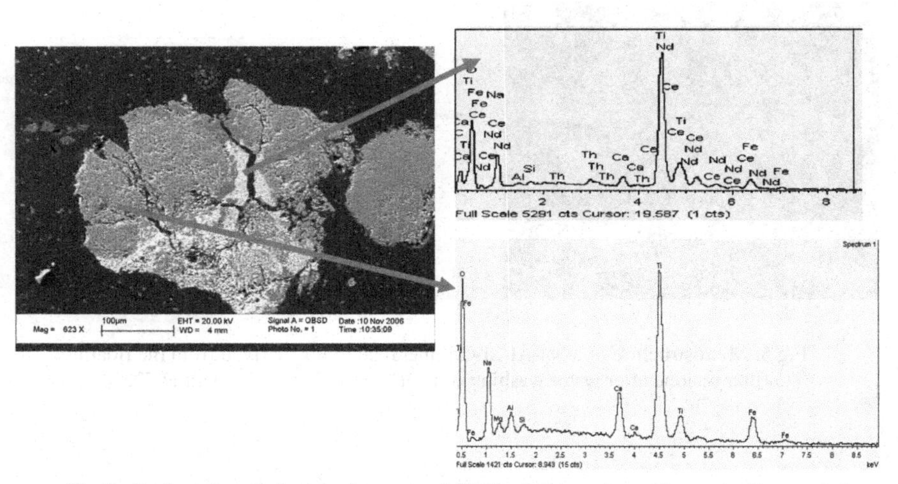

Fig 3: Backscattered electron image and EDX of the various phases in the roasted sample Bomar ilmenite with soda-ash (reaction 1) in air.

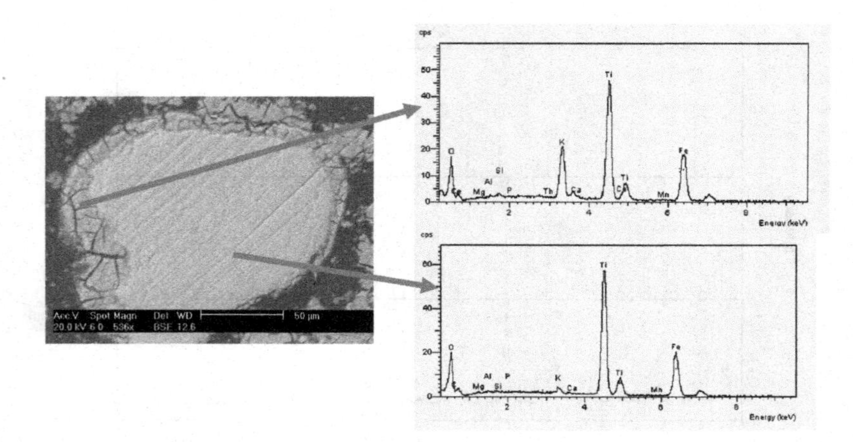

Fig 4: A characteristic backscattered image and the EDX spectra of a partially reacted ilmenite minerals with potash in air at 723 K for 30 minutes.

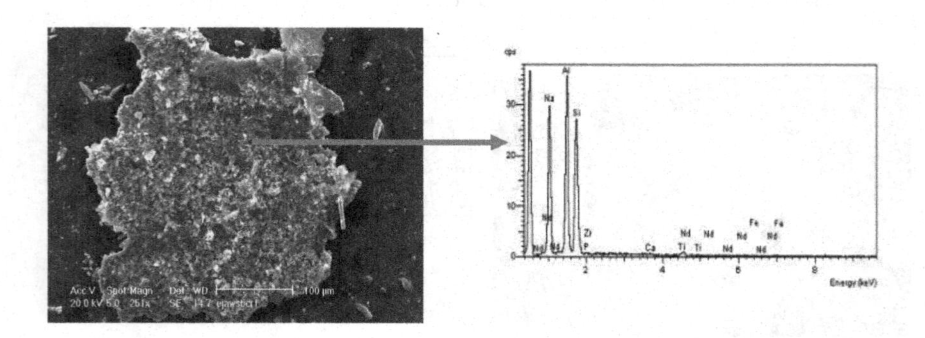

Fig 5: Microstructure of a Na-Al-Si-O mineral complex, derived from the floating filter residue after water washing of KOH-roasted ilmenite in air at 773K.

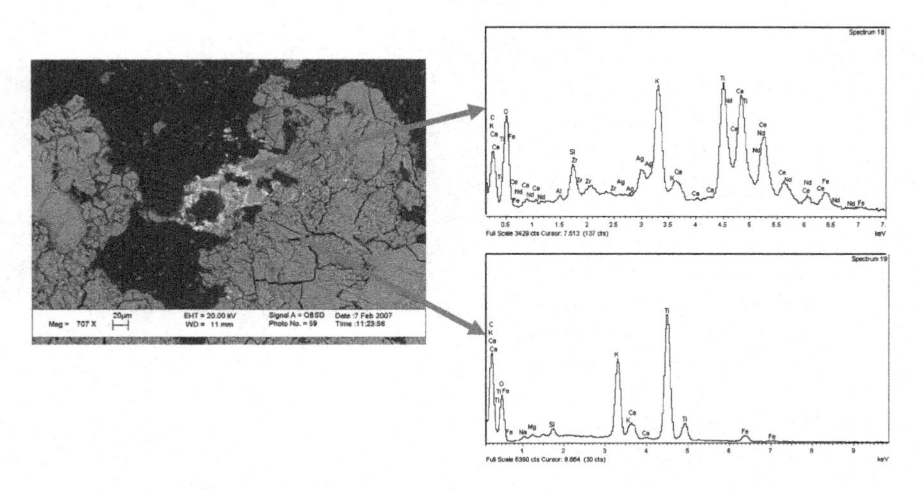

Fig 6: A backscattered image of the roasted and washed ilmenite with potassium hydroxide at 773K in air for 4 hours.

Energy Technology Perspectives
Edited by: Neale R. Neelameggham, Ramana G. Reddy, Cynthia K. Belt, and Edgar E. Vidal
TMS (The Minerals, Metals & Materials Society), 2009

Accelerated Electro-reduction of TiO_2 to metallic Ti in a $CaCl_2$ bath using an intermetallic inert anode

Xiaobing Yang, Abhishek Lahiri and Animesh Jha,

The Institute for Materials Research, Houldsworth Building, Clarendon Road, University of Leeds, Leeds LS2 9JT, UK

Abstract: In the FFC-Cambridge process, the cathodic dissociation of oxide and CO/CO_2 production on carbon anode is the basis for metal production in a $CaCl_2$ bath. By using an inert intermetallic anode, the CO_2 evolution can be eliminated altogether with acceleration in the electro-reduction kinetics. At present the process suffers from slow reduction kinetics of TiO_2 to Ti metal, which can be enhanced significantly by the incorporation of alkali species in the TiO_2 pellet at the cathode and in the $CaCl_2$ bath in the presence of an intermetallic inert anode. Nearly full metallization with greater than 99% of Ti metal containing 1500 ppm of oxygen is possible to achieve in less than 16 hours of electro-reduction. Analyses of phase composition and the microstructures morphology in fully metallized pellets revealed the presence of localized solidified Ti layer, indicating that fast reduction process also involved formation of liquid phase as middle stage reaction. In addition, in situ creation of porosity and maintain higher current level during electrolysis process were two key factors to fully reduction of TiO_2.

Introduction

During cathodic dissociation of TiO_2 in a $CaCl_2$ bath and anodic evolution of CO_2 on a carbon anode, which is the basis for the FFC Cambridge process [1], the rate of metallization slows down due to the formation of $CaTiO_3$ at the reaction interface. Calcium and alkaline earth perovskites are known for their poor electrical conductivity which adversely affects the overall electro-reduction. One of the ways to increase the porosity is via the addition of polyethylene precursors in the pellet or by directly starting from perovskite [2, 3]. The experimental evidences shown in the literature confirm although the speed of reaction increases, but a complete or near complete metallization is not possible by enhancing the in-situ porosity alone, as the insulating perovskite is very slow to decompose under the equilibrium that prevails in the cell. Plethoric evidences on carbon anode show with the current decrease gradually, and finally stabilize at a very low value between 0.2 and 0.4 A, which is a testimony of the presence of perovskite insulating barrier in the current path. Our aim in this paper is to demonstrate the mechanism of enhanced electro-reduction when alkali ions are present in the oxide pellet at cathode and in the bath. We also show for the first time the use of an intermetallic anode for Ti-metal production using the FFC Cambridge process. Evidences in support of enhanced current and nearly full metallization are presented and explained with the help of phase, microstructural and thermodynamic equilibrium calculations.

Experimental

The electrolysis experiments were performed in a molten salt mixture of $CaCl_2$-LiCl (180gms $CaCl_2$ and 20gms LiCl) inside an alumina crucible. The chloride salts were initially dried for 24 hours at 320°C by removing moisture, followed by slowly heating it to 900°C at a rate of 1°C min^{-1} in an atmosphere of argon, maintained at a rate of 500ml min^{-1}. We used two different types of anodes: a carbon anode for comparison, followed by replacement of carbon anode by an intermetallic Al-Ti-Cu anode [4] for enhanced current carrying capacity. The oxide pellet at the cathode was a mixture of TiO_2 and $KHCO_3$, prepared in a ratio of 1:0.5. Both the cathode and the anode were fixed on to a steel holder and fastened using a molybdenum wire, ensuring a good electrical contact. Once the mixture of $CaCl_2$-LiCl was molten, the electrodes were lowered down into the salt bath and the flow of argon gas was maintained at a rate as specified above. A constant voltage of 3.1V was applied through the electrodes and the current was continuously monitored using a computer-controlled software. After 15 hours of electrolysis the electrodes were extracted from the salt bath and the reduced pellet at the cathode was thoroughly washed using water. The microstructure of pellet was examined by using a scanning electron microscope (Cambridge Camscan 4 FEG SEM) operated at 20 kV. Chemical analysis of the phases present was then carried out using the "Link Oxford" energy dispersive X-ray spectroscopy (EDS) system on the scanning electron microscope. Wherever required, the phase identification was also carried out using X-ray diffraction for the analysis of reaction mechanism.

Results and discussion

A typical microstructure of electro-reduced titanium metal after 15 hours electrolysis is given in Figure 1. The inset EDX (energy dispersive X-ray) confirms the presence of pure titanium metal. Compared to the previously reported results [1], the dendritic structure is much coarse and can be seen clearly at the 200x magnification. The presence of coarse dendritic structure indicates that the reaction mechanism and kinetics may be different from those reported in the literature. Further evidence to support this point of view is shown by examining the microstructure in detail, as shown in Figures 2a and 3, where we see a clear evidence for the presence of two different morphologies of Ti metal – a sheet like morphology and the coarse dendrite underneath the sheet-like structure. The sheet-like morphologies are also surrounded by the regions of coarse dendrites. The sheet-like metal morphology is a direct evidence for the presence of a liquid phase which is present during the cathodic dissociation of oxide. In Figure 2b, we show the accompanying EDX analysis which confirms that the solidified sheet-like phases are pure titanium metal. The dramatic transition from a sheet-like metallic structure to dendritic morphology is evident in

Figure 3.

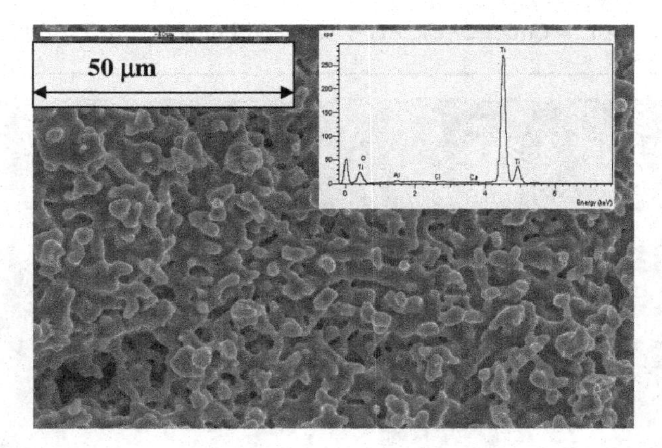

Figure 1: Microstructure of the fully metallized cathode along with a corrsponding EDX spectrum

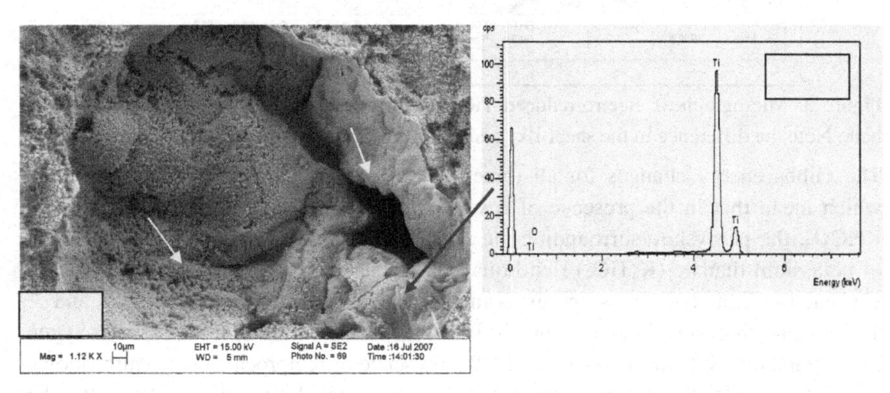

Figure 2: the microstructure of the fully metallized TiO_2 cathode, a) showing clearly the dentridic and solidified Ti metal morphologies, and b) EDX spectrum of the solidified region shows little evidence for oxygen

It should be mentioned here that during electrolysis, the decomposition of $KHCO_3$ in the pellet creates porosity via the liberation of steam and CO_2. As shown in Figure 2, the solidified sheet-like metallic structures were found to be associated with large pores, the formation of liquid phase in-situ is directly related with the rapid decomposition of $KHCO_3$ and subsequent reaction. From the phase equilibrium analysis presented in Figure 4, only possible liquid phase that is likely to form at 900°C is with K_4TiO_4. If this is true, in the presence of $CaCl_2$, the following reactions should take place:

$$CaTiO_3 + 2K_2O = K_4TiO_4 + CaO, \Delta G^o = -334349 \text{ J mol}^{-1} \text{ at T=900}^o\text{C} \quad [1]$$
$$CaTiO_3 + K_2O = K_2TiO_3 + CaO, \Delta G^o = -127211 \text{ J mol}^{-1} \text{ at T=900}^o\text{C} \quad [2]$$
$$K_2O + CaCl_2 = CaO + 2KCl, \Delta G^o = -346968 \text{ J mole}^{-1} \text{ at T=900}^o\text{C} \quad [3]$$

Figure 3: Micrograph of electro-reduced TiO_2:$KHCO_3$=1:1 pellet at 900°C in a $CaCl_2$-LiCl bath. Note the difference in the sheet-like and dendritic Ti metal, surrounded by pores.

The Gibbs energy changes for all three reactions at 900°C are large and negative, which mean that in the presence of freshly formed K_2O from the decomposition of $KHCO_3$, the perovskite surrounding the unreacted TiO_2 is removed by the formation of potassium titanate (K_4TiO_4) rich liquid via reaction 1, but not by the solid K_2TiO_3 via reaction 2, as the Gibbs energy is much smaller compared to reactions 1 and 3. This means that the liquid rich in K_4TiO_4 will become the main source of oxygen anion transport and will be mediated by the presence of reciprocal salt mixture, shown in reaction 3. This is possible because under the applied potential of 3.1 volts, the K_4TiO_4 liquid will continue to dissociate and generate K_2O which will then follow equilibrium in eq.3. Under these thermodynamic conditions, the perovskite phase will not be stable, and decompose readily which is why we observe acceleration in electro-reduction reaction.

The presence of porosity, which may arise as a result of the solidified microstructure and the gaseous constituents present during reaction, provides ingress of $CaCl_2$ liquid and further removes the dissociated oxygen ions from cathode and allows its diffusion mediation to anode via the $CaCl_2$-LiCl bath to the anode. Since the solubility and phase relationships for sub-oxides of Ti with K_2O containing liquid is not well known, it is difficult to explain how during the coarse of reaction when the 4+ Ti-ions reduce to lower valent states affect the overall metallic phase morphology.

It is however true that at a later stage the overall oxygen potential of the cell reduces, compared with the beginning of reduction. At present limited information is available as to how the coarse forms of dendrite structure nucleate underneath the sheet-like Ti metals in Figures 2a and 3. A clear distinction between the governing mechanisms may be essential for distinguishing and then comparing the residual oxygen concentrations present in the two morphologies. It might be possible that a gradual depletion of soluble oxygen from the sheet-like metallic matrix may lead to the dendritic growth. Alternatively it might be the presence of a liquid phase which feeds Ti metal via the cathodic dissociation of oxides and the subsequent discharge electrons occurs on the metallic surface: $Ti_xO_y = xTi^{4+}+yO^{2-}$ and $Ti^{4+}+4e = Ti$ at the cathode, and $yO^{2-} = 0.5yO_2+2y\ e^-$ at the anode. Here x=1 and y=2 are for TiO_2 and x will increase as the Magnelli phases and suboxides start forming.

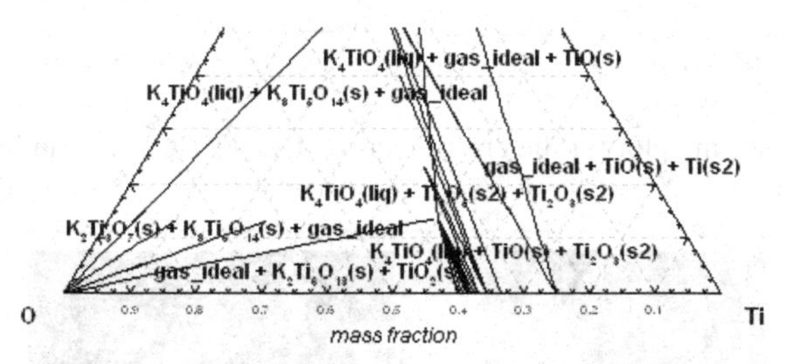

Figure 4: The K-Ti-O phase diagram at 1173K plotted using FACTSAGE thermodynamic software [4].

In order to make comparison and clarify the effect of current on the reaction mechanism, a carbon anode was also used during electrolysis in the presence of $KHCO_3$ in the pellet. However, only partial metallization was achieved in this case. Comparing the current-time diagrams for both inert and carbon anodes in Figures 5a and 5b, respectively, our observation was that on the use of metal anode there is an increase in the current whereas the current decreases rapidly for a carbon anode. As the experimental condition is same in both cases, the only reason for the partial reduction in the presence of carbon anode is due to relatively low current. Schwandt et al [5] and Alexander et al [6] have tried to relate the decrease in the current to the removal of oxygen from the cathode. In this study, we show that the low current is a manifestation of poor metallization and reaction products. To support this point of view, in Figure 6 we show the elemental maps of a cross-section of partially reduced TiO_2 sample which contained TiO_2:$KHCO_3$=1:0.2. Although high current was observed, as seen in Figure 5a with inert anode, the formation metallic layer was only confined to 500 μm. We also observe inside the pellet, the formation of KCl via reaction 3, which also competes with reaction 1 in terms of the magnitude of Gibbs energy change. It should be mentioned that no K_2TiO_3 was found in the partially reduced sample. This fact indicated that the reaction 1 and 3 dominate the overall equilibrium.

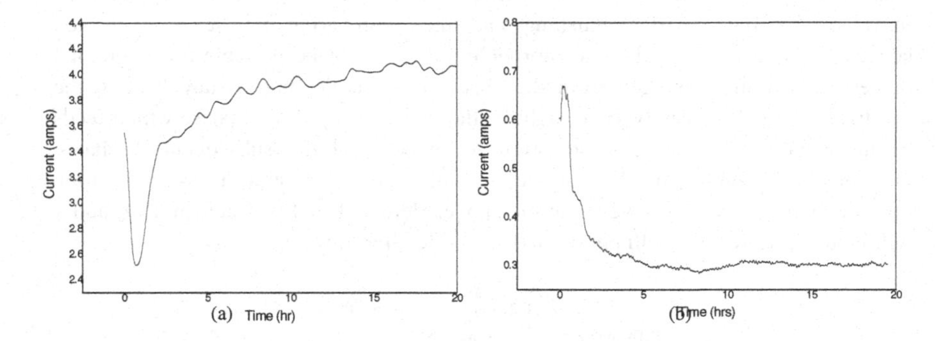

(a) (b)

Figure 5: Current time plot for the electrolysis of TiO_2+$KHCO_3$. 1) an inert anode and b) carbon anode.

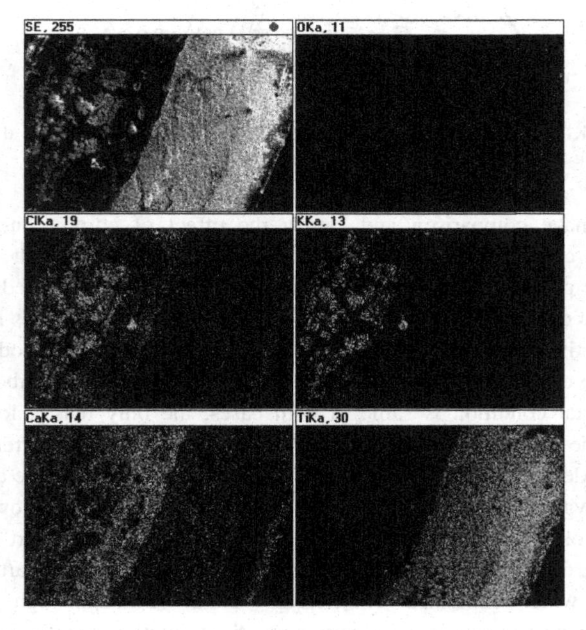

Figure 6: Elemental maps of the cross section of a partially reduced TiO_2 pellet with inert metallic anode. The right hand side of the micrograph was in contact with $CaCl_2$.

Conclusions

When inert anode was used, nearly full metallization of the TiO_2+$KHCO_3$ was achieved in less than 16 hours of electrolysis. The metal production removes the CO_2 evoluition altogether, and enhances the reaction kinetics, leading to much more energy efficient process. During electrolysis, the formation of liquid phase K_4TiO_4 in the pellet is important as this reaction helps to break down the insulating $CaTiO_3$ in the pellet and speeds up the electro-reduction. Carbon anode with alkali does not lead to complete reduction to poor current in the bath.

References

1. D.J.Fray, T.W. Farthing, Z.Chen, Nature, 407 (2000) pp. 361-363.
2. Jiang, K., et al., *"Perovskitization"-assisted electrochemical reduction of solid TiO2 in molten CaCl2.* Angewandte Chemie-International Edition, 2006. **45**(3): p.428-432.
3. Centeno-Sanchez, R.L., D.J. Fray, and G.Z. Chen, Journal of Materials Science, 2007. **42**: pp. 7494-7501.
4. X Yang, A Jha, 2005, 7th International Symposium on Molten Salt and Chemistry&Technology, 29August/2 Semptember 2005, Toulouse France.
5. C. Schwandt and D. J. Fray, *Electrochimica Acta,* **51** (1) (2005), p.66
6. D. T. L. Alexander, C. Schwandt, and D. J. Fray, *Acta Materialia* **54** (11), 2933 (2006).

Energy Technology Perspectives:

Conservation, Carbon Dioxide Reduction and Production from Alternative Sources

Energy Conservation in Metals Extraction and Materials Processing II

Organizers

Edgar Vidal
Cynthia Belt

Energy Technology Perspectives:

Conservation, Carbon Dioxide Reduction and Production from Alternative Sources

Energy Conservation and Technology

Session Chairs

Cynthia Belt
Marie Kistler

Energy Technology Perspectives
Edited by: Neale R. Neelameggham, Ramana G. Reddy, Cynthia K. Belt, and Edgar E. Vidal
TMS (The Minerals, Metals & Materials Society), 2009

MECHANISM AND APPLICATION OF CATALYTIC COMBUSTION OF PULVERIZED COAL

Zhancheng Guo[1], Zuohua Lin[2] and Chuanhong Li[2]

[1]**Key Laboratory of Ecological and Recycle Metallurgy, University of Science and Technology, Beijing, 100083, China**
[2]**United Energy Technologies (Beijing) Corpor., LTD, Beijing 100083, China**

Key words: Catalytic mechanism, CeO_2, Combustion, Pulverized coal, Application

Abstract

Catalytic mechanism of CeO_2 in combustion of coals was investigated in this paper. Catalytic combustion of four kinds of coals with different metamorphic grades by CeO_2 was examined, suggesting that catalytic function of CeO_2 increases with increase of metamorphic grade of the coals. One important reason for CeO_2 promoting combustion of the coals is that pyrolysis of anthracite is advanced by CeO_2, which not only raises the pyrolysis rate and pyrolysis conversion ratio, but increases contents of light components in pyrolysis gas; and both pore space and specific surface area of the pyrolysis char are enlarged by CeO_2, which is another reason. In order to realize in situ catalysis for reducing catalyst consumption, it is crucial to increase catalyst particle number and enhance the bonding force between catalyst and coal. In addition, remarkable energy saving effect was observed when a nanometer calcium-based rare earth catalyst obtained by hydro-thermal synthesis was applied to catalytic combustion of coal injection into blast furnace.

Introduction

Coal combustion is acknowledged as one of the most important coal applications. Since addition of catalysts could reduce the ignition temperature of pulverized coal and accelerate the coal combustion, catalytic combustion is generally accepted as an effective measure to realize fast combustion of coal. Furthermore, different catalysts added may produce different effects on ignition temperature of coal. It is reported activity of catalysts with the same cation follows this order: $OH^- > CO_3^{2-} > Cl^- > SO_4^{2-}$; while that of the catalysts which have the same anion follows: $Cs^+ > Rb^+ > K^+ > Na^+ > Ba^{2+} > Li^+ > Sr^{2+} > Ca^{2+} > Mg^{2+} > Be^{2+}$ [1-4].

Due to the special physical and chemical structure, CeO_2 plays an important role in the catalysis field. Many literatures point out that CeO_2 not only has great catalytic effect on energy conversion, succeeded in catalyzing combustion of gaseous fuels such as CH_4, CO, etc. but also can catalyze combustion of some liquid and solid fuels such as JP-10, black carbon, carbon deposit in the automobile exhausts and so on [5-9].

Pyrolysis, the first step of combustion, directly influences ignition temperature and combustion rate of coals. So a great variety of catalysts were applied to catalyze pyrolysis of coals. And plenty of studies suggest that the alkali and alkaline-earth metal compounds inherent in the coals show apparent catalytic effect on coal pyrolysis. Meanwhile, addition of alkali and alkaline-earth metals into coal combustion process also indicates obvious catalysis effect, accelerating coal pyrolysis, improving pyrolysis conversion ratio, increasing contents of pyrolysis gases and changing composition of the pyrolysis gases [10-15].

Although alkali and alkaline-earth metal compounds have good catalysis effect on coal combustion, their application in the catalysis field was limited because of their erosive attack to the combustion equipments. In contrast, CeO_2, as a substitution of the traditional coal catalysts, presents widely potential application due to its good catalysis ability, non-corrosive and non-toxic. However, investigation on CeO_2 catalyzing coal combustion is rarely reported. In this work, catalytic mechanism of CeO_2 in coal combustion was studied, in an attempt to provide theoretical foundation for developing new catalysis technologies.

Materials and Methods

Materials

Several typical coals including lignite, bituminous coal, anthracite and graphite were selected as materials in this paper. Particle sizes of them were all 150-100 μm. Proximate and ultimate analysis of the coal samples are given in table 1. CeO_2 used was chemically pure wit purity of >99% and size fraction below 150 μm, which was then classified into four size grades as 150-74 μm, 74-57 μm, 57-40 μm and -40 μm by distilled water, in order to check effect of catalyst particle size on catalytic performance. Effect of adhesive force between catalyst and pulverized coal on catalytic combustion was investigated by modifying surface electro-negativity of CeO_2 powders with the help of microwave and additives. To better mix the catalyst and pulverized coal, distilled water was firstly added and blended with them, then drying the mixture at 80 °C, finally mixing with a mortar for 10 min. And amount of catalyst addition is 2%, which is high enough to reduce errors.

Table1 Proximate analysis and ultimate analysis of coal samples

Coal samples	Proximate analysis wt %				Ultimate analysis wt %				
	M_{ad}	A_{ad}	V_{ad}	FC_{ad}	C_{ad}	H_{ad}	N_{ad}	S_{ad}	O^{*}_{ad}
Lignite	19.89	11.62	31.32	37.17	49.08	3.25	0.62	0.87	14.67
Bituminous	1.29	9.82	24.95	63.94	75.86	3.63	1.16	0.83	7.41
Anthracite	2.27	12.57	10.77	74.39	74.97	2.66	0.94	0.62	5.48

*By difference

Analytical Methods

The catalytic combustion experiments were carried out with an integrated thermal analyzer in the oxygen atmosphere with a heating rate of 15 °C/min at 100-1000 °C. Weight of the samples and temperature variation were measured on line; pyrolysis gas composition and tail gas constituent variation were monitored and recorded with a TGA-FTIR and a TGA-GC respectively; specific surface area of the pyrolysis char was measured by BET method.

In this paper, a layer combustion reactor was designed to further verify the catalysis effect of CeO_2 on the coal consumption. At the same time, this reactor, an alundum tube with the inside diameter of 40 mm, is also developed to avoid the errors resulted from the uneven composition of the coal samples, because slight amount of samples were required for the test performed with the integrated thermal analyzer. The combustion experiment was done in a tubular furnace with 1 g of pulverized coal as the material, and heating rate, oxygen flow rate and data record interval were 4 °C/min, 2 L/min and 30 s respectively. The catalysis behavior can be analyzed qualitatively by measuring surface temperature of the coal bed.

Data processing

Ignition temperature can be calculated by using TG and DTG data [16]; the combustion rate was obtained by the following formula,

$$v = [\Delta m / (1 - x)] / \Delta t m_0 \qquad (1)$$

Where, v, Δm, Δt, x and m_0 represent combustion rate, weight loss percent of the coal sample, time periods corresponding to the Δm, amount of catalyst addition and initial weight of the coal sample respectively.

Results and Discussion

Catalyzed Combustion of Four Different Kinds of Coal Samples by CeO_2

Four different kinds of coal samples, lignite, bituminous coal, anthracite and graphite, were all subjected to catalyzed combustion by 2% of CeO_2, and the results were analyzed with an integrated thermal analyzer. Fig.1 gives thermo-graphic curves of the four fuels during the catalyzed combustion. It can be easily seen that thermo-graphic time of the four fuels are all advanced by addition of CeO_2, compared with non-catalytic combustion. And the advanced periods for lignite, bituminous coal, anthracite and graphite were 80 s, 115 s, 200 s and 330 s respectively. In addition, peak width in the DTG curves of the four fuels were reduced by 15 s, 20 s, 120 s and 50 s for lignite, bituminous coal, anthracite and graphite respectively. Meanwhile, peak values in the DTG curves of them were increased from 24%/min, 41%/min, 25%/min and 35%/min respectively to 27%/min, 43%/min, 44%/min and 41%/min. These results show that CeO_2 addition promotes combustion of coal and graphite, not only reducing ignition temperatures of them, but shortening ignition time, making the combustion process more concentrated and combustion rate faster.

As can be observed from Fig.1, two peaks appeared in the DTG curves of both lignite and bituminous, which is caused by the special properties of the two coals. The special characteristics of lignite and bituminous result in two steps in the raw coal combustion process: the first step is the combustion of volatile constituents and part of the fixed carbon, and the second is purely combustion of the left fixed carbon. As shown in Fig.2 (a, b), the second peak of both lignite and bituminous were weakened by CeO_2, suggesting that combustion of the fixed carbon in the coal samples was catalyzed by CeO_2 to some extent.

Lignite, bituminous, anthracite and graphite are four typical representatives of pit coals with different metamorphic grades. According to the firing start time and firing delay time of the four metamorphous coals, it can be concluded that catalysis intensity of CeO_2 to the fuels increases with increasing of the metamorphic grade.

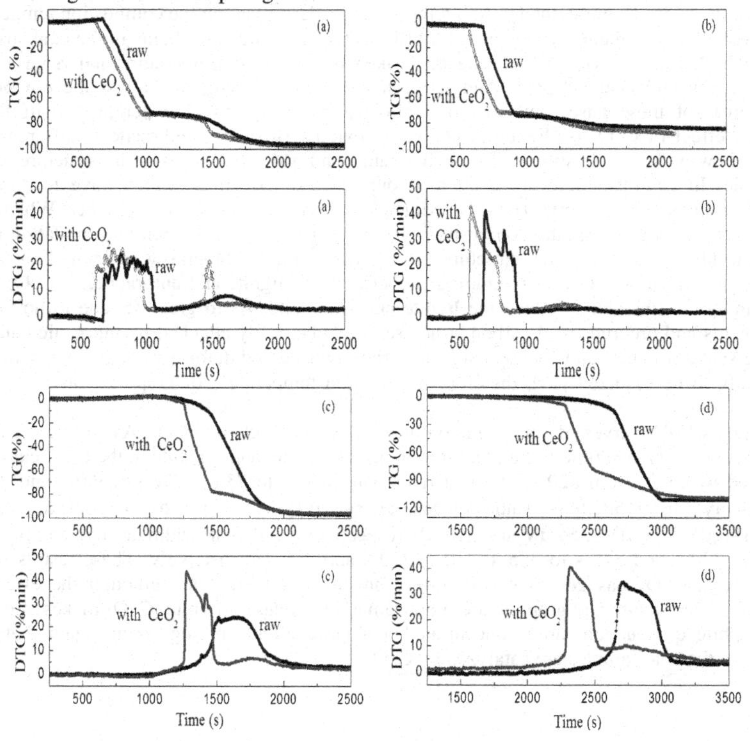

Fig.1 Comparison of TG and DTG curves of four coal samples with and without catalysis of CeO_2
(a) Lignite; (b) Bituminous; (c) Anthracite; (d) Graphite

Table 2 Combustion parameters of the coal samples with and without CeO_2 catalysis

Coal samples	Ignition temperature (°C)	Combustion rate range (%)	Combustion rate (%/min)
Lignite-raw	262	10.0~60.0	12.39
Lignite-CeO_2	259	10.0~60.0	13.64
Bituminous-raw	250	10.0~70.0	16.60
Bituminous-CeO_2	226	10.0~70.0	17.56
Anthracite-raw	458	10.0~50.0	12.31
Anthracite-CeO_2	408	10.0~50.0	21.16
Graphite-raw	714	10.0~70.0	14.46
Graphite-CeO_2	631	10.0~70.0	19.61

Table 2 gives the combustion parameters of the four fuels both before and after CeO_2 catalysis. It can be seen that ignition temperature of all the coal samples were reduced by CeO_2, but reduction extent differs from each other. Ignition temperature of lignite was reduced only a little, 3 °C; while the ignition temperature of bituminous, anthracite and graphite were reduced to a greater extent, they were 24 °C, 50 °C and 83 °C respectively. It is worthy to be noted that increase of the fixed carbon content of the coal samples and the extent of ignition temperature reduction by CeO_2 follow the same order, that is to say, the higher the content of fixed carbon in the coal samples, the greater the reduction of ignition temperature by CeO_2. And with the content of fixed carbon increasing, the ignition temperature dependence changes gradually from combustion of the volatile components to combustion of the fixed carbon. Therefore, data in the table 2 also show that CeO_2 has catalysis effect on the combustion of fixed carbon in the coal samples. In terms of lignite, the ignition temperature is dependent on the combustion temperature of the volatile constituents. There are a lot of branch chains and side chains in the coal structure with high thermal activity, which can be broken at very low temperature and release small molecule gas and coal tar [12]. And ignition temperature of lignite is just the combustion temperature of these gases and tar, so lignite has a low ignition temperature in itself, and catalysis effect of CeO_2 on lignite is not very obvious. However, anthracite is different from lignite. Few volatile constituents, branch chains and side chains exist in structure of the anthracite. In contrast, there are a great amount of aromatic rings which have low thermal activity and can hardly be pyrolyzed or oxidized, so ignition temperature is higher. When CeO_2 as catalyst was added into the combustion process, it benefits breakdown and oxidation of the chemical bonds in anthracite structure, as a result, the anthracite was ignited at a low temperature. Inner structure of bituminous is between the lignite and anthracite, so CeO_2 effect on bituminous is as well between the lignite and anthracite. As to graphite, due to no volatile components and ordered fixed carbon structure, it can be easily attacked by inorganic catalysts, causing space structure and backbone of the carbon twisted and deformed, bond energy lowered and easily to be oxidized [17], therefore the ignition temperature of graphite was lowered by CeO_2.

Fig.2 shows DTA curves of catalyzed combustion of four fuels by CeO_2. As can be seen from Fig.2, peak appearance time of lignite, bituminous, anthracite and graphite in the DTA curves are advanced by CeO_2 from 820 s, 790 s, 1610 s and 2810 s to 730 s, 720 s, 1360 s and 2450 s respectively. Meanwhile, peak values of them are raised from 160 μV, 205 μV, 130 μV and 175 μV to 190 μV, 220 μV, 255 μV and 210 μV respectively, and peak width are reduced from 340 s, 230 s, 370 s and 310 s to 325 s, 210 s, 250 s and 260 s respectively. These results further indicate that CeO_2 has good catalysis effect on coals and graphite, although the effects on bituminous, anthracite and graphite are more remarkable than on lignite. CeO_2 makes anthracite and graphite combustion more concentrated and more intensive, suggesting again that fixed carbons in the coal samples are catalyzed by CeO_2.

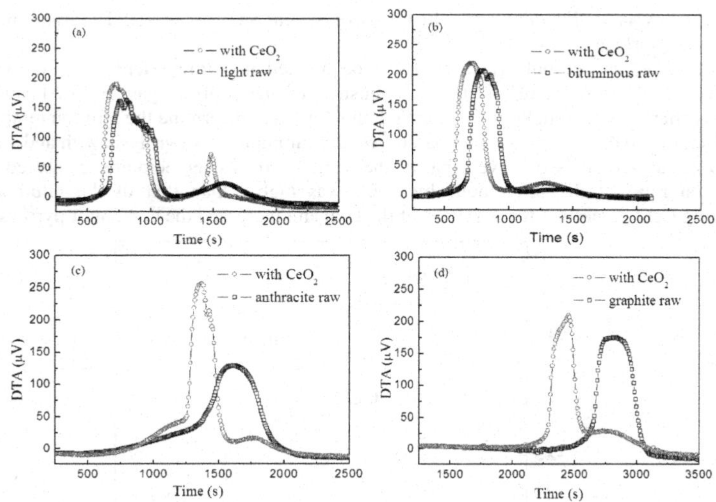

Fig.2 DTA curves of catalyzed combustion of four fuels by CeO_2

TG and DTA analysis did prove the good catalysis effect of CeO_2 on coal samples, but due to only slight amount of coal samples were used as material for each experiment, errors would be caused considering bad uniformity of the pulverized coal. To check CeO_2 catalysis effect on coal samples combustion to a large scale, a layer combustion reactor was designed to do the coal combustion test in a pipe furnace with 1 g pulverized coal as material. Ignition temperature and combustion rate of the anthracite were measured throughout the experiment, and Fig.3 shows variation of the coal layer temperature during the combustion process. As illustrated in Fig.3, the ignition temperature of anthracite was reduced from 540 °C to 455 °C by CeO_2 addition. At the same time, combustion rate of the anthracite, calculated according to the thermal-graphic curve, was increased from 1.37 %/min to 1.67 %/min after CeO_2 added. These results also reflect the catalysis of CeO_2 on combustion of anthracite, as well as on the fixed bed combustion of the anthracite.

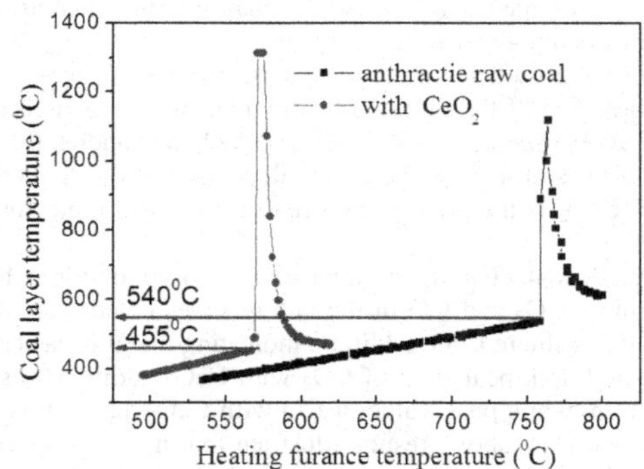

Fig.3 Effect of CeO_2 on ignition temperature of anthracite by layer firing

Catalysis Mechanism

TG, DTA and layer combustion analysis all indicated that ignition temperature was reduced and combustion rate of char was increased by CeO_2 addition. To disclose the catalysis mechanism, pyrolysis test was carried out with FTIR for pyrolysis gas analysis and BET measurement for the char specific surface area.

Combustion process of pulverized coal can be divided into three steps [18]: 1, pyrolysis of pulverized coal with gas released; 2, combustion of the prolysis gas; 3, combustion of the pyrolysis char. So the initial pyrolysis temperature and rate determine the ignition temperature of the pulverized coal. Fig.4 shows the TG curves of bituminous raw pyrolysis with and without 2% of CeO_2 catalyzed. It is observed that at the same pyrolysis temperature, catalyzed pyrolysis conversion ratio of the bituminous by CeO_2 was higher than that of the bituminous raw, suggesting CeO_2 benefited for pyrolysis of the coal and increased the content of pyrolysis gas.

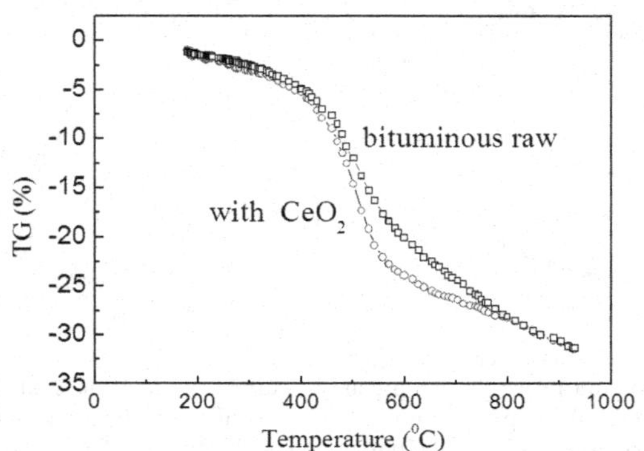

Fig.4 Effect of CeO_2 on pyrolysis rate of coals

If fast heated, except small amount of H_2, CO_2, CO, CH_4 etc., initial pyrolysis products of coal are mainly flammable light components which present as gases at elevated temperature, such as phenols, aliphatics, aromatics, carboxylates etc., and some long-chain hydrocarbons, CnHm with high ignition temperature. According to the previous literatures, lignite was pyrolyzed quickly at 600 °C, and pyrolysis conversion ratio was increased from 15% to 38% after addition of catalyst CoO. Meanwhile, contents of the flammable light components in the pyrolysis products, such as phenols, aliphatics, aromatics and carboxylates are raised by 150%, 80%, 100% and 200% respevtively. These results gave evidences that ignition temperature of the coal was decreased in that the catalyst not only promoted pyrolysis of the coal, but also increased the contents of the flammable light components in the pyrolysis products.

Specific surface area of pyrolyzed bituminous char below 500 °C was measured with a BET machine. Specific surface area of the bituminous raw char was measured as 264 m^2/g, and the average pore size as 68 Å. However, once the catalyst CeO_2 was added, the specific surface area was increased to 954 m^2/g, and average pore size decreased to 35 Å. Perhaps increase of the specific surface area by CeO_2 is the predominant reason for the combustion rate increase of the char.

Combustion of pulverized coal char in air atmosphere was monitored by TG-FTIR. It was observed that peak point of CO_2 and CO in the tailing gases of char combustion catalyzed by CeO_2 were earlier than that without CeO_2 addition, indicating that it is easier for char combustion with CeO_2 addition. In addition, peak area of CO_2 with CeO_2 addition was slightly bigger than that without CeO_2 addition, while peak value of CO with CeO_2 addition is slightly smaller than that without CeO_2 addition. The above results illustrate that not only specific surface area and flammability of the pyrolysis char were increased by CeO_2, but percentage of the combustion product CO_2 was also improved. Furthermore, structure of coal ash, the char combustion product, was changed greatly; in detail, the coal ash with CeO_2 blended had more developed void pore space structure and smaller particles, as shown in Fig.5.

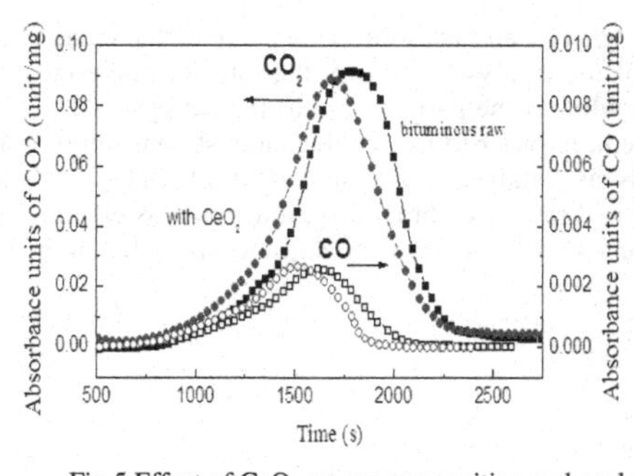

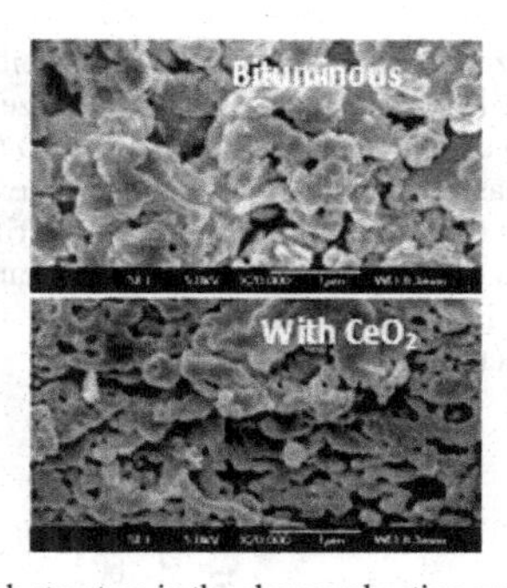

Fig.5 Effect of CeO_2 on gas composition and coal ash structure in the char combustion products

The catalysis mechanisms of inorganics on coal combustion have been intensively studied [19-22], to sum up, there are two. One is oxygen transfer theory, believing the catalyst works on the oxygen in the combustion process; the other is electron transfer theory, indicating the catalyst reacts on the solid fuels and accelerates the electron transfer in the fuels. On the basis of the two mechanisms, catalysis effect of CeO_2 on anthracite combustion was explained. It is believed that the catalyzed pyrolysis in the early stage could be described as the electron transfer theory, considering the characteristics of catalyzed pyrolysis and combustion of anthracite, as illustrated in Fig.6A. The catalyst transfers electrons to the chemical bonds to make them easier to break down and form small molecules [10,11], thus accelerating the pyrolysis process. These flammable small molecules burn at low temperature on surface of the catalyst, transferring heat to the char, and the model is given in Fig.6B. The char formed in the pyrolysis process is good at adsorbing oxygen because of its porous structure with big specific surface area [12,18]. Huge amounts of oxygen were thus adsorbed on the char, and then a great variety of active oxygen was formed when the adsorbed oxygen was contacted with the catalyst which has special inner structure. The catalyst passed the active oxygen to the char and was reduced at the same time by the char, while the reduced catalyst was oxidized again by the oxygen, thus forming a circle. In this circle, catalyst acts as carrier of oxygen, realizing catalyzed combustion of the char. The oxygen transfer theory had a good expression in the late stage of the catalyzed combustion process.

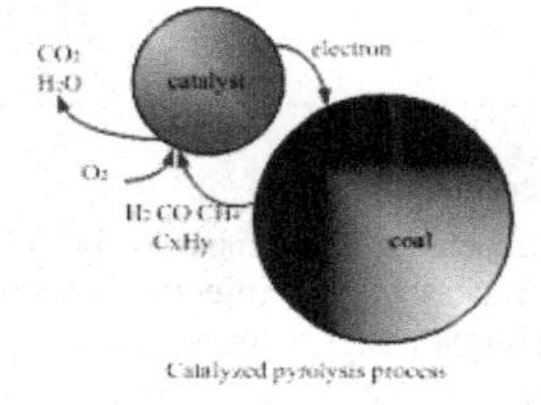

A B

Fig.6 Schematic diagram of catalyzed pyrolysis and combustion process

Catalyst Activity Analysis

Effect of Catalyst Particle Size. Given a certain composition, the catalyst activity depends on the number of active atoms or molecules on the surface of the catalyst particle. Because smaller particles correspond to more active atoms or molecules on the surface, the smaller the catalyst particle size is, the higher the catalyst activity is, as described in Fig.7. Moreover, since coal combustion is a deflagrating reaction, smaller catalyst particle means more active particles and better catalysis effect, for a specified catalyst addition amount. Therefore, in terms of catalyzed combustion of pulverized coal, impact of the catalyst is not depended on the addition amount of the catalyst, but on number of the active particles in the catalyst. Taking a 50 μm of catalyst particle size as example, granted the specific gravity and addition percentage of the catalyst of 2 g/cm^3 and 1% respectively, if particle size of the catalyst is 10 μm, the ratio of pulverized coal number to catalyst particle number is 1: 0.6; while if the particle size of the catalyst is 1 μm, the ratio increases sharply to 1: 600. To that extent, in order to reduce the catalyst consumption, it is an effective way to produce nanomaterials as catalyst. Catalysis effect of CeO_2 with four different size grades on ignition temperature and rate of anthracite combustion was examined, as shown in Fig.8, as well indicating that the smaller the catalyst particle size, the higher the catalytic activity.

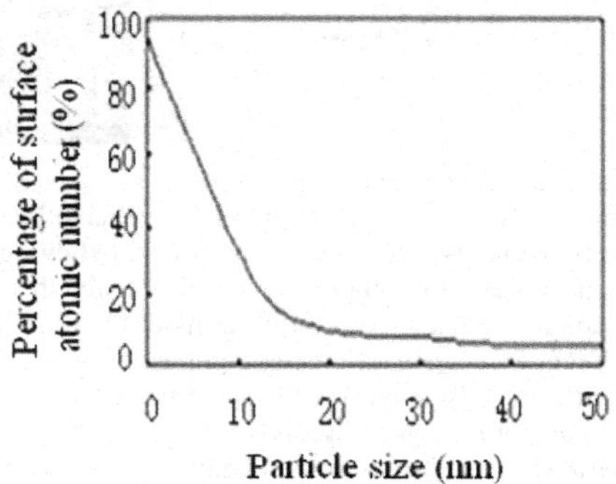

Fig.7 Relationship between surface atomic number and particle size

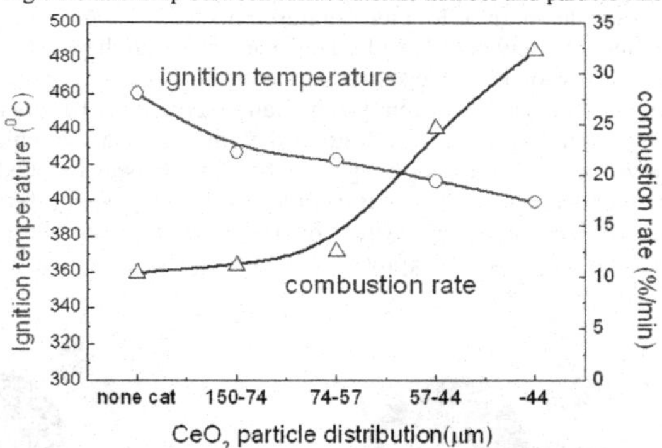

Fig.8 Effect of catalyst particle size on ignition temperature and combustion rate of anthracite

Surface Modification Of Catalyst. The combustion process of coal injection into blast furnace is a gas-solid two-phase flow reaction, so only if the catalyst adheres tightly to the pulverized coal, dose the catalyst work efficiently. Because surface of the coal is lipophilic and electronegative, in order to make catalyst attached to the pulverized coal and enable the interface reaction among coal, catalyst and oxidant, it is necessary of surface modification for the catalyst. Although it is difficult to determine the two-phase flow reaction accurately in the lab, qualitative description to the effect of surface modification to the catalyst on the reaction is still available by means of the fixed-bed measurement. The DTA curves of CeO_2 with different surface electronegativity catalyzing combustion of the pulverized coal are given in Fig.9, suggesting that the higher the electronpositivity of the catalyst, the better the catalysis effect.

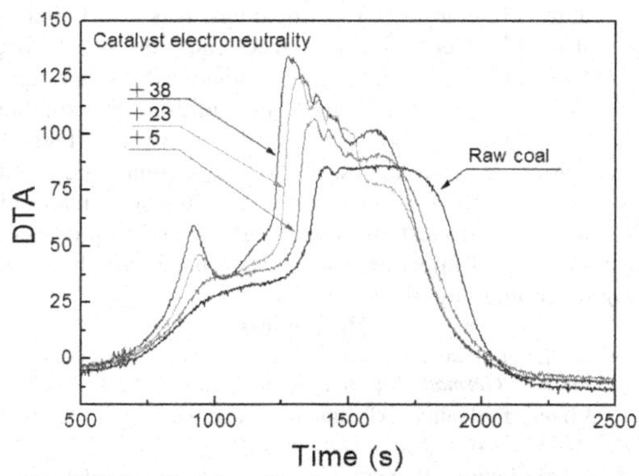

Fig.9 Effect of catalyst electric property on coal combustion

Application of Catalyzed Combustion of Pulverized Coal

Catalyzed combustion of pulverized coal is mainly applied in power plant boiler, industrial stoves of pulverized coal fuel, coal injection for blast furnace ironmaking and so on. While the most appropriate application is perhaps the coal injection for blast furnace, just in that the residence time of the injected pulverized coal in combustion area is very short. Catalyzed combustion can not only raise combustion rate of the pulverized coal and lower carbon content in the blast furnace dust, but also benefits for smooth working of the blast furnace and increasing the coal injection amount. With support of the national "863" program in China, a nanometer calcium-based rare earth combustion improver for blast furnace coal injection was produced by a technology of hydrothermal synthesis-crystallizing separation-surface modification. The combustion improver was added into the coal pulverizer in proportion with raw coal in the milling process, and the experimental results are listed in table 3.

Table 3 Test results of coal injection into blast furnace with combustion improver

Blast furnace volume	200m^3		750m^3		2600m^3	
Key parameters	Base period	Test period	Base period	Test period	Base period	Test period
Coal type	Anthracite	Anthracite	Mixed coal	Mixed coal	Mixed coal	Mixed coal
Addition of combustion improver (%)		0.5%		0.5%		0.5%
Air temperature (°C)	1153	1154	1207	1207	1048	1012
Oxygen enrichment percentage (%)	1.88	1.89	2.64	2.61	~2%	~2%
Sinter ore, TFe (%)	59.61	59.70	57.13	56.95	54.82	54.64
Gravity dust, C (%)	49.6	36.0			~25%	~18%
Cloth dust, C (%)	46.2	41.8	26.6	20.9	~18%	~12%
Coal ratio (kg/thm)	160	166	144	139	133	134
Coke ratio (kg/thm)	386	364	357	351	434	418
Comprehensive coke ratio (kg/thm)	514	497	472	462	540	525
Energy saving effect (before calibration)	17 kg/thm		10 kg/thm		15 kg/thm	
Energy saving effect (after calibration)	15 kg/thm		12 kg/thm		23 kg/thm	

Conclusions

Different catalysis effects were observed when CeO_2 was used for catalyzing combustion of coal samples with four different metamorphic grades; higher degree of metamorphism resulted in better catalysis effect, suggesting that it is advantageous for CeO_2 to catalyze the combustion of coal which contains carbon structure with higher thermal stability and higher aromaticity. As to the reasons for CeO_2 catalyzing combustion of the coals, it is concluded that there are two. One is catalyzed pyrolysis of coal by CeO_2, not only accelerated the pyrolysis process, but increased the content of light components in the pyrolysis products; the other reason is specific surface area of the pyrolysis char is greatly increased. In addition, the smaller the particle size of catalyst, and the higher the surface electronpositivity, catalysis effect of the catalyst on coal combustion is more obvious. So it is necessary to prepare nanometer catalyst and make sure electronpositivity of the catalyst for reducing the catalyst consumption in the catalyzed combustion process. The industrial test show that the ore coke ratio was decreased remarkably by catalyzed combustion for coal injection into blast furnace, which is expected to be a feasible way for energy saving in ironmaking plant.

References

[1] K. Hedden and A. Wilhelm, "Catalytic Effects of Inorganic Substances on Reactivity and Ignition Temperature of Solid Fuels," *Germany Chemical Engineering*, 3(1980), 142-147.

[2] H. LI and Q. LIN. "Effect of Alkaline, Alkaline-earth and Transition Metals on Catalytic Oxidation of Coals," *Journal of DALIAN University of Technology*, 29(3)(1989), 289-294.

[3] H.F. Chen et al., "Catalytic Ignition Property of Coal II The Effects of Different Catalysts," *Journal of Fuel Chemistry and Technology*, 21(3)(1993), 180-184.

[4] Z.H. Wu et al., "Catalytic Effects on the Ignition Temperature of Coal," *Fuel*, 77(8)(1998), 891-893.

[5] M. Issa et al., "Oxidation of Carbon by CeO_2: Effect of the Contact Between Carbon and Catalyst Particles," *Fuel*, 87(2007), 740-750.

[6] A. Setiabudi et al., "CeO_2 Catalysed Soot Oxidation the Role of Active Oxygen to Accelerate the Oxidation Conversion," *Appl. Catal., B: Environmental*, 51(2004), 9-19.

[7] B.V. Devener and S.L. Anderson, "Breakdown and Combustion of JP-10 Fuel Catalyzed by Nanoparticulate CeO_2 and Fe_2O_3," *Energy Fuels*, 20(2006),1886-1894.

[8] F. Arena et al., "Probing the Factors Affecting Structure and Activity of the Au/CeO_2 System in Total and Preferential Oxidation of CO," *Appl. Catal., B: Environmental*, 66(2006), 81-91.

[9] S.Q. Li and X.L. Wang, "The Ba-hexaaluminate Doped with CeO_2 Nanoparticles for Catalytic Combustion of Methane," *Catal. Commun.*, 8(2007), 410-415.

[10] J.B. Yang and N.S. Cai, "A TG-FTIR Study on Catalytic Pyrolysis of Coal," *Journal of Fuel Chemistry and Technology*, 34(6)(2006), 650-654.

[11] Q.R. Liu et al., "Effect of Inorganic Matter on Reactivity and Kinetics of Coal Pyrolysis," *Fuel*, 83(6)(2004), 713-718.

[12] Ke-chang XIE. *Coal Structure and Its Reactivity* (Beijing, BJ: Science Press, 2002), 350-358.

[13] S.Y. Lin et al., "Comparison of Pyrolysis Products Between Coal, Coal/CaO and Coal/$Ca(OH)_2$ Materials," *Energy & Fuels*, 17(2003), 602-607.

[14] N.A. Oztas and Y.Yurum, "Pyrolysis of Turkish Zonguldak Bituminous Coal. Part 1. Effect of Mineral Matter," *Fuel*, 79(10)(2000), 1221-1227.

[15] K.S. Sunil et al., "Influence of Alkali on Pyrolysis of Coals," *Fuel*, 67(1988), 1683-1684.

[16] W.D. FAN et al., "Thermogravimetric study of the effect of oxygen concentration on combustion characteristics of natural gas soot," *Journal of Fuel Chemistry and Technology*, 33(5)(2005), 550-555.

[17] Tao Zhu. *Coal Chemistry* (Beijing, BJ: Metallurgy Industry Press, 1984), 198-200.

[18] D. Paolo et al., "Investigation of the Combustion of Particles of Coal," *Fuel*, 75(1996), 1083-1088.

[19] W.M. Douglas, "Mechanisms of the Alkali Metal Catalyzed Gasification of Carbon," *Fuel*, 62(1983), 170-175.

[20] J.W. Patrick and A. Walker, "A Study of the Copper-catalyzed Combustion of Carbon," *Carbon*, 12(1974), 507-515.

[21] D. W. Mckee, "The Copper-catalysed Oxidation of Graphite," *Carbon*, 8(1970), 131-139.

[22] F.J. Long and K.W. Sykes, "The Catalysis of the Oxidation of Carbon," *Journal de Chimie Physique*, 47(1950), 361-367.

Energy Technology Perspectives
Edited by: Neale R. Neelameggham, Ramana G. Reddy, Cynthia K. Belt, and Edgar E. Vidal
TMS (The Minerals, Metals & Materials Society), 2009

Improving Energy Efficiency in a Modern Aluminum Casting Operation

C. Edward Eckert[1], Mark Osborne[2], Ray D. Peterson[3]

[1] Apogee Technology, Inc.; 1600 Hulton Road, Verona, PA 15147; USA,
[2] General Motors Powertrain; 823 Joslyn Ave., Pontiac, MI 48340; USA,
[3] Aleris International; 397 Black Hollow, Rockwood, TN 37854; USA

Keywords: Aluminum, Melting, Isothermal Melting, Energy Efficiency, Secondary Aluminum

Abstract

The theoretical melting energy requirement for a typical hypoeutectic aluminum-silicon alloy is approximately 520 BTU/lb. It has been demonstrated, however, that even a state of the art secondary processing-automated lost foam casting operation can exceed this value by at least an order of magnitude when the actual thermal energy input from melting to solidification is monitored. Metal transfer and holding operations constitutes over 65% of this expenditure. The authors present relative benchmark energy expenditure information by unit operation for an off-site melting/lost foam casting line with a daily throughput in excess of 100,000 lbs. Efficiency improvements through optimization of the current process, and anticipated energy values at the culmination of a U.S. Department of Energy sponsored project to develop, integrate and demonstrate an advanced melting, transportation, and dispensation system will be cited.

Introduction

The size of the U.S. aluminum remelt pool was approximated to be 14.21 billion kg (31.27 billion pounds) by a recent study[1], with the engineered castings (foundry) sector accounting for 3.25 billion kg (7.16 billion pounds) or 23% of that total. This study considered overall process yield and all major contributors to scrap. The values cited, therefore, are more meaningful than simply tabulating annual pounds shipped by sector.

The foundry sector consists of Sand Castings, PM Castings and Die Castings categories appearing in Figure 1. All other categories comprise the wrought alloy sector. Wrought alloy production enjoys an energetic benefit from both scale and a relatively continuous operation. Many foundry operations, however, operate in batch mode at less than design metal throughput with characteristically longer melt holding periods. Since *any* detention of molten metal beyond melting represents an overhead use of energy, melt holding (secondary) energy expenditures can appreciably inflate melting energy requirements in the foundry.

2006 Aluminun Industry Wide Melt/Re-melt Pool Assessment														
Category (million lbs)	RCS	Sheet	Plate	Foil	Extrusions	Die Forgings	Hand Forgings	Rings	Wire & Cable	Rod & Bar	Sand Castings	PM Castings	Die Castings	**Total**
Category Total Melted	5808	8015	887	2055	5965	341	29	41	495	470	1228	3058	2875	31287
Gross melt loss %	6	4	4	4	4	4	4	4	3	3	6	6	6	
Skim/dross	329	308	34	79	229	13	1	2	14	14	69	173	163	1429
Total Cast	5479	7707	853	1976	5736	328	28	39	480	457	1158	2885	2712	29838
Ingot/cast losses	261	367	41	94	273	16	1	2	24	22	0	0	0	1100
Scalper chips	248	316	77	9	91	15	1	2	4	0	0	0	0	764
Started ingot/billet/cast weight	4970	7024	735	1873	5372	298	26	36	452	435	1158	2885	2712	27974
Process losses	1404	2088	228	478	1170	75	5	7	73	57	521	1154	271	7531
In-plant recovery %	82	80	85	80	85	85	90	90	90	92	55	60	90	
Total Shipments	4075	5619	625	1498	4566	253	23	32	407	400	637	1731	2441	22307
Finished part recovery %	70	85	30	98	90	80	30	30	98	95	85	85	90	
Skeleton, ears, defectives	1223													1223
Mach. chips, turnings & borings			438			50.6	16	22			96	260	244	1127
Trim, skeleton, defectives		843		30										873
Ends, defectives					457				8	20				485
Elwin Rooy and Associates, March 2007														

Figure 1 – Estimation of the Annual Remelt Pool Size for Aluminum by Category

Based on the average values of 1.2 kW-hr/kg (1,862 BTU/lb), and 0.02 kW/kg (31 BTU/hr-lb) to melt and hold aluminum in the foundry, a 10 hour melt holding period during drawdown will add 0.2 kW-hr/kg (310 BTU/lb) for a total energy investment of 1.4 kW-hr/kg (2,172 BTU/lb). These values assume that holding furnaces are closed and no free metal surface exists. In the frequently encountered situations where the top of a holding furnace is open to permit access for metal withdrawal purposes, specific heat loss rates can be as high 23.6 kW/m^2 (7,483 $BTU/hr\text{-}ft^2$). Any reduction of uptime, such as weekend or off-shift idling, adds additional holding cost.

Multiple metal transfer events further escalates the energy expenditure, since the heating methods used during such events are typically inefficient, i.e.: direct flame impingement burners to heat ladles operating at 8% thermal efficiency. The tacit energy losses associated with the irrecoverable metal loss from direct heating can approach 45 kW-hr/kg (70,000 BTU/lb). It is therefore reasonable to expect that the *actual* in-plant energy expenditure, (including melt loss) required to melt, prepare, hold, and dispense typical foundry alloys to the casting process could very easily exceed 2 kW-hr/kg (3,100 BTU/lb). Conversely, the *theoretical* melting energy requirement for a typical hypoeutectic foundry alloy (i.e.: 356) is approximately 0.34 kW-hr/kg (520 BTU/lb).

Melt preparation operations in the foundry therefore present compelling opportunities for energy conservation. An in-plant energy expenditure reduction of 1 kW-hr/kg (1550 BTU/lb) applied to the 3.25 billion kg (7.16 billion lb) 2006 remelt pool would reduce energy consumption in the engineered castings sector by 3.25 billion kW-hr (11.1 trillion BTU) per year. These energy savings, alone, could melt an additional 9.7 billion kg of aluminum.

Accordingly, a U.S. Department of Energy sponsored project was initiated in 2005 with the objective of reducing melt preparation and holding energy expenditures in foundry operations to levels approaching theoretical limits.[2] Both melt preparation and molten metal holding energy requirements are considered. The approach proposed for this project was to supply fully processed high quality molten aluminum *directly* to the casting process, using an off-site melter operating at the highest possible energy efficiency.

Further, using recently developed self-heated and transportable molten metal containment vessels with direct metal dispensation capability would eliminate the need for traditional holding furnaces. The energy expenditures associated with maintaining a molten metal inventory and additional processing of molten aluminum at the casting facility are obviated using this approach. Using conservative estimates, an annual tacit domestic energy savings of 18.25 billion kW-h (62.3 trillion BTU) by the year 2020 in the engineered castings sector was approximated at project conception.

This paper overviews the project and examines baseline energy consumption data that typifies the conventional system and provides some insight into the advanced system.

Summary of Current System

The Aleris International metal preparation facility in Saginaw MI (AIS) and General Motors Powertrain Saginaw Metal Castings Operation (SMCO) are the participants in this project, with Apogee Technology, Inc. as the primary participant. A distance of approximately 4 km (2.48 mi) separates AIS from SMCO. Molten specification grade 319 alloy is transported from AIS to SMCO using unheated 13,600 kg (30,000 lb) capacity over the road ladles of contemporary design. Approximately 4 deliveries per day are made.

The overall metal preparation, delivery, and dispensation scheme of this system is shown in Figure 2.

Specifics of this system have been described elsewhere.[3,4,5,6] In general, however, a mixture of RSI, primary metal, cast scrap, and foundry run-around is melted and alloyed to specification at AIS in a large, somewhat conventional, reverberatory melter. The metal is then pumped to a transfer trough, which flows into a cascade transfer top drop delivery system for over the road 14,000 kg (30,000 lb) preheated ladles. These ladles are maintained with top gas heaters until delivery is required by SMCO.

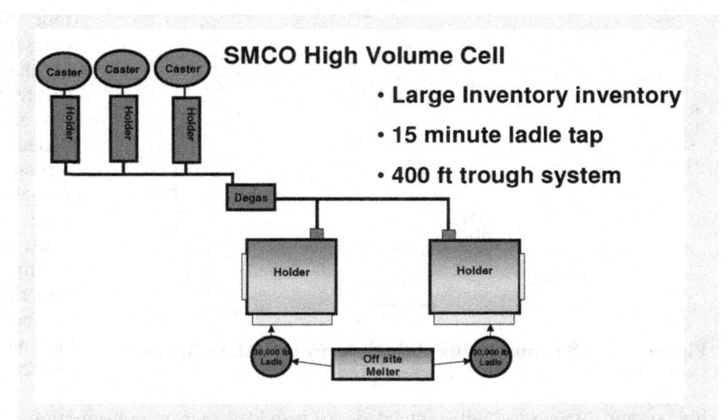

SMCO High Volume Cell

• 15 minute ladle tap
• 400 ft trough system

Figure 2 – Schematic Representation of Molten Metal Flow in System

At SMCO, these ladles are then bottom tapped into large 68,000 kg (150,000 lb) reverberatory holding furnaces. The metal subsequently passes through radiant heated troughs into a dual degas station, and then to a separate ladling furnaces of 10,000 kg (22,000 lb) capacity. At each ladle furnace the metal is recirculated and degassed. Metal is then either pumped or ladled into a mold according to the process. In excess of 78,000 kg (171,600 lb) of molten metal inventory is maintained in-line with this scheme at SMCO. Such capacity is required to provide a ballast of metal to ensure constant flow conditions, and to allow for sufficient residence time for heating and metal treatment. Real time tailored alloy chemistry adjustment or alloy change is not possible with the current system.

Energy Tracking System

The first task of this project was dedicated to establishing and maintaining a comprehensive energy tracking system. This system provides detailed temporal energy expenditure information essentially from initial charge melting to the pouring cup. A mass balance is also embedded within the tracking system to account for energetic losses associated with melt loss. For data collection and analysis purposes, the overall metal flow scheme was considered as 5 unit operations:

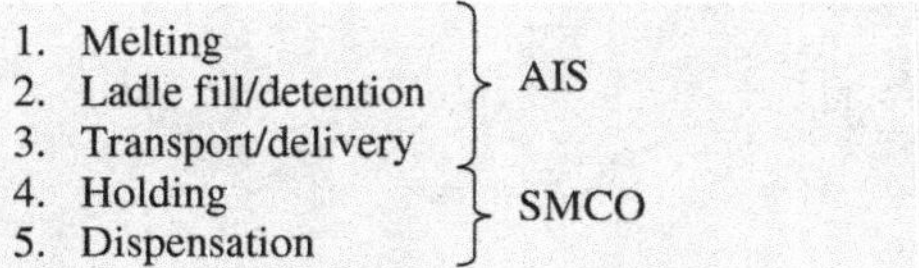

1. Melting
2. Ladle fill/detention
3. Transport/delivery
4. Holding
5. Dispensation
AIS
SMCO

Energy input to the ladle and contents occurs using gas burners at AIS. Detention of the ladle until needed by SMCO also occurs at AIS. Since ladle – furnace metal transfer at SMCO is brief; the transport/delivery operation is categorically identified at AIS with the sensible heat content considered by energy tracking as in input/asset to SMCO. Ladle preheat energy is amortized over the throughput.

Figure 3 illustrates the components of the energy and mass balances that are tracked at AIS.

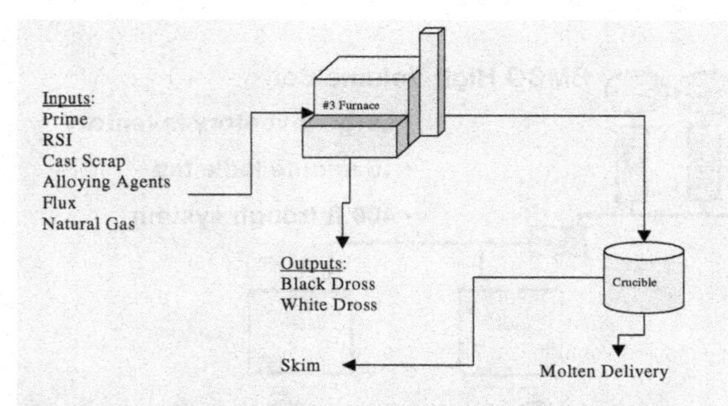

Figure 3 – AIS Components of the Energy and Mass Balance

Both AIS and SMCO represent modern, state of the art facilities. Accordingly, a resident data acquisition system existed at SMCO, able to monitor a number of energetically meaningful process parameters. A robust energy tracking system was developed from this backbone and has been operating at AIS and SMCO since June 2006.

This system was originally intended to provide energy consumption data for project benchmarking purposes, but has proven to be an effective energy management tool with long-term value. It was necessary for this study to develop an overlay system capable of querying various registers and consolidating the information. This work was undertaken jointly between GM Worldwide Facilities, SMCO, and Consumers Energy (the local power authority). The approach used is a system-wide energy balance that includes the energetic content of a mass balance.

Energy to recover aluminum from black dross is considered, as well as the tacit energy cost of white dross (unrecoverable aluminum). Figure 4 represents in schematic form the energy extraction points. At Aleris, specific melting energy, furnace holding energy, and ladle holding energy values are tracked, as well as the energy equivalents represented by dross.

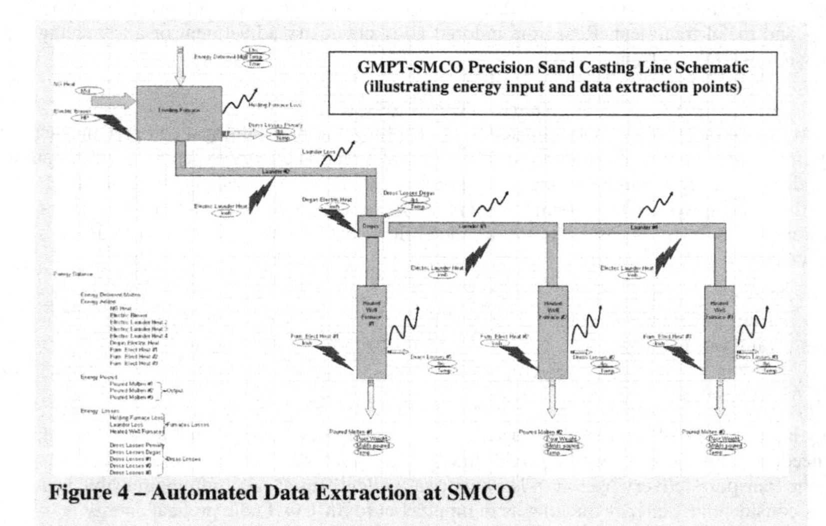

GMPT-SMCO Precision Sand Casting Line Schematic
(illustrating energy input and data extraction points)

Figure 4 – Automated Data Extraction at SMCO

Data extraction for the SMCO system include: molten metal deliveries (mass and temperature), leveling furnace (temperature, natural gas, and electric energy), degas furnace (temperature and electric energy), launder system (temperature and electric energy), pour furnaces (temperature and electric energy), and the daily molds poured.

Real time information is available for the various data extraction points in the system. Figure 5 is a template used for this purpose. Additionally, however, this information is complied on a weekly basis for generation of summary reports. A comprehensive energy survey requires mass balance information regarding skim generation, metal recovery, and melt loss. Since off-line skim assays, for example, are needed to generate this data, a time window was designated over which information is integrated.

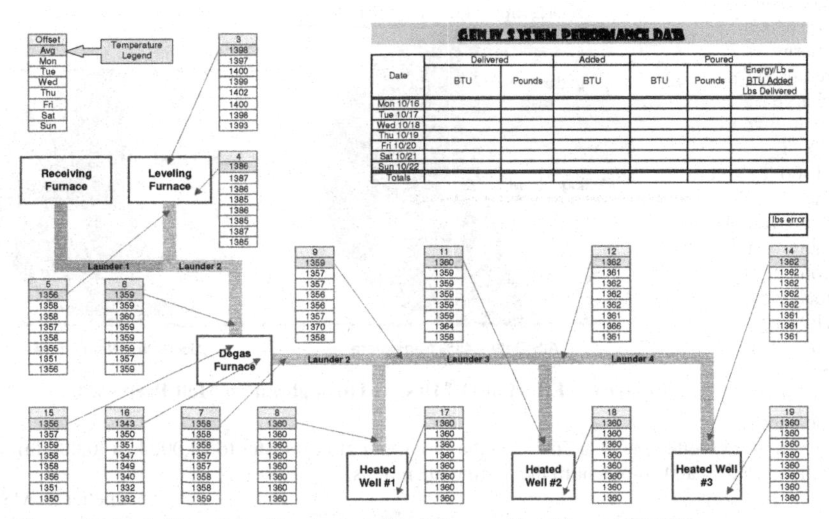

Figure 5 – Data Template Characteristic of Energy Tracking at SMCO

Benchmarking the Current System

System wide benchmarking data was collected and analyzed over 6 quarters of operation, with 2 single – quarter intervals of stringent data collection. Data of this nature can be stratified in a number of ways. These include source energy, operating mode, location, unit operation, and the order of operation. All source energy is natural gas from melting at AIS through holding at SMCO. Once metal flows from the 68,000 kg (150,000 lb) holding furnace at SMCO, however, source energy becomes exclusively electric. Electric energy itself is derived from a primary source (i.e.: coal) at prevailing power-plant conversion efficiencies. Energy expenditures viewed in the context of national consumption, must consider the *tacit* source energy equivalent, which was 9780 BTU/kW-h, implying a power-plant thermal efficiency of 35% in 2003[7]. Our analysis will be confined to in-plant process energy only, except for melt loss energies – 45.1 kW-h/kg (70,000 BTU/lb), which were not major contributors to energy expenditures at SMCO.

At the lowest level of examination, it can be said that any additional energy input beyond that which is required to produce specification grade alloy at a temperature suitable for casting is parasitic. This characteristic would be typical of a zero inventory make-to-order process. The other extreme is a process or its components that are operated at production rates far displaced from the design or "nameplate" throughput. Examples include situations where the throughput capabilities of the constituent unit operations are not coordinated, or an otherwise well designed process being operated at a large displacement from optimum.

Figure 6 demonstrates the effect of operation at attenuated throughput. Values shown are for AIS only and reflect energy partitioning of three unit operations performed within AIS: direct melting, metal holding in the melter, and metal holding after transfer to a ladle. The latter includes transport/delivery to SMCO. The former consists of energy required for charge meltdown to 760^{0}C (1400^{0}F) and the irrecoverable aluminum content of melt loss. Holding energy is being measured as the energy used to hold molten metal at 760^{0}C in the furnace or after transfer to 14,000 kg (30,000 lb) delivery ladles and shipment to the casting operations.

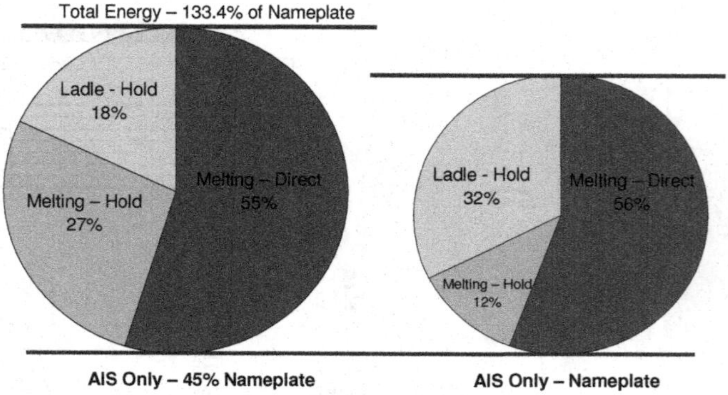

Total Energy – 133.4% of Nameplate

Ladle - Hold 18%; Melting – Hold 27%; Melting – Direct 55% — AIS Only – 45% Nameplate

Ladle - Hold 32%; Melting – Hold 12%; Melting – Direct 56% — AIS Only – Nameplate

Figure 6 – Effect of Off Design Throughput on Melt Preparation

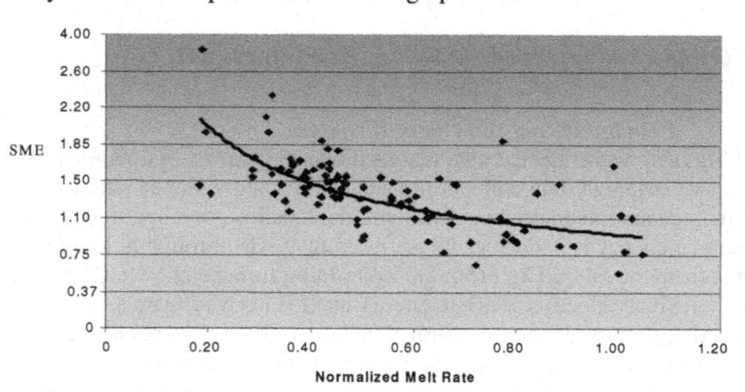

Figure 7 – AIS Melter Energy Consumption as a Function of Throughput

The melter at AIS is a 250,000 lb capacity gas fired reverberatory unit with charge preheating capability and a firing rate of approximately 50 MM BTU/hr. Interestingly, the energy *fraction* ascribed to direct melting is essentially identical in both cases, however, the *quantity* of energy used in melting at design throughput is 74.8% of the value at an attenuated 45% throughput. Aleris is desirous of achieving optimal operating conditions for maximum energy efficiency. Accordingly, melter efficiency has been measured at various throughput values. Figure 7 illustrates the effect of throughput on specific melting energy (energy used/quantity melted).

Both specific melting energy (SME) and throughput are normalized relative to melter design values in Figure 7. Data from approximately 100 melter heats are considered. Although significant higher order variables exist between individual heats, reasonable agreement ($R^2 >$ 0.85) can be seen between SME and throughput. The energetic difference between design throughput and 45% of this value is about 35%, which is good agreement with Figure 6. Reasons for reduced SME include the maintenance of containment heat during contiguous operation, the avoidance of flat bath operation, increase of burner efficiency at the design firing rate, and efficient charge heating with higher throughput. Excess firing of a melter beyond design values can result in an increase in SME due to high melt loss and the development of inhibitive bath surface oxide layers.

SMCO represents *32.9% of the total metal energy input* from melting to solidification under typical operating conditions. This value is inclusive of melt loss (only 4%), shown as a separate category, but represents holding and dispensation unit operations. Data was collected over the three operational categories of leveling furnace, pouring furnace, and launder system. Metal holding requirements understandably dominate SMCO's energy input. Although the particular holding furnace used is a state of the art unit with a congruous efficiency, holding large quantities of metal in such furnaces is energy intensive. A reduction of holding furnace temperature through the use of a launder (trough) system capable of increasing metal temperature, in-line, represents an energy saving opportunity with the existing leveling furnace. The energy expenditure distribution by operation for SMCO appears in Figure 8.

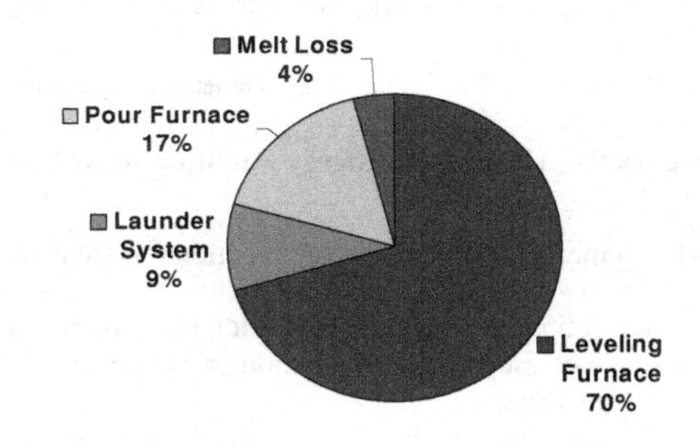

Figure 8 – SMCO Precision Sand Line Energy Expenditures

Although approximately one-third of the total energy expenditure measured in the combined AIS-SMCO system is attributable to SMCO, significant improvements can be made at SMCO by operating at design throughput. The primary reason for this is that all energy expended at SMCO is for melt holding and is therefore overhead by definition. Unlike melting, significant changes in mass specific energy consumption do not appear to occur. SMCO's energy expenditure is therefore a linear function of metal residence time. Metal residence time is inversely proportional to throughput.

Normalized to the maximum total off-design (45% attenuated throughput) system energy expenditure, AIS and SMCO accounts for 67.1% and 32.9% respectively of the total energy input. At nameplate throughput, the *total system* energy expenditure decreases to 64.1% of this value. The AIS contribution (also normalized to the maximum system energy expenditure) is 50.2% and SMCO constitutes 14.0%. AIS/SMCO energy partitioning relative to the total energy expenditure at nameplate throughput (64.1% of total energy at attenuated throughput) is 78.2% and 21.8% respectively.

Figure 9 clearly illustrates that throughput has the greatest relative impact on SMCO energy consumption.

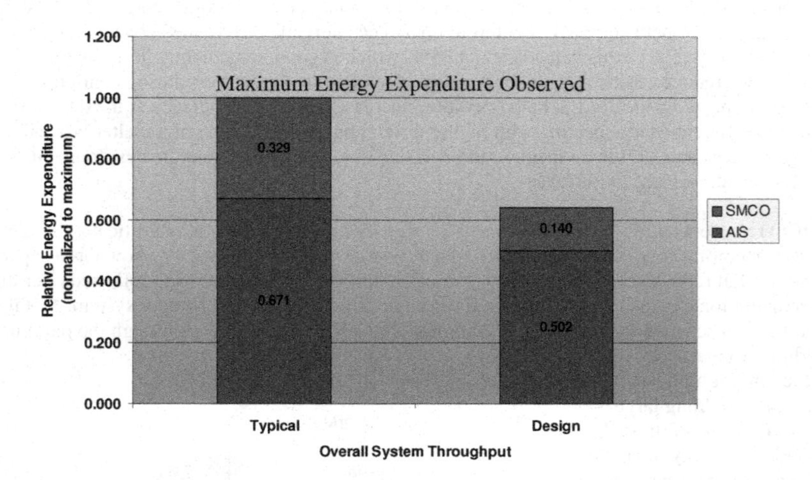

Figure 9 – AIS/SMCO Energy Partitioning at Design and Attenuated Throughput

Five points emerge in summary to conventional system benchmarking:

1. A 55% throughput attenuation (45% of nameplate value) results in a 33.4% increase in melt preparation process energy primarily due to melter efficiency increase.
2. AIS direct melting and holding relative energy partitioning remains essentially constant at 55 – 56% between nameplate and attenuated throughput conditions.
3. The SMCO contribution to total system energy decreases by 57.5% when system throughput is at the nameplate value. SMCO's impact (all holding) on sub optimal performance is greater than that of AIS.
4. Optimum operation of the current system in the most efficient mode results in a maximum one-third reduction of total energy expenditure (no new technology).
5. Although not particularly apparent from the data presented, the total system energy requirements at any throughput conditions are several multiples of theoretical melting requirements with a reasonable metal detention time. On this basis, the overall process thermal efficiency can be less than 10%.

Advanced System Description and Performance Projections

The advanced system that is being developed for AIS/SMCO demonstration and use is the integration of 3 emerging technologies. They include: the Isothermal Melting process (ITM), electric ladles (TeL), and conductive trough (CT). All of these processes are based on electrical power as the source energy. Further, continuity of operation and minimization of metal residence time are underlying design tenets. The ITM process, for example, is displacement driven flow with 2 independent heating systems – the first for metal holding purposes and the second for use during melting. Transport lag for control purposes is minimized, and automated remote monitoring provides Internet accessible system health status. Specific details of the ITM, TeL, and CT processes have been given elsewhere.[8,9,10,11]

Figure 10 illustrates a 400 kg/hr (880 lb/hr) melt rate ITM. The control panel in the foreground was added for specific monitoring activities and is not typically a component used in ITM operations. Approximately 630 kg (1400 lb) of metal is contained within this melter. Metal outflow (displacement driven) is seen in the front of the melter with solids charging on the opposite side. The control screen seen in the background is multi-layered and Internet accessible. An ITM may be gas blanketed as seen here.

Figure 10 – 400 kg/hr ITM (0.36 kW-h/kg)

Efficiency

Trough System	Date Measured	Isothermal Holding Power	Shell Temperatures	Lid Temperatures
RT	11/07	535 – 790 w/ft	102°F +/- 13.6^{O}F (1σ)	130°F +/- 20.1^{O}F (1σ)
CT- Gen1	08/07	> 1200 w/ft	185°F +/- 19.8^{O}F (1σ)	257°F +/- 59.5^{O}F (1σ)
CT- Gen2	11/07	698 – 975 w/ft	132°F +/- 12.9^{O}F (1σ)	219°F +/- 40.8^{O}F (1σ)
CT- Gen2	4/08	761 w/ft	109°F +/- 13.7^{O}F (1σ)	211°F +/- 31.5^{O}F (1σ)
CT- Gen3	4/08	460 - 526 w/ft	105°F +/- 18.4^{O}F (1σ)	108°F +/- 23.5^{O}F (1σ)

Performance

Trough System	Power kW	ΔT @ 19,000 lb/hr	Recovered Metal Heating Power (kW)	Recovered Metal Heating Power (BTU/hr-ft)
CT Gen 2	47.6	25.4	36.8	6,280 BTU/hr-ft

Figure 11 – Performance Characteristics of the CT (Gen 3 latest)

The total energy investment by SMCO is approximately 33% of the total system input, however, this is a "hold only" system and no melting is performed. Requirements of the casting process necessitate the use of high holding furnace temperatures to accommodate an approximate 40°F decrease in the trough (launder) system. The use of CTs will provide for sufficient addition of in-line make up heat that permits a *reduction* of holding furnace temperature and corresponding energy savings. During a 12-month operational evaluation, the thermal performance of the CT, among other things, was measured. This information is summarized in Figure 11. "RT" is the best commercially available conventional trough and "CT – Gen 1 and 2" represent developmental versions of the CT. A 6.1 m (20 foot) section of this trough is capable of imparting almost 37 net kW to a flowing metal stream – sufficient to raise metal temperature 25°F at 8630 kg/hr (19,000 lb/hr).

A 3180 kg/hr (7,000 lb/hr) ITM will be located in an enclosure at AIS. Based on gravity driven charge induced displacement, molten metal will flow from the ITM into a small heated docking station and will be periodically transferred into an 11.3 MT (25,000 lb) capacity TeL receiving vessel. Once full, the TeL will be transported from AIS to SMCO and received at a second docking station. Metal will flow directly from the TeL to the docking station in a controlled and isothermal manner consistent with the casting rate requirement. Metal treatment will also occur in the docking station.

A length of CT will continuously transfer metal from the docking station to the casting unit at a temperature several degrees lower than required by the casting unit. Since the CT is capable of imparting in-line sensible heat, the final delivered metal temperature will be adjusted by in-line closed loop control. The entire in-line molten metal inventory will be only several thousand pounds in excess of the TeL capacity, which has 100% drawdown. Molten metal is essentially provided on a demand basis without significant detention.

The energetic performance expectations for this system are shown in Figure 12. Differentiation is by unit operation or function. These values are compared with benchmarks that represent typical industry performance at the 95th percentile. The performance of the existing AIS/SMCO system is at the high end of the ranges cited in Figure 12.

Unit Operation	Industry 95th Percentile	Project Goal	Enabler
Melting	0.9 – 1.6 kW-h/kg 1400 – 2500 BTU/lb	0.353 kW-h/kg 552 BTU/lb	Isothermal Melting (ITM)
Holding	16 – 42 w/kg 25 – 65 BTU/lb-hr	5.5 w/kg 8.6 BTU/lb/hr	BSPP Holders
Transport	> 32 w/kg 50 BTU/lb-hr	4.5 w/kg 7 BTU/lb/hr	Electric Ladles (TeL)
Melting to Solidification (Includes all melt loss)	0.9 – 5.15 kW-h/kg 1400 – 8,000 BTU/lb	[illegible]	ITM + TeL + Conductive Trough

Figure 12 – Comparison of Anticipated and Actual Energy

Figure 13 depicts the expected performance of the advanced system distributed over 4 energy categories, expressed in w-hr/kg. Note that less energy will be used from initial charge melting to the pouring cup than most conventional melting processes alone.

Energy attributable to melt loss appears to be disproportionately high relative to total because of low overall energy expenditure and high tacit energy content of melt loss (45.1 kW-h/kg). Notably, only 179.2 kW-hr/kg (278 BTU/lb) will be required for melt detention. Very low system capacitance and total drawdown containment vessels are partially responsible for this anticipated performance.

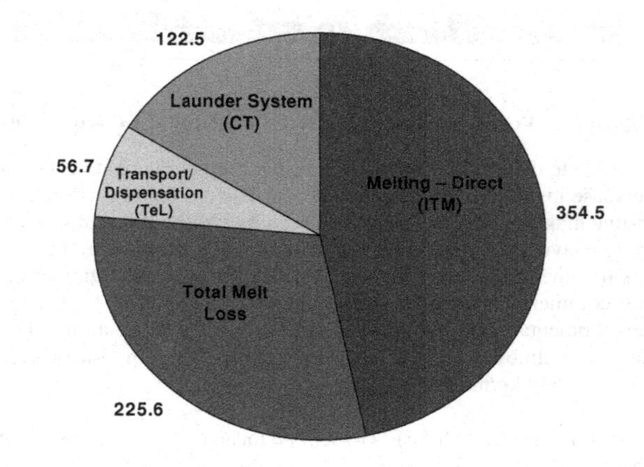

Figure 13 – Energy Apportionment of Advanced System (w-hr/kg)

Efficient heating also dramatically influences overall thermal efficiency. Electricity was selected as source energy for all unit operations of the system, using resistance as the energy conversion method. The thermal efficiency of this conversion process is essentially 100%. By contrast, unrecuperated natural gas combustion processes typically operate at 20% to 30% thermal efficiency. Energetic demands of these systems, such as wall losses for metal containment, are multiplied by a factor of 3 to 5 to represent the actual burnerhead energy input. Electrical – thermal energy conversion does not result in the in-plant generation of products of combustion (POCs), including greenhouse gases.

The analog to Figure 9 is presented in Figure 14, wherein the advanced system is included for comparison purposes. Due to a systemic de-emphasis on holding energy of the advanced system, SMCO based energy apportionment is only 19% of the total energy required by this system. Again, no melting occurs at SMCO, and all energy expenditures at this location are parasitic by strict definition.

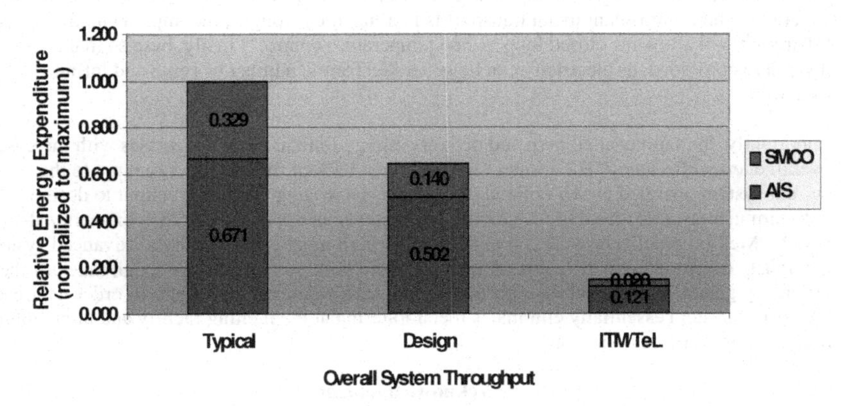

Figure 14 – AIS/SMCO Energy Apportionment Illustrating Current and Advanced Systems

The following points summarize the advanced system:

1. Reduced system capacitance decreases metal detention, increases energy efficiency and lowers response time, approaching a high throughput make-to-order system.
2. Greater proportionate energy expenditures at AIS correlate to higher overall system efficiency; with AIS/Total energy ratios of: Attenuated throughput – 67.1%, Design Throughput – 78.2%, and Advanced System – 81.2%.
3. Electric resistance energy conversion operates at near theoretical efficiency is used exclusively in the advanced system.
4. Two characteristics of the advanced system drive energy savings: minimal parasitic loss and high source energy conversion efficiency.

Summary

At fruition, the proposed system will significantly improve melting energy efficiency, recovery, and melter floor space utilization at secondary aluminum alloy processing operations. Additionally, it will eliminate holding furnaces at shape casting facilities, provide in-line temperature control, and alloy chemistry flexibility, and raise the quality of molten metal to near aircraft castings standards. Since the Aleris/SMCO operation is typical (if not state of the art) of the majority of domestic high volume foundries, proliferation of the technology to other facilities will not be problematic. The ITM/TeL process essentially allows a foundry to out source melting and eliminates the need for large holding furnaces. This reduces labor costs (among other things), which frequently constitutes 50% of the cost to produce castings.

A paradigm shift will occur in the industry as a consequence of the proposed system. Geographically optimized central melting operations will melt aluminum with a specific melting energy of 0.353 kW-h/kg (552 BTU/lb-hr), and without in-plant process emissions.

Molten aluminum will be transferred from the melter in a turbulent free manner into heated vessels, thereby dramatically reducing transfer related melt loss and maintaining metal quality. At the foundry, these heated ladles will function as holding furnaces, since they have integral heating, metal treatment, and controlled dispensation systems. The ladles are expected to have a specific holding energy requirement of less than 5.5 w/kg (8.5 BTU/lb-hr), which is less than 25% of the current average holding furnace energy cost of 29 kW/kg (45 BTU/lb-hr). In situations where intra-plant metal transport is facilitated by troughs, the implementation of the CT trough will allow for closed loop in-line temperature control. Finally, base source energy diversity is provided, as electricity can be generated from a number of finite and infinite sources.

Importantly, the ratio of melt prep and delivery energy to total energy increases with increased overall energy efficiency. The former represents just 67% of the total energy input to the current system operated at sub-optimal throughput. Increasing metal throughput to design optimum increases this ratio to 0.782, which represents an upper limit of the conventional system. Melt prep/delivery is 81.2% of the total system input energy with the advanced system. Obviously energy efficiency is remarkably improved with this system. Although much of the efficiency gain is due to a substantial reduction in melting energy, the make-to-order nature of the project system essentially eliminates metal loitering at the melting facility and large volume holding at the casting operations.

Acknowledgements

A portion of this work is being supported under the US Department of Energy projects DE-FC07-01ID14021 and DE-FG36-05GO15047. The authors gratefully acknowledge Doug Kaempf, Glen Strahs and Bradley Ring, US Department of Energy – Office of Industrial Technology; Gary Merritt and Bill Dobish of Aleris International; John Sullivan of General Motors Worldwide Facilities; Professor David Schwam of Case Western University; and Mike Kinosz, Scott Moline, George Ducsay, and Tom Meyer of Apogee for their insight, support, guidance, and advice provided for this work.

---

[1] Elwin Rooy and Associates, March, 2007
[2] DE-FG36-05GO15047 "Energy Efficient Melting & Direct Delivery of High Quality Molten Aluminum"
[3] Osborne, M.; Eckert, C.E.: Meyer, T.N.: Kinosz, M.; "Preventative Metal Treatment Through Advanced Melting System Design"; Shape Casting: 2nd International Symposium; TMS; 2007
[4] Rooy, E.L.; "High Intensity Resistance Heating Offers Power and Efficiency for Industrial Heating Processes"; Industrial Heating, February 2008
[5] Osborne, M; "Advanced Aluminum Melting and Dispensation – Foundry Applications"; Ohio Technology Showcase – Casting Section"; Cleveland, OH; September 28, 2005.
[6] "Isothermal Melting: Reaching for the Peak of Efficiency in Aluminum Melting Operations"; Energy Matters-Spring 2008; Washington D.C.; U.S. Department of Energy, EERE-OIT http://apps1.eere.energy.gov/industry/bestpractices/energymatters/articles.cfm/article_id=271
[7] Choate, W.T.; Green, J.A.S.; "U.S. Energy Requirements for Aluminum Production: Historical Perspective, Theoretical Limits and New Opportunities"; U.S. Department of Energy – EERE, Washington, D.C.; 2003; Appendix – C, Table D-1
[8] "The Isothermal Melting Process – A Success Story", (Washington, DC: U.S. Department of Energy Industrial Technology Program, Friday Apr 16, 2004), http://www.oit.doe.gov/cfm/fullarticle.cfm/id=822
[9] "The Isothermal Melting Process"; Aluminum Now; Vol. 6; No. 4; July/August 2004; Washington, DC; The Aluminum Association; 26-29.
[10] New Technology Spotlight – Isothermal Melting; Light Metals Age; August 2005; 70
[11] US Patents: 5,850,072; 5,850,073; 5,894,541; 5,963,580; 5,968,223; 6,069,910; 6,066,289; 6,444,165; 6,872,294

Energy Technology Perspectives
Edited by: Neale R. Neelameggham, Ramana G. Reddy, Cynthia K. Belt, and Edgar E. Vidal
TMS (The Minerals, Metals & Materials Society), 2009

OXYFUEL – ENERGY EFFICIENT MELTING

Thomas Niehoff[1], Dave Stoffel[2],

[1]Linde Gas, 85716 Unterschleissheim, Carl-von-Linde-Str. 25, Germany
[2]Linde North America, Inc., 2100 Western Court, Suite 100, Lisle, IL 60532, USA

Keywords: Energy, Oxy-fuel, Furnaces, Combustion Systems, Process Modifications

Abstract

Energy in the form of natural gas, oil and electricity is expensive and will continue to be a rare resource in the future. Recycling of metals, instead of primary production, is a logical and crucial step towards energy conservation. Energy consumption can be impacted by using advanced combustion systems for metals recycling. Air-fuel combustion has been the conventional method to melt and recycle metals. Global competitive pressures, "greenhouse" gas emissions, and high energy prices force melt shop operations to optimize energy usage and reduce overall operating costs. Oxy-fuel combustion and process technology can be applied to metal recycling operations to meet these objectives. Linde has converted several hundred furnaces from air-fuel to oxy-fuel technology and has extensive experience in optimizing the use of oxygen in furnaces. Additionally, Linde has the process knowledge necessary to avoid costly start-up issues that can accompany the transition to oxy-fuel technology. This paper will describe the benefits and potential issues for conversion to oxy-fuel.

Introduction

The furnace combustion system is the heart of a fossil fuel fired melting process and must be properly matched to the furnace to work as one cohesive unit. Furnaces vary widely in size, shape and features. Rotary furnaces are barrel-shaped and can be designed for tilting or non-tilting operation. In a rotary furnace, the products of combustion create a single or double pass through the furnace prior to exiting. Reverberatory furnaces are box-shaped and can also be designed for tilting or non-tilting operation. Reverberatory furnaces are constructed of single or multiple chambers, with or without metal bath movement, and designed for continuous or batch processing. Different furnaces are used for different feedstock materials and products.

Furnace combustion systems also vary in many ways. Combustion systems can be of air-fuel or oxy-fuel based technologies, or a combination of the two, resulting in an air-oxy-fuel system. Each of these types of systems has different combustion characteristics which can produce very different results. Here we will describe specific differences in melting processes and their associated energy requirements.

Furnaces

Rotary furnaces are typically used for fast melting of metals and drosses. Rotary furnaces are not normally used for metal treatment or holding. The material is charged in intervals and melted as quickly as possible. The furnace constantly rotates during metal processing, creating dust and smoke that is carried over and exhausted with the products of combustion. A gas extraction fan pulls the off gases out of the furnace or hood into the gas cleaning plant.

Reverberatory furnaces are typically used for larger scale melting operations, where metal treatment or holding of metal may be required. Reverberatory furnaces cannot rotate to achieve homogeneous bath conditions. A liquid metal pump, an electro magnetic stirrer or gas bubbles can be used to generate bath movement and homogenize the molten metal in terms of temperature, chemistry and density.

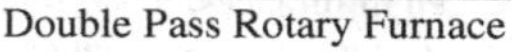

Double Pass Rotary Furnace

Direct Charge Reverberatory Furnace

Side Well Charged Reverberatory Furnace

Figure 1. Furnaces Types

Each melting process and furnace type requires a different combustion system. A melting process with organically contaminated (e.g. oil, paint, plastic) feedstock material ideally needs a combustion process that burns the gasified combustibles in the furnace chamber and uses their heat to save primary energy (natural gas, oil, etc).

A rotary furnace combustion system needs to be powerful and flexible to quickly melt the material down and then carefully reduced the firing rate as to not overheat the metal. When combustibles are charged with the feedstock material in a rotary furnace, it is important to increase the retention time of the gases in the furnace to allow a full burnout of the combustibles. Reduction of gas volumes by cutting out the nitrogen associated with air-fuel combustion and using extra oxygen via lance injection will enhance the process and save energy. For the various steps (see Figure 2) of the melting process, different flue gas volumes are generated and different firing rates and flame shapes are required.

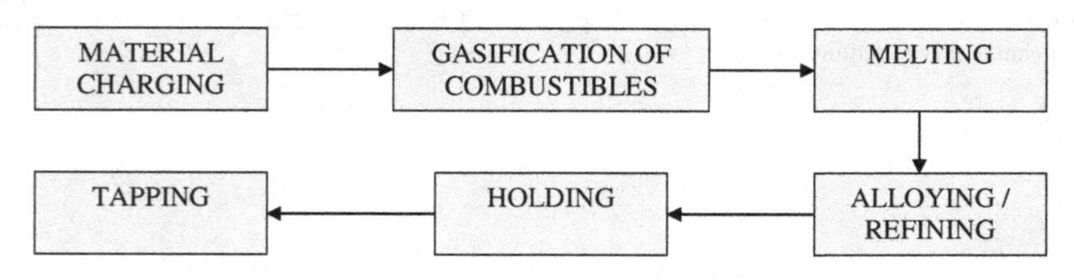

Figure 2. Melting Process Steps

Reverberatory furnaces can be operated as batch or continuous melting processes. Subject to the type of materials being processed and features of the combustion chamber(s), there are different requirements to the combustion system. Some reverberatory or hearth-type furnace melting processes require a long and luminous burner flame, with highly convective furnace gas conditions to achieve a good melting result. Other furnaces and process types require a soft, lazy burner flame with reduced luminosity to avoid overheating of the metal.

For holding intervals and slow melting characteristics of clean feedstock material, air-fuel combustion systems and regenerative air-fuel combustion systems have demonstrated positive melting results and economic melting. Fast melting and flexible operating needs for contaminated charge materials are best achieved by oxy-fuel and air-oxy-fuel combustion systems.

Combustion Systems

Air-fuel combustion systems generate the highest amount of flue gases and allow for the shortest retention time of the gases in a furnace. This is due to the nitrogen contained in air (78 vol.-%), which is carried through the process without contribution. Flue gas ducts, extraction fans and dust removal systems must be sized accordingly.

Air-oxy-fuel combustion systems are flexible in their way to adjust for specific needs of a melting process or furnace. The flame length, luminosity and convective heat can be adjusted by the air to oxygen ratio as desired. An extra oxygen lance can be used to burn out combustibles in the furnace chamber.

Bright and radiant oxy-fuel combustion systems are optimum for fast melting in rotary furnaces. There they achieve economic and efficient melt results. Oxy-fuel combustion systems can be equipped with an extra lance to support and enable combustion of charged organics.

Oxy-fuel combustion systems can be specifically tailored to achieve the best results in terms of energy efficiency, emissions and melting cost. Flames of oxy-fuel combustion systems can vary widely to give best results to your melting operation. Oxy-fuel flames can be bright and luminous, round, flat, long, short, flameless, low temperature and high temperature. Linde has a wide range of combustion technologies to offer to cover all aspects for efficient and cost effective melting of metals. Linde has decades of experience converting conventionally operated melting processes to oxy-fuel or air-oxy-fuel.

Linde's Non Ferrous Combustion Technologies

WASTOX® combustion technology incorporates oxygen lancing technology with burner combustion management to effectively combust emissions (i.e. volatile organic compounds) within the furnace. By monitoring flue gas compositions, WASTOX® burns off carbon based materials within the charge and uses this energy to reduce overall burner heat input to the furnace. This patented technology has been successfully installed on single and double pass rotary furnaces throughout the world.

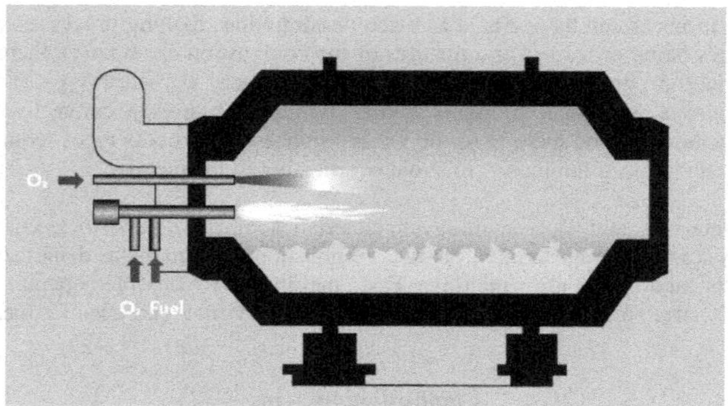

Figure 3: WASTOX® principle.

Figure 4: WASTOX® Installation in a double pass rotary furnace.

FLEXFLAME® combustion technology tailors the oxy-fuel burner flame length to the specific requirements of the melting operation. The ability to adjust the flame length allows the operator to optimize the heating characteristics of the flame for each of the various melting process steps, without having to alter the burner firing rate or nozzle geometry. FLEXFLAME® technology has been installed on both reverberatory and rotary furnaces to increase productivity, reduce overall operating costs and furnace emissions.

Low Temperature Oxy-Fuel (LTOF) combustion technology is designed to boost melting capacity within a reverbatory "box-type" furnace. The uniquely designed burner allows for the recirculation of flue gases into the burner mixing zone, which dilutes the oxygen concentration and slows down the combustion reaction, thus creating a lower flame temperature comparable to air-fuel technology. The LTOF flame characteristics result in uniform furnace temperatures, elimination of furnace hot spots, reduced fuel consumption, reduced emissions and improved metal yields.

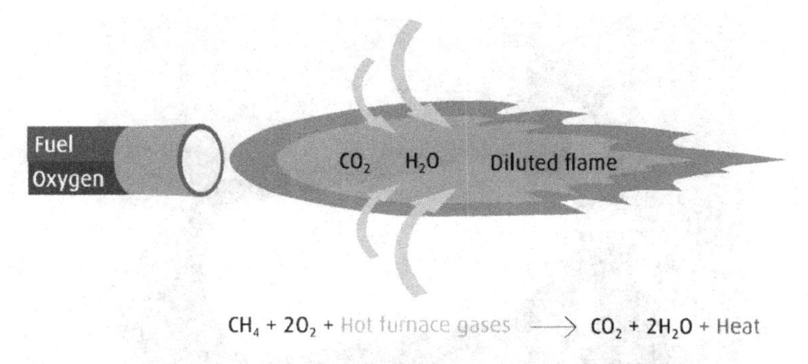

Figure 5: Low Temperature Oxyfuel (LTOF) Principle

AIROX® combustion technology combines the benefits of air-fuel technology with oxy-fuel technology to create a versatile system that can operate between 21% oxygen (air-fuel) to 100% oxygen (oxy-fuel) combustion. This flexibility in operating range allows the operator to adjust the flame characteristics for each melting process step while managing overall operating costs through the optimization of energy consumption.

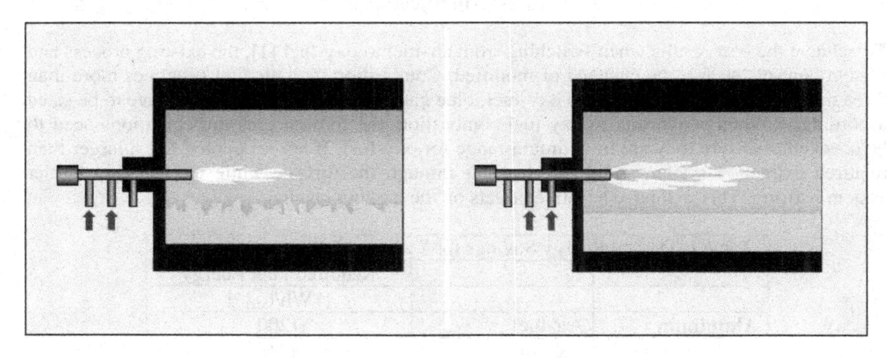

Figure 6: AIROX® Principle (right side oxy-fuel, left side air-fuel).

Linde worked with Hertwich Engineering GmbH and CORUS Aluminium Voerde GMBH to develop the Universal Rotary Tiltable Furnace (URTF) to meet the demands of the secondary aluminum industry. The URTF provides outstanding process control to increase productivity and metal yield while minimizing energy consumption.

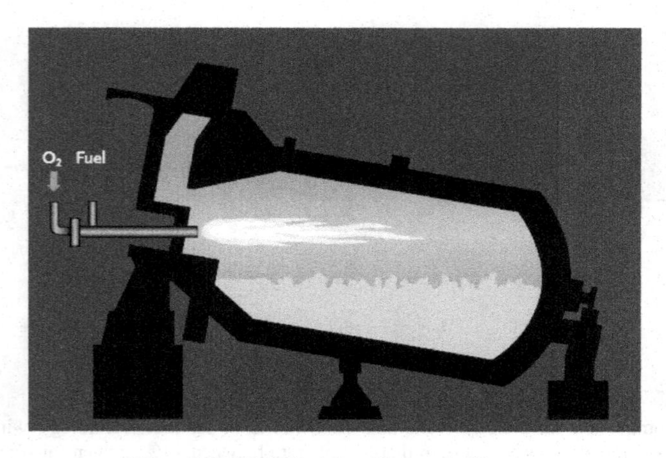

Figure 7: URTF Principle with oxy-fuel burner.

Process Modifications

To achieve the best results when switching from air-fuel to oxy-fuel [1], the existing process and system controls need to be changed or modified. Combustion with air-fuel generates more than three times the flue gas compared to oxy-fuel. Flue gas ducts and extraction fans have to be sized accordingly. When converting to oxy-fuel combustion, the furnace pressure conditions need to be checked to ensure they are in optimum range for oxy-fuel. If not accounted for, a larger than required extraction system could pull leak air through the furnace while the oxy-fuel burner system is firing. This will have negative effects on the melting results.

Table I: Typical Energy Savings for Various Melting Processes [1]

		Required Fuel Energy [kWh/t_{Metal}]
Aluminum	Air-fuel	1200
	Oxy-fuel	660
Copper	Air-fuel	1100
	Ox-yfuel	440
Iron	Air-fuel	900
(cupola)	Oxy-fuel (+10%)	720

Oxy-fuel burners with a bright and luminous flame have higher flame temperatures compared to air-fuel flames. When installing oxy-fuel technology onto a furnace, it is advisable to install thermocouples to protect the refractory material from overheating.

When choosing the oxy-fuel combustion system, extra care should be taken to properly fit the size and shape of the flame(s) into the furnace chamber. Short oxy-fuel flames will create very localized heating and high temperature gradients in the furnace. Long oxy-fuel flames can impinge upon the refractory or mechanical parts of the furnace and damage them. The flame's length needs to be adjusted to the specific furnace hearth dimensions and optimized for charge material and process conditions. This includes a variation of flame characteristics for the different steps of a melting process with just one burner type or several burner types. Linde operates R&D centers with scientists to develop the best combustion technology for various melting processes. CFD computer modeling of the furnace and burner profile assist with proper oxy-fuel burner placement and use. Trained and skilled combustion application engineers assist the customer in obtaining the best, most efficient results for their melting operation.

Summary

Converting air-fuel combustion and melting operations to oxy-fuel technology will save energy, increase productivity and reduce overall operating cost. Typically, air-fuel technology gets converted to an oxy-fuel or air-oxy-fuel type of combustion process. The conventional process interlocks, safety interlocks and process controls need to be adjusted to the new process specifics to ensure safe operation and desired results. Linde offers a range of combustion technologies to improve existing and new melting processes. Additionally, Linde optimizes the process conditions and controls to achieve maximum benefits for your operation. More than a hundred furnace conversions and decades of experience in oxy-fuel combustion and process know-how, make Linde industry experts.

References

1. T. Niehoff "Oxyfuel – Solutions for Energy and Environmental Conservation", TMS 2008, Feb. 2008, New Orleans, LA

Energy Technology Perspectives
Edited by: Neale R. Neelameggham, Ramana G. Reddy, Cynthia K. Belt, and Edgar E. Vidal
TMS (The Minerals, Metals & Materials Society), 2009

ENERGY SAVINGS AND PRODUCTIVITY INCREASES AT AN ALUMINUM SLUG PLANT DUE TO BOTTOM GAS PURGING

Klaus Gamweger[1], Peter Bauer[2]
[1]RHI AG, Technology Center Leoben, Magnesitstraße 2, A-8700 Leoben, Austria
[2]NEUMAN Aluminium Austria GmbH, Werkstraße 1, A-3182 Marktl im Traisental, Austria

Keywords: Gas purging, porous plugs, aluminum slugs, melting rate

Abstract

Gas purging systems are well established in nonferrous metallurgy at multiple steps during metal production, including melting, converting, alloying, and metal cleaning. Since the kinetics of all of these stages are positively influenced by using gas purging systems the overall process benefits are substantial. In the aluminum industry, inert or reaction gases are blown into the melt using porous plugs to achieve improved metal grades, higher productivity, and more efficient energy utilization. To enable the most effective application of the process gases a complete package, termed AL KIN, was developed by RHI that includes porous plugs, refractory expertise, gas supply technology, and gas control equipment. This paper discusses the technological and economic advantages of this system and the specific benefits of fuel reduction, production time savings, decreased process gas consumption, and improved refractory service life and maintenance are illustrated using an aluminum slug plant in Austria.

Introduction

In addition to high quality production, increasing productivity is the main goal in many aluminum cast houses to maintain or even strengthen their market position. The Neuman Aluminium Group, based in Austria, North America, and China, is the largest slug-producer for tubes, cans, and automotive parts worldwide.
Recently, a comprehensive improvement program was initiated at the Neuman Aluminium Austria slug plant in Marktl - the main factory and technological centre of the group. The principal goal of the optimization project was to increase the productivity of the strip-casting production. The investment included a gas purging system, ingot preheating equipment, a new insulated launder system, and optimized melting cycles. However, the introduction of the gas purging system was pivotal to the project and has accounted for approximately 50% of the benefits achieved. This paper describes the AL KIN gas purging technology and details the process benefits resulting from the improvements at Neuman Aluminium Austria.

AL KIN gas purging system

Porous Plugs and Refractories

The porous plugs for aluminum furnaces are manufactured from chromium-alumina with an optimized porosity and pore size distribution for this application. This guarantees not only the formation of very small bubbles, which is extremely important for the degassing properties of the system, but also enhances resistance against aluminum infiltration up to a bath depth of 1400 mm even when no gas is being purged through the plug.

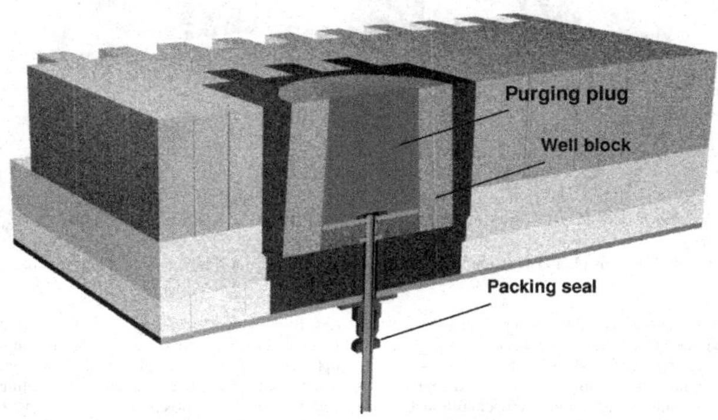

Figure 1. Schematic of a purging plug set installed in a brick-lined tilting furnace.

The fired material is distinguished by a high mechanical strength as well as corrosion resistance against all process gases.

The porous plugs are available with a maximum length of 330 mm and the dimensions are selected according the furnace size and lining thickness. The porous refractory is enclosed in a stainless steel casing manufactured from an appropriate steel grade to withstand chlorine attack or high temperatures. The well block is produced from a high alumina low cement castable with the addition of a non-wetting agent to guarantee infiltration resistance against aluminum. A porous plug set (i.e., plug and well block) installed in the hearth of a brick-lined tilting furnace is illustrated in Figure 1. If the plugs are installed in a stationary furnace with limited space underneath, the piping is incorporated into the permanent lining with the gas connection on the furnace sidewall. If chlorine is also used as a purging gas the connection pipe and steel shell must be tightly sealed using a special packing seal to eliminate any leakage risk.

After defining the requirements and targets of a specific gas stirring system the optimal plug number, position, and gas flow rates are calculated. For this engineering work, performed at the RHI Technology Center Leoben (Austria), the Fluent computational fluid dynamics software is used. Depending on the process (e.g., melting or holding furnace), the required goals, as well as the on-site conditions (e.g., furnace dimensions, gas supply, and gas pressure) the gas bubble distribution or bath agitation is optimized.

Gas Control Station

In addition to the plug sets, the piping and gas control station are also included in the package. The gas control station is central to the AL KIN purging system. The standard equipment has two gas inlets (typically for nitrogen and chlorine), a gas mixing station, and an outgoing pipe for each porous plug. In service experience has shown that maintaining a constant gas flow irrespective of the molten metal bath depth is of utmost importance to run the stirring system effectively. The gas pressure must be carefully monitored in real time by the software and adjusted to produce a consistent stirring action. A minimum pressure of 6 bar for the inlet gas is required to provide sufficient flexibility to maintain the mass flow constant.

Gas Control Software

The software controls each plug individually to achieve a constant gas flow rate. This software is run independently of the on-site furnace computer system and enables constant monitoring and adjustment of the gas flow rates as required during the treatment phases of the process. Depending on the particular requirements, different gas mixes and different flow rates for each individual plug can be programmed.

AL KIN Installation and Process Benefits in a Melting Furnace

A 5.8-tonne melting furnace at Neuman Aluminium Austria was equipped with 6 purging plug sets (Figure 2).

Figure 2. Porous plug installation in a 5.8-tonne melting furnace at Neuman Aluminium Austria.

Typically, the furnace is charged with 3-tonne ingots and 2.5–2.8 tons of internally generated punching scrap and aluminum fines. The standard production range is 3103, 3207, and 3102 alloys. During melting and alloying it was routinely necessary to open the furnace door and stir the bath, which led to a temperature loss of 40 °C each time. Due to the agitation effect of the plugs (Figure 3) this is no longer necessary. The nitrogen gas flow rates of 9 Nl/min during melting and 6 Nl/min during holding were optimized during the installation phase and are now fixed for each batch. As a of result gas purging the melting rate has increased significantly by between 33 and 38% with a major positive influence on the productivity (Table I).

Figure 3. Porous plug gas purging in a 5.8-tonne melting furnace.

Prior to plug installation, the feed lying on the furnace floor had to be moved using a large steel rack mounted on a forklift truck. This frequently resulted in major damage to the refractory lining and on one occasion the furnace was put out of operation due to significant roof damage. Eliminating this mechanical stirring method has also had additional cost benefits as prior to purging plug installation the steel rack had to be replaced every month.

The excellent mixing effect of the porous plugs has also resulted in a homogenous temperature distribution from the top of the melt to the floor. Therefore, the bath surface is no longer overheated to provide energy transfer from the burner into the liquid metal. Previous investigations have shown that the temperature gradient in a 1100 mm deep bath is often more than 50 °C. With the use of porous plugs this thermal stratification can be eliminated within only 5 minutes. This not only saves considerable amounts of natural gas but also decreases refractory wear in the highly stressed three-phase zone at the metal, dross, air interface, termed the belly band zone. As a result repair cycles in this area have been extended.

Additional temperature losses of approximately 100 °C in total took place in the 20 m launder system from the furnace to the casting station and in the impeller degassing box. The launders were redesigned and insulated and the impeller is no longer necessary due to efficient degassing with the porous plugs in the furnace. Thereby, maintenance costs for the rotary lances have also been eliminated.

Due to all these improvements the metal temperature in the furnace has been decreased significantly, which has not only reduced the natural gas consumption but has also had a substantial influence on the metal loss due to combustion and dross formation (see Table I). These losses increase exponentially with the temperature from 750 °C upwards, which directly influences the production tonnage.

Production figures	Without plugs	With 6 plugs
Melting time (min)	90–105	60–65
Metal loss due to dross formation (%)	5.4	1.2
Metal loss due to combustion (%)	1.00–1.23	< 0.5
Specific natural gas consumption (%)		– 15.1
Production increase per year (%)		16.0

Table I. Summary of production figures with and without the use of 6 porous plugs in the melting furnace at Neuman Aluminium Austria.

Conclusion

The presented case study at Neuman Aluminium Austria demonstrates that the introduction of an appropriate gas purging system can improve the productivity of an aluminum melting furnace significantly. All kinetic-dependent processes increased due to the introduction of inert gas bubbles into the melt. The melting rate of solid charged metal - in this case mainly ingots and punching scrap - was increased by up to 38%. As a result of this substantial improvement and homogenization of the bath temperature, the specific gas consumption has decreased by approximately 15%, which also has a major positive impact on CO2 emissions. In addition, reduced opening of the furnace door and appropriate launder insulation has enabled the process temperature to be decreased, which has led to a reduction of metal loss due to lower dross formation and reduced metal combustion. In summary, the result of this improvement project was a yearly production increase of 16% at lower costs.

Energy Technology Perspectives
Edited by: Neale R. Neelameggham, Ramana G. Reddy, Cynthia K. Belt, and Edgar E. Vidal
TMS (The Minerals, Metals & Materials Society), 2009

ENERGY CONSERVATION & PRODUCTIVITY IMPROVEMENT MEASURES IN ELECTRIC ARC FURNACES

Ajit Kumar Jaiswal[1]

[1]Steel Authority of India Ltd. – Centre for Eng. & Technology, 4th Floor RDCIS Lab Bldg., Ranchi, Jharkhand, 834002, India

Keywords: EAF, Power Consumption, Oxy-fuel burners, Foamy Slag, Electric, Hot Metal, Scrap preheating, Hot DRI

Abstract

Electric Arc Furnace (EAF) being a power-intensive equipment, EAF based industries are highly dependent on scarce electric energy. This is the reason Government of India is highly concerned about this and laid emphasis on the conservation of electric energy.

This paper highlights factors contributing to reduction in specific electric energy consumption in EAF and suggests measures not only for power saving, but also for productivity improvement. To name a few, these are proper sealing of furnace, introduction of oxy-fuel burners, foamy slag practice, increased usage of hot metal/DRI, post-combustion of CO gas for preheating of scrap, improved selection & design criteria of electrics, power demand management, and process automation. Few cases also have been cited, where the above measures have helped in reaping significant benefits.

Introduction

For a long time, Government of India had focused on increase in energy supply, in particular, through electricity generation. Now the emphasis of Government of India has shifted from electricity generation to its conservation for following two reasons:

- Power plants have become very cost-intensive
- By saving energy, more of it can be delivered at lower cost to the consumers

Power generating units, transmission & distribution agencies as well as consumers all have roles to play in the area of energy conservation. By energy conservation, we mean curtailing energy requirement by reducing energy wastages and taking up measures to improve energy efficiency of electric equipment/appliances. In almost all cases, this would be economically beneficial.

India's cost-effective energy conservation potential has been estimated by the Planning Commission at 23 per cent of total commercial energy generated [1]. A national movement for energy conservation can significantly reduce the need for fresh investment in energy supply systems in coming years. It is imperative that we make all-out effort to realize this potential. Whether a household or a factory, a small shop or a large commercial building, a farmer or an office worker, every user can and must make this effort for his own benefit and for the nation's benefit as well.

In this paper, we will not look out for wastages of energy in the area of generation, transmission & distribution, but we will confine our discussions on energy conservation measures in a small but prominent fraction of consumers that is EAF. During 2006-07, mini steel plants contributed 25.20 Mt out of total of 50.85 Mt of steel produced in India. , in which EAF's share was 9.8 Mt. EAFs, where a large number of A.C. and D.C. drives, furnace duty transformers and other electrical & technological equipment are operating for long duration everyday. These are obviously power intensive and consume a good share of electricity.

Reduction in energy conservation is of paramount importance for EAF operators, next come productivity and cost of production. Several developments have taken place worldwide in EAFs over the period from last decades and as a result, some of the parameters like Tap-to-tap time, Specific electrical energy input and Specific electrode consumption have improved considerably, which has been illustrated with the help of the following diagram: [2]

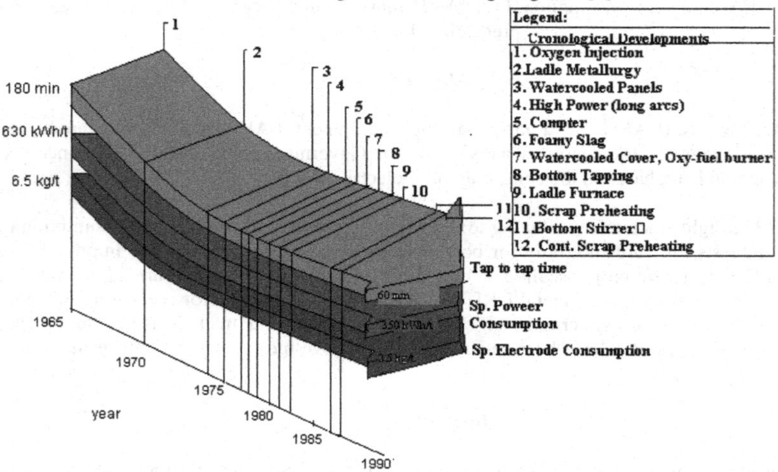

Figure 1. Progressive development and related improvement in some of the parameters of EAF

Before start discussing, let us have a look at a typical energy balance of an EAF in the following figure: [3]

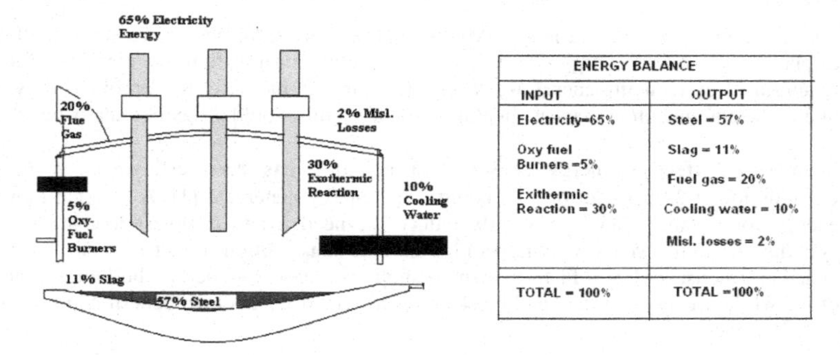

ENERGY BALANCE	
INPUT	OUTPUT
Electricity=65%	Steel = 57%
Oxy fuel Burners =5%	Slag = 11%
Exithermic Reaction = 30%	Fuel gas = 20%
	Cooling water = 10%
	Misl. losses = 2%
TOTAL = 100%	TOTAL =100%

Figure 2. A typical Energy Balance in EAF

Ways & Means of Energy Conservation in EAFs

Some of the important measures helpful in improving specific energy consumption in EAFs have been described in the following lines with citing of cases where these measures have helped in improving not only specific energy consumption, but many more parameters of EAFs.

Introduction of Oxy-fuel burners

By lancing oxygen into the EAF bath, electrical energy gets supplemented by chemical energy. In the initial stage of oxygen lancing, steel makers were confronted with adverse effects of refractory life, electrode consumption and yield loss. With the advances of system engineering and automation, it is now possible to eliminate the adverse effects. Very efficient oxy-fuel burners are now available with the following facilities and capable to be tailored for any variety of operating conditions: [4]

- advanced supersonic oxygen lance,
- various fuel injecting burners concentric to the supersonic nozzle,
- water cooling system, and
- PLC & automation systems.

A typical arrangement of Oxy-fuel burner in an EAF is shown in the following figure: [4]

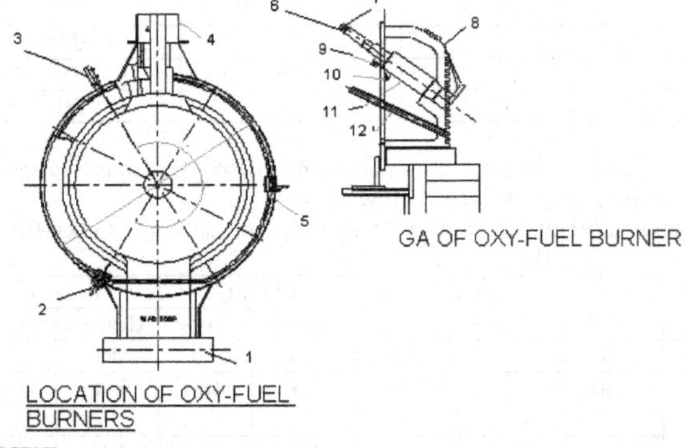

LEGEND

1.	Removal Hatch for Slag Disposal	8.	Water Cooled Copper Block Hard faced with Refractory Mass
2,3,5.	Oxy-Fuel (for carbon) Burners	9.	Water Inlet
4.	Tapping Slide Gate	10.	Water Outlet
6.	Propane Gas Injection	11.	Nozzle Assembly
7.	Oxygen Gas Injection	12.	Carbon Injection

Figure 3. A typical general arrangement of Oxy-Fuel Burner & its location in an EAF

This type of oxy-fuel burner has a circumferential burner that surrounds the supersonic nozzle. Supersonic nozzle is needed to accelerate the gas faster and its efficient utilization. The efficiency depends upon: [4]

- turbulent mixing between the furnace environment and jet,
- diameter of the exit nozzle, and
- design of the nozzle.

The burner provides additional melting capabilities in its vicinity to rapidly melt the scrap and to enable the supersonic metallurgical oxygen stream to begin the de-carbonization quickly. It contributes in: [4, 5]

- improving productivity,
- lowering cost of production,
- eliminating need of manual lancing,
- lowering power consumption,
- speeding up bath de-carbonization which in turn helps in removal of P, Si & S,
- enhancing foamy slag practice,
- reducing chances of violent boiling caused by severely over oxidized regions of the bath which is absent on account of uniformity of bath temperatures,
- reducing power-on time, and
- useful post combustion of CO.

Oxy-fuel burners were introduced in some four plants say Plants: A, B, C & D , which have tap weights and charge –mix constitutions as follows: [4]

Table I. Tap weights & Charge-mix Constitutions of Plants A, B, C & D.

Parameter	Plant A	Plant B	Plant C	Plant D
Tap Weight (t)	35	50	40	40
Charge-mix:				
DRI %	40	55	0	25
Pig Iron %	10	0	25	25
Hot Metal%	0	40	50	0
Scrap %	50	5	25	50

Values of some of the parameters before and after commissioning of oxy-fuel burners in the above four plants were very encouraging as can be seen below:

Table II. Parameters Improvements in Plants: A, B, C & D after Commissioning of Oxy-fuel Burners

Parameter	Plant A		Plant B		Plant C		Plant D	
	Before	After	Before	After	Before	After	Before	After
Total O (Nm3/t)	24	34	20	34	30	39	19	37
Power Consumption (kWh/t)	560	480	610	380	500	460	570	310
Tap-to-Tap Time (Min)	120	83	143	68	122	95	123	84

The decrease in specific power consumption is maximized in the plant D, where there is maximum jump in oxygen intensification. However Plant B has achieved a modest 30 % reduction with 70% increase in oxygen consumption. Hence, it can be concluded that the benefit of oxygen intensification in power consumption will be maximum when the existing level of oxygen usage is low. Also it can be seen that tap-to-tap time has decreased considerably after installation of oxy-fuel burners.

These improvements may be the reason that the global trend is for more and more use of oxygen. It has risen to 35 Nm^3/t and expected to go up to 40 Nm^3/t [4].

But, most of the Indian mini steel plants/foundries have not gone for oxy-fuel burners for their EAFs so far only for the reasons that oxy-fuel burners require heavy investment. Indian liquid fuel has high S content and most of these units do not have captive oxygen plant as big steel plants have [6].

Foamy Slag Practice [5, 8]

Another measure to reduce electrical energy is foamy slag practice. Slag foaming is advantageous in the sense that electrode arcs get submerged in surrounding foamy slag. Thickness of foamy slag should be greater than 1.5 times of the length of the arc to ensure sufficient submerging. This should form as early as possible. Oxy-fuel burners in this respect are very helpful. It should start injecting oxygen and carbon nearly at a rate of 15-30 kg/min just after the start of heat. Slag will start foaming shortly thereafter from the bath reactions (approximately 4 minutes). Submerged arc results in reduction in radiation heat loss and has a potential of reducing power consumption by 10%.

Quality of lime plays a significant role in slag foaming. It can increase slag basicity to around 2.0, maintain slag FeO content around 15-20%, stabilize the slag, and maintain a high quality slag throughout the heat. Lime having smaller concentration of inherent silica and other impurities should be charged with some MgO in order to improve its quality.

Supplementing Steel Scrap with Hot Metal in Charge-Mix

Another measure can be supplementing steel scrap with the use of hot metal. Hot metal can provide sensible energy to the melt and thereby reducing energy consumption. Supersonic oxy-fuel burners, if also used can help in de-carburising hot metal, that too very uniformly. If the plant is having blast furnaces, charging of hot metal at around 1200^0 C is going to help in:

- reducing electric energy consumption,
- productivity in sense of reduced tap-to-tap time, and
- also cleanliness of EAF tapped liquid steel as hot metal has low residual element.

It has been experienced that productivity increases with increase of hot metal up to 40% in the charge-mix. After that it reduces with increase of hot metal percentage as de-carburising becomes the rate limiting process for a given oxygen injection capacity [5].

Companies which already had a Mini Blast Furnace or have gone fresh for a Mini Blast Furnace, they can only practice hot metal charging into their EAFs. In case of India, such companies are just a few. To name, these are Jindal Steel & Power Ltd, Usha Martin, Visveswarya Iron & Steel, Adhunik Metalliks Ltd, Ispat Industries Ltd, whose installed production capacities range from 10 Mtpa to 3 Mtpa.

A Case of Combined Effect of All The Above Three Measures [5]

Let us take an example of another plant, where 3 pyro jet burners for simultaneous oxygen intensification & carbon injection were installed into a 100 t EAF. Later hot metal was introduced in the charge-mix up to 40-50%. Each pyro jet was equipped to inject oxygen and carbon at flow rates of 2000 cum/hr. and 60 kg/min. respectively. These burners were found very favorable in an attempt to increase higher and higher proportions of hot metal in the charge as these burners make the foam slag very stable by avoiding local slag over-oxidation and balancing the slag chemistry and temperature in the jet impact area.

Improvements were observed after introduction of pyro-jets & foamy slag practice, which further bettered after use of hot metal. These can be seen in the following table:

Table III. Improvements After introduction of Pyro Jets and Then Hot Metal in Charge

Parameter	Before start of any practice	After start of pyro-jets	After start of pyro-jets & 40-50% hot metal charging
Power -on time incld. lancing time (min)	51	42	36

Tap-to-tap time (min.)	59	49	44
Power consum. (kWh/t)	364	281	177
Electrode consum. (kg./t)	1.28	0.92	0.67
Oxygen consum. (Nm3/t)	45	46.9	47.1
Fuel consum.(Nm3/t)	-	1.2	0.78
FeO in slag (%)	25	22	21

Besides, reduced power on time and specific power consumption, improvement in cost of production and productivity are also significant as tap-to-tap time reduced due to exceptional de-carburising capability of pyro-jets even for 40-50% hot metal charge.

Scrap Preheating/Hot Direct Reduced Iron (DRI) Charging

As can be seen in the figure No. 2.0, 20% of energy input is going away with the waste gas. By using this gas in preheating the scrap being charged to the EAF, some of the energy can be recovered and can reduce the energy required to melt the scrap. Heat transfer to scrap is a function of scrap size and time. Scrap can be preheated up to 315 to 450^{0}C. Not only the reduction in energy input, other additional advantages were also experienced, which are as follows: [7]

Reduction in Energy Consumption	:	by 40 to 60 kWh/t
Reduction in Electrode consumption	:	by 0.3 to 0.36 kg/t
Reduction in Refractory consumption	:	0.9 to 1.4 kg/t
Reduction in Tap-to-tap time	:	by 5 to 8 min.
Removal of moisture from scrap	:	100%

Problems faced with scrap preheating are scrap sticking to bucket, poor controllability due to cycling of off-gas temperature, and flow rate through different operating phases of EAF. Another problem is with logistics, if tap-to-tap time is less than 70 minutes. Modern methods of scrap preheating are CONSTEEL Process and Fuchs Double Shaft System. Schematic diagrams of both the processes are shown in the Figures nos. 4 & 5 respectively. CONSTEEL Process is a method, which involves sized scrap conveyed inside a tunnel like shell. Hot gases and gases from natural gas burners in the tunnel preheat the continuously charged scrap. In Fuchs Double Shaft System, there are two furnaces, each with a shaft and one common electrode mast and set of electrodes. When one furnace is in the refining mode, its hot off-gases are directed to pass through ductwork to heat the scrap in another furnace and its shaft, then the process is reversed between furnaces & shafts after a cycle of tap-to-tap. But, so far only a small number of steel makers have adopted preheating of scrap in India. Its application will increase only after a more cost-effective process is evolved [7].

DRI needs more power at a rate of 12 kWh/t and 11 kWh/t respectively, if it has to reduce its 1% iron oxide and to neutralize its acid gangue. Even then, continuously charging of DRI is widely practiced for its lower cost compared to scrap, higher productivity, decreased electrode breakage & oxidation, less frequency of roof opening meaning less thermal shock, and less CO in furnace. For sake of energy conservation, the effluent gas of EAF cannot be used to preheat DRI as the gas would re-oxidize the metallic content of the DRI. A new trend of continuous feeding of hot DRI at 500 – 600 ^{0}C is being seen for the units, which has access to DRI unit. This has saved power by 105 kWh/t. There can be two ways of reducing power consumption in case of DRI charging:

a) Decrease in power consumption in the tune of 18 kWh/t has been experienced for 1% decrease in metallization of DRI
b) Practice of hot DRI charging has been able to reduce power consumption by 20 kWh/t of liquid steel for each 100^{0}c increase in composite solid charge temperature.

[9]

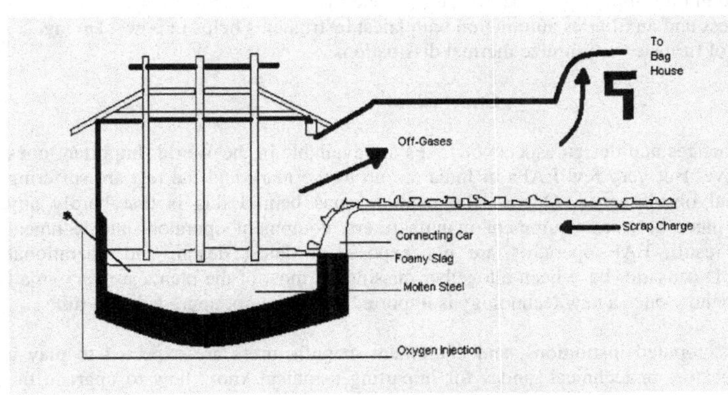

Fig 4.0. Schematic of CONSTEEL Process

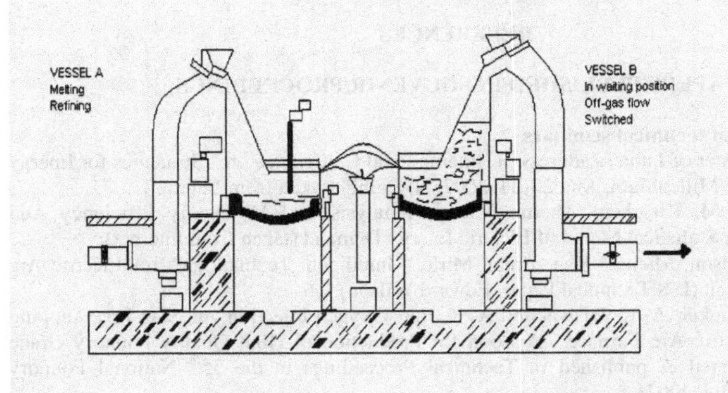

Fig 5.0:Schematic of Fusch's Double Shaft Furnace System

Aspects to Be Considered in Design of An EAF for Minimizing Energy Losses [8]

Following aspects need to be considered in designing or selecting equipment for an EAF:
1) With the usage of ultra high power transformers, the production rate has more than doubled along with a decrease in power consumption to the tune of 40 kWh/t. But, the furnace lining and roof have to be water-cooled.
2) The distribution transformers and distribution boards need to be located at the load centers in order to minimize the distribution losses due to cables.
3) In place of off-load tap changers, now all modern EAFs are having on-load tap changers, with whose help, secondary voltage of furnace transformers are varied on line without breaking the circuit for smooth voltage regulation. It contributes in:

- uninterrupted arcing,
- reducing melting time by 4-5 minutes, and
- keeping uniform temperature gradient in bath.

Tap positions can be smoothly controlled through PLC controller and Tap positions can be

monitored on HMI screen.
4) Efficient electrode movement with the help of dedicated PLC controllers contributes in terms of arc stability and in turn improves the performance of the furnace to a great extent.
5) Complete process and auxiliaries automation with latest instruments helps in power saving.
6) Proper sealing of furnace to minimize thermal dissipations.

Conclusion

Very effective practices and design aspects of EAFs are available in the World. Important ones are described above. But very few EAFs in India are up to the mark and the rest are suffering from technological obsolescence. One of the main reasons behind this is that hardly any interaction takes place between equipment manufacturers, equipment operators and technical institutes. As a result, EAF operators are not exposed to latest design and operational technologies. R& D activities have been altogether missing, as most of the plants are very small to afford it. If somehow once a new technology is imported, these are not upgraded with time.

Electricity Boards, reputed institutions, and equipment manufacturers are expected to play a major role as educators or technical guides for imparting technical know how to upgrade the knowledge level of the users. This in turn will help the national cause of `Energy Conservation'.

REFERENCES

PAPERS PUBLISHED IN SOUVENIR/PROCEEDINGS

Papers presented at technical seminars
1. Prime Minister of India' s address in `International Conference on " Strategies for Energy Conservation in the Millennium, Aug 23-24, 2002, N. Delhi" (taken from Internet).
2. H. Pfeifer, M. Kirschen "Thermodynamic Analysis of EAF energy Efficiency And Comparison with A Statistical Model of Electric Energy Demand (taken from Internet)
3. Edward Wilson, Michael Kan, Anjan Mirle " Intelligent Technologies for Electric Arc Furnace Optimization (ISS Technical Paper, Edward Wilson) 1-6
4. Pradip K Thakur, A. K. Jaiswal and A. K. Jha "Oxygen Injection and Slag Free Tapping System in An Electric Arc Furnace : A Boon for Production of High Quality Foundry Grade Steel (Paper presented & published in Technical Proceedings in the 55th National Foundry Congress, Agra, India, 2007)
5. Xu Xiaohong, Ruan Xiaojiang, Zhang Guowei, Deng Xing Michael Grant and Chen Tao "High Efficiency Production Practice of a 100 t DC EBT EAF At Xing Cheng Steel Works" (Paper published in AISTech 2006 Proceedings – Volume II), 413-421
6. Non implemented case study – Energy Conservation vide www.energymanagementtraing.com (taken from Internet)
7. Robert J. Schmitt, CMP, Pittsburgh, "Commentary: Electric Arc Furnace Scrap-Preheating" (Published by The EPRI Center for Materials Production in Electrical Power Research Institute in 4/97)
8. S. K. Srivastava, Dr. K. N. Jha, S.C. Prasad, A.K. Gupta "Energy Saving Technologies in Electric Arc Furnace (Presented & published in Technical Proceedings in a National Seminar on Energy Efficiency in Foundries & Secondary Steel Sectors organized by IIF, Ranchi Chapter, 17-18 th August, 2007)

JOURNAL
9. B.V.R. Raja, N. Pal, P.L.Talwar, N.P. Jayaswal, Alloy Steel Plants, Steel Authority of India Ltd., India "Technologies for cost reduction in EAFs" (published in Steelworld, April 2005)

Energy Technology Perspectives
Edited by: Neale R. Neelameggham, Ramana G. Reddy, Cynthia K. Belt, and Edgar E. Vidal
TMS (The Minerals, Metals & Materials Society), 2009

Evaluating Aluminum Melting Furnace Transient Energy Efficiency

Edward M. Williams[1], Donald L. Stewart[1] and Ken Overfield[2]

[1] Alcoa Technical Center, 100 Technical Drive, Alcoa Center, Pennsylvania 15069 USA
[2] GREA Engineering, 900 Gay Street, Knoxville, Tennessee 37902

Keywords: Melting, Energy

Abstract

Recent increases in energy cost have lead to a renewed focus on energy efficiency during aluminum melting operations to reduce fuel usage. The goals of a batch melting operation are to melt the charge using the minimum required energy, to melt the metal to the required temperature within the required timeframe, and to avoid generating unwanted contaminants by overheating the charge.

The heat transfer efficiency of a conventional hydrocarbon fired aluminum melting furnace varies over the course of the melt cycle, depending on the conditions in the furnace. The purpose of this work was to develop a method of determining the transient heat transfer efficiency throughout the furnace cycle and to take advantage of this knowledge to optimize the melting process through furnace controls and proper production operations. This furnace survey methodology has been developed and used to evaluate a number of furnaces.

Description of a Typical Batch Melting Furnace Cycle

A typical batch melting furnace cycle will consist of three stages:

1. Charging – Solid metal charged through an open roof, solid or molten charged through an open door, and/or molten poured through a furnace blister
2. Melting – From the time the door or roof is closed and the burners are turned to high fire until the charge is all liquid and has reached a temperature suitable for alloying (typically ~1350 °F / 730 °C)
3. Holding – From the end of the melting stage until transfer or casting is complete

In some cases, further charging can take place after the melting cycle has started. This can happen when the entire solid charge will not fit in the furnace in one load, or if additional molten metal is added while the charge is melting.

The bulk of the energy used in the melter is consumed during the melting cycle. The thermodynamic minimum energy requirements for melting pure aluminum are:

- Energy required to heat pure aluminum from room temperature to the melting point at 1220°F (660°C) is 288 BTU/lb (670 kJ/kg)
- Energy required to change state of pure aluminum from solid to liquid (heat of fusion) is 171 BTU/lb (397 kJ/kg)
- Energy required to heat molten from melting point at 1220°F (660°C) to 1350°F (730°C) is 36 BTU/lb (83 kJ/kg)
- Total energy required to melt a solid charge from room temperature to 1350°F (730°C) is 495 BTU/lb (1151 kJ/kg)

This means that if the furnace was 100% efficient at putting the input fuel energy into the charge, it would take 495 BTU of fuel energy to melt each pound of cold charge. Combustion fired furnaces operate well below 100% efficient due to inefficiencies in heat transfer from the flame to the charge and the energy required to heat the combustion air or oxygen. This efficiency loss occurs because a significant portion of the energy input must be used to heat up the non-burning portion of the combustion air and because the hot combustion gases that do not transfer their energy to the charge go up the furnace stack carrying a significant portion of the energy away. A typical cold combustion air furnace will have energy efficiency in the 25-30% range. Furnaces with heat recovery from the flue gas will typically be 35 to 45% efficient, depending on the system used and how efficiently it is operating.

In order to minimize the energy required for melting, the changes in the heat transfer efficiency during melting need to be understood. Immediately following the charging operation, at the start of the melting stage, the furnace will be filled with relatively cool solid charge, as shown in Figure 1. Heat transfer to the charge will occur primarily by radiation from the roof, walls and directly from the flame. There will also be some convection heat transfer from the hot combustion gases blowing in the furnace. Radiation heat transfer is driven by the temperature difference between the two surfaces, as well as the total surface area exposed. The convection heat transfer is also driven by the temperature difference between the combustion products and the charge.

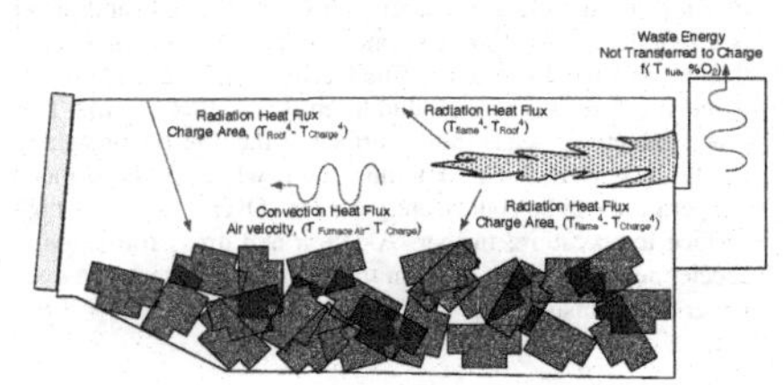

Figure 1. Start of the melting stage

At the start of the melting stage, the temperature differences between the charge and the flame as well as the charge surface area are at a maximum, so the heat transfer is at the highest efficiency during this part of the cycle. At this point, the heat transfer from the flame to the roof and walls will also be at a maximum, since the refractory is also at its coolest immediately following the charging operation. The bulk of the energy not transferred to the charge or to the refractory ends up going up the flue in the form of hot combustion gas. The amount of waste energy going up the stack is a function of the stack gas temperature and the gas composition, which can be characterized by the % oxygen in the flue gas. Higher flue temperatures and higher % oxygen generally represents higher heat losses up the stack.

As the cycle goes on, the charge will begin to melt as shown in Figure 2. At this point, the bulk of the charge will have reached the alloy solidus temperature (1220 °F for pure metal). Some fraction of the charge will have started converting to liquid. The heat transfer will be reduced from where it was at the start of the melting stage, due to the higher bulk temperature of the charge and the reduced charge surface area. It is also likely that the roof and walls will be approaching their maximum temperature. At this point in the cycle, the smaller differences in temperature and the reduced charge surface area will cause the heat transfer efficiency to be reduced. This leads to more of the energy input going up the flue as waste heat. Typically the flue temperature has climbed from the cycle start and is now approaching its maximum.

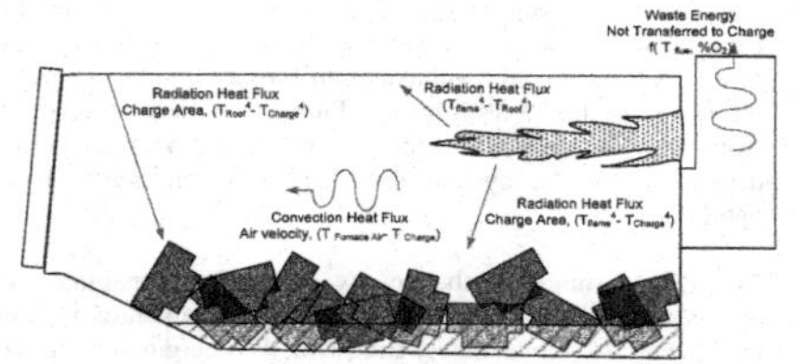

Figure 2. Melting begins

A little further along in the melting stage, 'flat bath' is reached. This is shown in Figure 3. Flat bath is the point at which the bulk of the charge is molten and most of the remaining solids are submerged or can be submerged. There will be a dross layer on the surface of the molten that was generated during the melting cycle. At this point, heat transfer from the flame and from the roof and walls will be driven by the surface temperature of the dross and the conduction heat transfer through the dross layer. This conduction heat transfer is a function of the thermal conductivity of the dross and the thickness of the dross layer. The thermal conductivity of dross is much lower than that of aluminum ($k_{Al} = 28 \times k_{Al2O3}$). This low conductivity leads to the surface of the dross approaching the furnace wall and roof temperature, while the bulk metal temperature can be significantly cooler. Over time at this high surface temperature, further oxidation and dross formation is accelerated, leading to an even thicker layer of dross and even lower heat transfer.

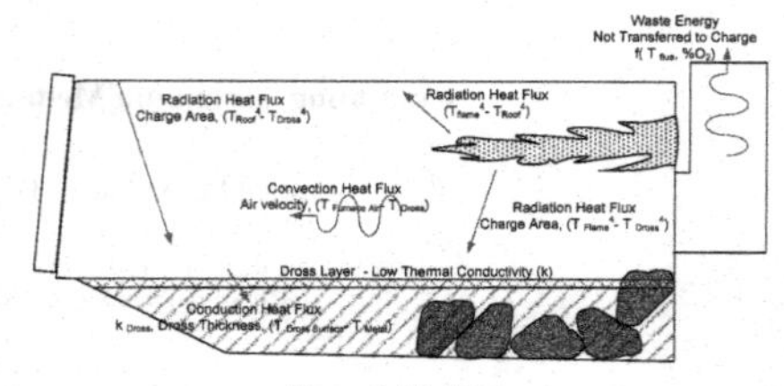

Figure 3. Flat Bath

Once flat bath is achieved, it is necessary to skim off the insulating dross layer to improve the heat transfer to the charge and finish melting in the most efficient manner. When the skim is removed, the heat transfer to the charge should once more be driven by the radiation and convection directly to the liquid metal surface, where conduction carries the heat to the solids until everything is molten.

Transient Survey Methodology

In order to determine the transient energy efficiency, an energy balance around the furnace is needed. The energy balance of input to outputs is:

$$\dot{E}_{in} = \dot{E}_{Flue} + \dot{E}_{Wall\,/\,Hearth} + \dot{E}_{Useful} + \dot{E}_{Door\ Open}$$

Looking at each individual part of the energy balance,

$$\dot{E}_{in}(t) = Fuel\ Input(t) \cdot Fuel\ Heating\ Value$$

Transient energy input to the furnace is determined by measuring and logging the rate of fuel input to the furnace, using either mass or volume flow, and multiplying by the fuel heating value. The energy input will vary over the course of the furnace cycle as the furnace burners turn down.

$$\dot{E}_{Flue}(t) = f(\,Flue\,Temp(t), Flue\,Composition(t), Flue\ Flow(t))$$

The biggest loss over the course of the melting cycle is the energy that is carried up the flue stack in the form of hot combustion products. This loss can be determined at any point during the cycle from determining the heat content (enthalpy) of the flue gas relative to ambient temperature and the mass flow of flue gas up the stack. To determine the heat content, the temperature and composition of the flue gas must be known. Assuming that all combustion takes place in the furnace (a must for efficient furnace operation), the mass flow rate and make-up of the combustion products can be determined from the fuel type and input rate. If the % oxygen in the flue gas is measured continuously, the excess ambient air in the flue gas can be determined. The accuracy of the excess air calculation will be poor during times where there is a large infiltration of ambient air into the furnace, such as when the door is open. For the details of these calculations for a variety of fuel compositions, see reference[1].

$$\dot{E}_{Wall\,/\,Hearth} \approx Constant$$

The energy loss through the walls, roof, and hearth for a properly insulated furnace is roughly constant over a typical melt cycle. The heat transfer from the furnace interior to the outer surface is a function of the difference between the interior and exterior temperatures and the thickness and thermal conductivities of the wall materials. Due to the large mass of wall refractory and the low thermal conductivity of the insulation, the heat loss through the walls will not vary greatly over the melt cycle unless the furnace interior is allowed to cool for a long period of time. Observations made at production furnaces have show very little change in the furnace outer surface skin heat flux over the melt cycle. This value can be considered a constant and can be measured using heat flux sensors, or by measuring the furnace skin temperature and calculating the heat loss.

$$\dot{E}_{Door\ Open} \approx f(Door\ Open\ Time)$$

The door open losses are transient energy losses that occur when the furnace door is opened and heat from the charge and the refractory is lost due to radiation and convection to the atmosphere. These losses can be significant if the door is left open for long periods of time during the melting or holding portions of the cycle. It is very difficult to get a measurement of the actual losses that occur during the door open periods. For this method, the door open losses are considered to be zero and the calculations made around the door open times are known to be circumspect.

$$\dot{E}_{Useful} = \dot{E}_{in} - \dot{E}_{Flue} - \dot{E}_{Wall\,/\,Hearth} - \dot{E}_{Door\ Open}$$

The useful energy is the portion of the energy input that actually goes into the load to heat and melt the charge. It is difficult to directly measure the transient useful energy into the charge, since the temperature of the charges is heterogeneous during the melting cycle. Therefore it is necessary to calculate the useful energy in by taking the energy input and subtracting all of the known losses.

With this data, the transient energy efficiency can then be calculated and plotted over the melt cycle:

$$Efficiency(t) = \frac{\dot{E}_{in}(t) - \dot{E}_{Useful}(t)}{\dot{E}_{in}(t)}$$

Table I lists the variables that need to be measured on a continuous basis in order to perform these calculations and run the furnace transient energy survey.

Table I Variable required for furnace survey

Required Variables
Fuel Input Rate
Combustion Air Flow
Flue Temperature
Flue % Oxygen
Bath Temperature

Production Furnace Data

An example of data collected from a production furnace as part of a survey is shown in Figure 4. This plot shows the transient bath temperature, the flue temperature and the burner firing rate for a melting cycle, starting from the point the furnace door is shut after charging. The melting cycle starts with the burners on full fire and a steady increase in the flue temperature is observed for the first two hours. At 125 minutes, the furnace door is opened, and molten metal is added to the charge. Following the molten addition, the furnace goes back to 100% firing until the flue over temperature set point is reached at which point the burners start cutting back and controlling to the flue set point to avoid overheating the flue. This continues until about 240 minutes into the cycle, when the furnace door is opened again and the furnace is skimmed. The furnace then goes back to high fire until the bath thermocouple shows that the charge is melted and the burner switches to bath temperature control. At this point the charge is melted and ready to transfer.

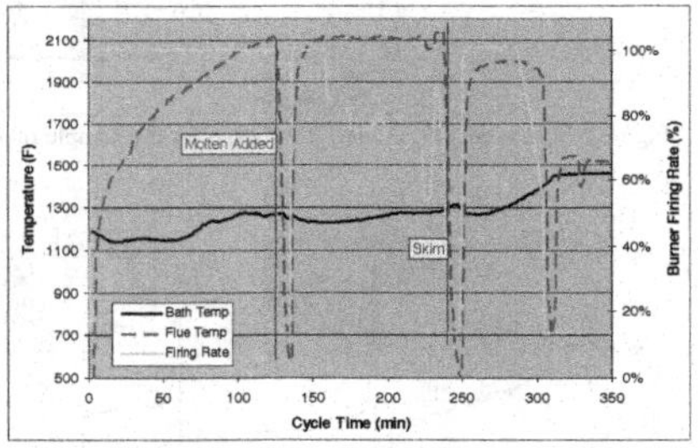

Figure 4. Example of transient furnace data

Figure 5 shows the results of performing the transient energy calculations on the data collected from this charge. The figure shows transient energy into the furnace and the energy losses up the flue and through the walls. The difference between the energy in and the energy loss in this plot is the useful energy delivered to the charge plus the door open losses that occur during and just after the time periods that the door was open.

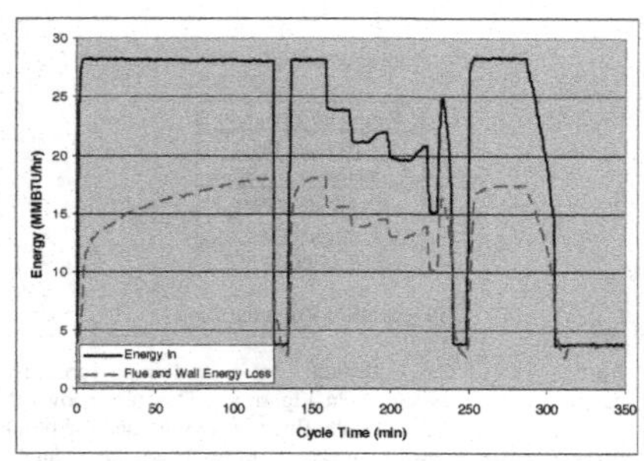

Figure 5. Energy input and loss for the sample melt cycle

Figures 6 and 7 show the useful energy input to the charge and the energy efficiency over the course of the cycle.

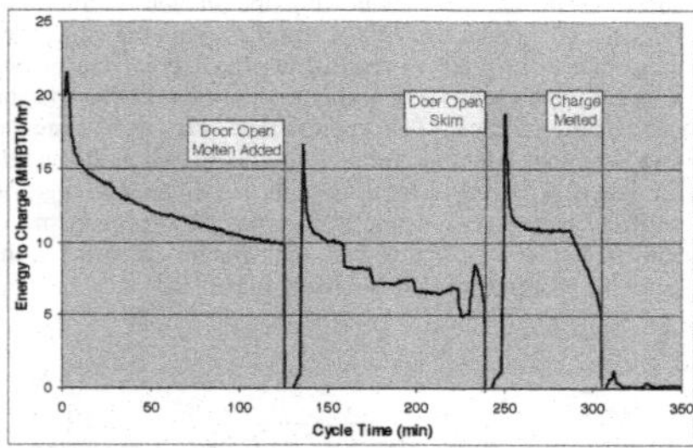

Figure 6. Transient energy into the charge for the sample melt cycle

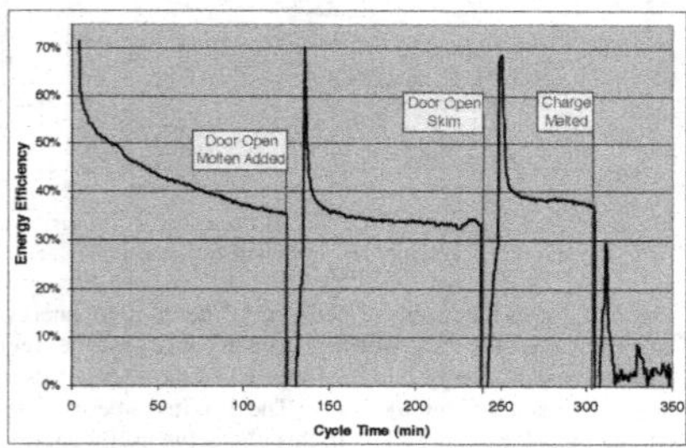

Figure 7. Energy efficiency for the sample melt cycle

As the theory predicts, the highest heat transfer to the charge occurs at the start of the cycle. This efficiency drops gradually over the first couple of hours as the charge temperature increases and the radiation and convection heat transfer efficiency is reduced. When the furnace door is opened at 125 minutes into the melting cycle to add molten metal, the energy efficiency goes negative as cold ambient air infiltrates the furnace and the furnace is losing more energy than what is being input with the burners on low fire. After the furnace door closes and the burners start firing again, the efficiency appears to briefly spike up as less energy is lost up the furnace stack relative to the input. This is due to the furnace walls absorbing more of the energy from the flame after they cooled with the door open. This is not actual additional energy going to the charge but is making up for the losses that occurred with the door open. Very soon after the door is closed, the refractory heats back up and the efficiency returns to the trend it was following before the door was opened. As the burner turns back based on the flue maximum set point temperature, energy input to the charge is reduced, but the efficiency stays about the same leveling out at about 33% of the input energy going to the charge. The rate of energy input to the charge drops to about 7 MMBTU/hr. After the furnace is skimmed at 235 minutes and the insulating blanket of dross is removed, the rate of energy input to the charge increases to 11 MMBTU/hr and the energy efficiency increases to 38% as the charge is able to absorb more of the heat input.

The energy efficiency of this charge over the melting cycle looks higher than the 25-30% that would be expected for this type of cold combustion air furnace. Due to energy losses during charging, holding and the transient door losses during the molten addition and skimming operation, the actual overall efficiency for this cycle from start of charging to the end of transfer was 29.8%.

This methodology has been applied to several production furnaces at a variety of Alcoa locations and has provided similar results. It has been used to modify furnace operations both from the standpoint of both production activities and burner control.

Conclusions

A method for determining the transient energy efficiency of a batch melting furnace has been developed. This method uses continuous data collected from the furnace fuel input and flue temperature to track the energy inputs and losses to determine the energy input to the charge. This methodology has been proven through surveys on production furnaces. Using the knowledge gathered from these transient furnace surveys, good decisions can be made about proper furnace operations from a production and burner control standpoint.

Acknowledgements

The authors would like to acknowledge the support of Donald Stanko, John Hartill and Kim Archibald from the Alcoa Technical Center in carrying out the plant evaluations performed for this work.

References

1. Reed, Richard J., *North American Combustion Handbook,* 3rd ed., North American Mfg. Co., Cleveland, 1986.
2. Incropera, F. P., "Thermodynamic Implications of Improved Furnace Performance", *Aluminum Industry Conservation Workshop V*, 1980, pp. 73-89.

3. Wiesner, J., "Furnace Firing for Peak Efficiency", *Aluminum Energy Conservation Workshop V*, 1980, pp. 207-228.
4. Weaver, Clark, "The Evolution of Furnace Operations in Alcan Smelter Castshops", *7th Australian Asian Pacific Conference Aluminum Cast House Technology*, 2001, pp.57-63.

Energy Technology Perspectives
Edited by: Neale R. Neelameggham, Ramana G. Reddy, Cynthia K. Belt, and Edgar E. Vidal
TMS (The Minerals, Metals & Materials Society), 2009

BILLIONS OF DOLLARS TO BE SAVED WITH RELIABILITY EXCELLENCE

Darrin Wikoff, CMRP, Life Cycle Engineering

TMS (The Minerals, Metals, & Materials Society)
184 Thorn Hill Rd.; Warrendale, PA 15086-7514, USA
Life Cycle Engineering, 4360 Corporate Rd, Charleston, SC, 29405

Keywords: Energy, Reliability

Abstract

The data, compiled by the Department of Energy (DOE), indicates that the average industrial plant could reduce its total energy cost by 14.8% by implementing effective Life Cycle Asset Management processes. The actual savings varies by industrial classification, but in all cases the potential is substantial and could have a marked impact on the operating profit of the company.

Improvements geared towards improving equipment reliability have distinctive linkages to environmental performance, such as reducing the amount of product and raw material waste through routine monitoring of system parameters through predictive technologies, and preventing interruptions to production cycles with a focus on Overall Equipment Effectiveness (OEE).

Alcoa successfully reduced solvent disposal costs by more than 40% and GE reduced greenhouse gas emissions by more than 250,000 metric tons. This presentation will show you how companies have successfully reduced energy and disposal costs through a focused effort on manufacturing process reliability.

Introduction

Smelter in the Park

The quaint town of Portland, Victoria protrudes outward towards Antarctica from Australia's southeastern coastline. Portland is a quiet rural community surrounded by agriculture that draws tourists by the thousands to the southernmost edge of Victoria by way of breathtaking cliff top views of the majestic Southern Ocean, the intrigue of shipwreck coves along the Great Ocean Road, and what most will call the eighth wonder of the modern world, the "Smelter in the Park". Integrated within Portland's beautiful landscape is a 356,000 ton/year aluminum smelter that stands as a symbol of collaboration between nature and industry. Nestled atop a ridge which faces "Sail Rock", a popular destination of local and visiting fisherman, the smelter is engulfed in a mix of brush and wetland preserves that are home to many of Australia's feathered and four-legged natives. Also within its interior you'll find parks and natural landscapes abundant where man and machines coexist with the native inhabitants.

The smelter, which employs more than 800 workers and supports thousands indirectly, was commissioned in 1986 after seven years of negotiation between business leaders and environmental protection agencies. As a monument of cooperation between environmentalists and industrialist, the need for success was paramount. If the "Smelter in the Park" failed, heavy industry like aluminum smelters and oil refineries, both of which play a significant role in national exports, would struggle to grow their business in order to meet consumer demand from Australia. The endeavor itself was extraordinary because it was implemented in a culture that is based on a healthy work / life balance. Life itself is a balance of convenience and conservation to protect the natural environment in Australia. The smelter was designed with conservation in mind, focused on alternative energy sources, such as wind and wave power. Waste recovery systems were added resulting in less than 20 cubic meters to landfill annually. Life cycle asset management strategies were utilized to preserve equipment health to maintain manufacturing and environmental performance. After more than twenty years of cohabitation, the smelter and the "park" are still prosperous and continue to set new expectations for sustainability within Australia and other regions of the world.

Practices aimed at improving reliability can pay big dividends

The burgeoning problem of waste generated by manufacturing processes is a source of growing concern among corporate enterprises and government markets, requiring American manufacturers to assume greater responsibility. However, justification for improvements within brownfield projects is often difficult. When we examine the chart of accounts recorded by the U.S. Department of Commerce, "Utilities" – purchaser's cost of power generation and supply, natural gas distribution, water consumption and sewage treatment – is hidden or masked by larger intermediate expenditures. In reality, utilities expenditures across the U.S. are less than 2% of the total intermediate cost of modern manufacturing. If we dig deeper, we quickly realize that this nominal percentage equates to more than $49 trillion annually. U.S. manufacturing has the opportunity to save billions by reducing utility costs by as little as 0.1%!

There's a social and fiscal need to change the way businesses looks at environmental performance. Some companies have chosen to allocate capital to the re-engineering of manufacturing systems in order to reduce waste generation, while others have invested in more energy efficient components, such as electric motors, air compressors and large centrifugal fans or pumps. However, organizations like Alcoa, General Dynamics and 3M Company have realized that Reliability Excellence is a far greater opportunity for sustained performance improvement. Studies conducted by the Department of Energy (DOE) indicate that a reliability-based asset management strategy yields a reduction in energy consumption, alone, by 33% on average across industries. Table 1.0 illustrates the annual savings potential based on the DOE studies.

Table 1.0

Company	Type of Plant	Energy Savings (kWh/Year)	Savings (% Reduction)	Annual Cost Savings*	Payback on Investment (Years)
General Dynamics	Metal Fabrication	451.778	36%	$68.000	1.5
3M Company	Laboratory	10,821,000	6%	$823,000	1.9
Peabody Coal	Coal Processing	108,826	20%	$6,230	2.5
Stroh Brewery	Beer Brewing	473,000	52%	$19,000	0.1
City of Milford	Municipal Sewage	36,096	17%	$2,960	5.4
Louisiana Pacific	Strand Board	2,431,800	50%	$85,100	1.0
Nisshinbo California	Textiles	1,600,000	59%	$100,954	1.3
Alumax	Aluminum Smelter	3,350,000	12%	$103,736	0.0
City of Long Beach	Waste Incineration	3,661,200	34%	$329,508	0.8
Bethlehem Steel	Integrated Steel	15,500,000	50%	$542,500	2.1
Total/Average		**38,663,734**	**33%**	**$2,166,330**	**1.5**

*Savings are calculated based on purchaser's cost. Average equates to $0.055 per kWh/year.

Reliability is not just a measure of maintenance or your organization's ability to preserve asset integrity. Reliability Excellence is a culture in which the systems, business processes, structures and disciplines exist to preserve the integrity of manufacturing performance. Narrowly focusing on raising environmental performance by preventing component fatigue as with valves, pump seals and piping systems will only create a greater need for reaction and quite frankly will result in higher maintenance costs with little risk management security. Gaining control of the process is a much greater goal for driving environmental performance. Process reliability deals with the capability of your organization to effectively detect variation in the manufacturing process and efficiently eliminate the source before increases in energy consumption or waste generation occur. As an example, Alcoa, a world-wide leader in alumina refining and the manufacturing of aluminum products, has successfully reduced energy consumption as a result of training operators in better standards of loading, starting, and operating manufacturing equipment. Through Operator Care, standards of practice are both identified and reinforced so that even minor variations, which have the ability to generate waste, are prevented.

The path to sustainability powered by Reliability Excellence begins with a non-technical solution first. The reason being that for more than fifty years our businesses have looked at energy costs and waste disposal costs as a financial obligation only, one that defined our acceptable levels of risk. Today, the business arena has changed and the magnitude of impact relating to environmental performance is no longer acceptable. Environmental performance, today, is synonymous with Sustainability. Either we learn as an industry to cohabitate or we will be extinct. The visionaries behind the "Smelter in the Park" figured this out in 1986. Business leaders today should look at environmental performance as a cultural and communicable responsibility. Begin by educating your organization on your company's environmental performance to create awareness regarding the collective responsibility to improve. Keep in mind that there are two primary messages that need to be conveyed: 1) Why is this type of improvement important to the business? and 2) How will improving environmental performance impact the employees? Next, assess the ability of your reliability systems, business processes and structures to support energy conservation and eliminate unnecessary waste generation, remembering that the first step is to stabilize equipment reliability and process control. A model for Reliability Excellence and the areas of focus during both the assessment and improvement process is shown in the following illustration.

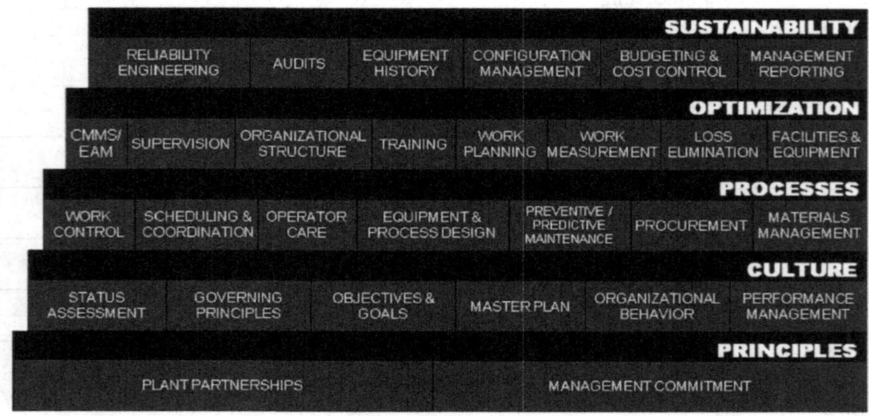

© 2008 Life Cycle Engineering, Inc.

The model for achieving Reliability Excellence is based on establishing the principles of reliability that support your business needs, in this case environmental performance. Then develop a culture of reliability in order to support the business principles through appropriate behaviors. Standards of practice must also be created through business processes in order to establish daily practices that reflect the principles and culture of Reliability Excellence. Optimizing performance within the newly established practices can then begin via elements such as supervisory practices, skills training and work measurement systems that highlight opportunities for improvement. Finally, your organization will be able to build a sustainability model to further drive environmental performance through the disciplines of Reliability Engineering and continuous improvement techniques like audits and daily management reporting systems. Once the gaps in your organization's reliability practices have been assessed, it is imperative that leadership create an ability to improve the current situation through a focused improvement team structure supported by active and participative sponsors for change.

Eastman Kodak provides us with a common example of reliability-based environmental performance improvement. From 1999 through 2006, Eastman Kodak utilized the Lean Manufacturing methodology of kaizen – a Japanese term for improvement that looks at a specific problem through all functional aspects of the business - in order to focus reliability improvements on energy reductions within their manufacturing processes. They began each kaizen event by defining the problem statement as "what do we use energy to do?" In the end, Eastman Kodak found that there were two main arenas in which energy reductions would produce significant savings, 1) those steps within the manufacturing process that were used to generate heat, and 2) the magnitude of energy used to turn motors and mechanical drive systems, be that air, steam or electric drives. Over the seven year period, Eastman Kodak applied Reliability Centered Maintenance strategies which resulted in energy savings of nearly $15 million, as reported by "The Lean and Energy Toolkit" published by the Environmental Protection Agency (EPA).

Reliability Excellence is the foundational step necessary to stabilize performance and remove variation. Practices aimed at improving reliability can pay big dividends in striving toward sustainability success. Regularly monitored and well-maintained equipment can save energy, increase uptime, drive profitability, and advance core sustainability objectives. Armed with robust reliability systems we will ensure sustainability of both manufacturing and the environment while saving billions that would have otherwise been lost on energy and waste.

Energy Technology Perspectives:

Conservation, Carbon Dioxide Reduction and Production from Alternative Sources

Extraction Processes/ Refractories/Modeling and Analysis

Session Chairs

Edgar Vidal
Mark Cooksey

Energy Technology Perspectives
Edited by: Neale R. Neelameggham, Ramana G. Reddy, Cynthia K. Belt, and Edgar E. Vidal
TMS (The Minerals, Metals & Materials Society), 2009

ENERGY AND SUSTAINABLE DEVELOPMENT IN HYDROMETALLURGY – AN EMERGING PERSPECTIVE

K. Sarveswara Rao
Hydro- & Electro-Metallurgy Department
Institute of Minerals & Materials Technology (CSIR)
Bhubaneswar 751 013, Orissa, India.

Keywords: Aqueous processing, complex ores, particle size distribution, water salinity, energy savings

Abstract

Any innovative approach made in non-ferrous metallurgical industry centers upon developing new processes, saving materials, improving production quality, and born-again materials. Aqueous processing is commonly used to treat lean grade or more complex ore bodies. Its sustenance largely depends on energy and environmental compliance including water and wastewater management. The present paper focuses the importance of using particle size distribution and surface area measurements during ammoniacal dissolution of a Cu-Pb-Zn sulphide bulk concentrate and with an overview of the effect of water salinity on the process chemistry and residue mineralogy. Accordingly, an effort is made here to discuss the emerging perspective and highlight the recent successes and trends in terms of energy savings in non-ferrous hydrometallurgy.

Introduction

The nature and structure of the global energy industry is likely to change significantly in the 21st Century. Rabago et al [1] have comprehensively discussed how the mining and minerals industries can adapt to the challenges and opportunities posed by trends and drivers in global energy use. According to the authors, the drivers of sustainability include water, efficient energy use, climate modification, transportation technologies and energy security. Sustainability and competitiveness for the minerals industry will depend on natural capitalism, resource efficiency, new processes, saving materials, improving production quality, born-again materials and the solutions economy. Based on this motivation, the theme of the present paper is centered upon applications as relevant to energy and sustainable development in non-ferrous hydrometallurgy.

Many sustained efforts are needed to maintain the pace of industrial development of any country, more so with extractive metallurgy. In this context, aqueous processing is commonly used to treat lean grade or more complex ore bodies. The present author [2] has previously described some recent energy saving measures practiced via process intensification, the emerging importance of an interdisciplinary nature of hydrometallurgy, and processes for treating complex materials. There is a strong nexus between energy and process intensification, and between process intensification and the move to sustainable processes. For example, in leaching processes the attainment of autogenous steady-state temperature of autoclave, agitation/mixing of slurry, are the parameters identified that influence the energy saving measures. In solution separation and purification, there are variations in energy consumption based not only on the origin of the material leached, but also variations based on the particular circumstances of the plant installation, for example, costs of pumping pregnant liquor miles away from the leach site to the SX/EW plant. Process intensification of copper extraction using emulsion liquid membranes and the ionic liquid synthesis in a microstructured reactor for process intensification, are the typical examples for future potential areas for energy conservation in hydrometallurgical processing of complex materials.

In hydrometallurgical processes, energy, water and environment would play a major role. The present paper also describes the effect of particle size vis-à-vis surface area during mineral dissolution, as compared to energy saving in comminution; and the use of high salinity water as the source of process water and the effect of salinity on the process chemistry and residue mineralogy. Accordingly, an effort is made in the present study to discuss the emerging perspective and to highlight the recent successes and trends in the energy savings in hydrometallurgy.

Particle size distribution and surface area measurements during oxidative ammonia leaching of bulk concentrate

Energy use tends to increase with declining ore grade and with finer mineral grain size. So, hydrometallurgical processes aim at optimum particle size and to balance the energy use. For leaching of bulk concentrates, the particle size distribution of both feed materials and leach residues is very important for gaining an insight into the extent of fine grinding and micro-fracturing of host rock particles for mineral liberation and/or exposure to leaching medium. With this background, data were generated on particle size distribution and surface area measurements during ammonia leaching of sulphides.

For this work, single sulphide concentrates such as chalcopyrite ($CuFeS_2$), sphalerite (ZnS), galena (PbS), their mixtures, and a Cu-Zn-Pb bulk concentrate were used to carry out oxidative ammonia leaching experiments in a 2 L capacity Parr autoclave, USA make (Model 4542) as described [3]. The surface areas of starting materials and leach residues were measured with the high speed surface area analyzer (Model 2200, Micromeritics make). Surface morphology data were also generated on the starting material and some partial leach residues with SEM (Model JSM 35 CF, JEOL Japan make) and details available elsewhere [4]. The data generated are shown in Table I. While the starting materials are all having a uniform particle size of -63+45μm and 1.0 m^2/g surface area before leaching, the size distribution of leach residues is different and same is true with active surface area values.

The observed increase in surface area of bulk concentrate leach residue from 1.0 to about 40 m^2/g, is attributed to the formation of goethite from the iron that is released during dissolution of chalcopyrite and also the liberation of fine silicate minerals present therein. SEM data indicate that original sulphide particles are massive with uniform surface features whereas the leach residues have shown characteristic features such as development of triangular pits, micro-channeling and porous/cavernous textures. A significant increase in the fraction of fines, presumably, the particles of silicate minerals released from the matrix during the leaching, has resulted in increased surface area of leach residues. Such increase in surface area is also related to dissolution of metal sulphides. Such a correlation has been reported previously in literature for leaching of chrysocolla. A similar correlation has been found to be true for the present study also.

Table I: Size distribution of sulphide leach residues and active surface area

Sample No.	Sample	Particle weight distribution, %			Surface area of whole sample, m^2/g
		Coarse -63+45μm	Middle -45+38μm	Fine -38+36.5μm	
1.	Chalcopyrite, Cp*	19.7 (100)	26.7	53.6	15.23
2.	Cp-Sphalerite, Sp* 1:3	17.6 (100)	16.1	66.3	6.27
3.	Cp- Galena, Gn* 9:11	4.6 (100)	7.6	87.8	27.46
4.	Cp-Sp-Gn * 20:57:23	6.6 (100)	6.3	87.1	18.65
5.	Bulk Concentrate*	29.7 (100)	18.6	51.7	40.10
6.	Bulk Concentrate**	27.0 (100)	22.8	50.2	41.00

* Oxidative ammonia leaching, 135^0C and 2 h.
** Oxidative ammonia-ammonium sulfate leaching, 135^0C and 2 h.
( ) Particle weight distribution of untreated feed materials shown in parenthesis.

Effect of water salinity on the process chemistry and residue mineralogy / Water and wastewater management

Most metallurgical operations must also deal with fugitive emissions and water discharges. The nonferrous metals industry (NFMI) has also put a considerable effort into ensuring that aqueous discharges from mines, mills, smelters, and refineries meet stringent standards. The most common technology used is simple alkaline precipitation of heavy metals, usually with lime or sulfide, and subsequent recycling of the precipitates to smelters or solid waste treatment plants. These precipitates or sludges are often excellent feeds for smelters that can accept them as the moist, variable feeds. Due to importance of environmental hydrometallurgy for treating waste water and resource recovery, the differences in the ore composition and process-water salinity need attention for they do not complicate those apparently simple processes. A change in the salinity of the process water could have significant effects on solubilities, activities/activity coefficients, leaching kinetics, extent of reactions and process mineralogy.

Whittington et al. [5] have examined the effect of water salinity on the reactions occurring during pressure acid leaching of a blend of laterite ores, as relevant to residue volume/mass, overall acid consumption and nickel extraction. However, the influence of water salinity on the process chemistry and residue mineralogy needs to be quantified and it appears that more detailed studies in this area should be carried out by researchers in the future. Sea water can be considered as the source of process water in view of better material of construction available on date to take care of corrosion problem of the equipment. A good amount of basic research is still to be done in this chloride metallurgy area.

Process Innovations Versus Comparisons of Energy Consumption

Global Mining Initiative and the International Council on Mining and Metals (ICMM) has concerned with energy consumption and energy efficiency of sustainable development for mineral extraction. As rightly pointed out by Marsden [6], none of the published work delves on the range of processes currently in use, or emerging, for copper extraction today on the relative energy efficiency compared with conventional processes using mineral processing and pyrometallurgy. One reason could be the big gap in between the areas of process development and the energy audit. In the generic energy model developed by Marsden, all sources of energy consumption involved in copper extraction were considered from the ore in the ground to the final product to the market in the following manner.

Depending on copper ore grade, seven generic processing routes were selected to compare energy consumption for copper extraction both by commonly used processing routes and emerging technologies. The total energy consumption estimated with various head grades of 0.125% Cu and 1.0% Cu ore are shown here for the sake of comparison, respectively.

1. Mining, ROM stockpile leaching, SX and EW (33668 kJ/lb Cu, 9405 kJ/lb)
2. Mining, three-stage crushing, heap leaching, SX and EW (43976 kJ/lb Cu, 10694 kJ/lb)
3. Primary crushing, SAG milling with pebble crushing, ball milling, flotation, smelting and electrorefining (83036 kJ/lb Cu, 20194 kJ/lb)
4. Mining, primary and secondary crushing, HPGR tertiary crushing, ball milling, flotation, smelting and electrorefining (68068 kJ/lb Cu, 18323 kJ/lb)
5. Mining, primary crushing, SAG milling with pebble crushing, ball milling, flotation, high-temperature pressure leaching, SX and EW (77189 kJ/lb Cu, 15303 kJ/lb)
6. Mining, primary crushing, SAG milling with pebble crushing, ball milling, flotation, medium-temperature pressure leaching, direct EW and SX/EW for low grade process stream and EW bleed (77144 kJ/lb Cu, 14624kJ/lb)

7. Mining, primary and secondary crushing, HPGR tertiary crushing, ball milling, flotation, medium-temperature pressure leaching, direct EW and SX/EW for low grade process stream and EW bleed (62253 kJ/lb Cu, 12763 kJ/lb)

The basic approaches involved for process commercialization are summarized [7]. A process usually reaches commercial status mainly because of its ability to meet certain criteria like technical ease, economy, reduced unit operations and energy savings, as compared to other processes. There may be several borderline cases where the decision on process selection really becomes difficult. The problem has become more acute in recent times, mainly due to advances in technology involving newer, more energy-efficient, and environment friendly processes. It is noted that project development programme for any process route will contain many activities, such as ore characterization, pre-concentration tests, amenability tests, preliminary cost estimate, continuous mini-plant operation, feasibility study, pilot plant testing, and finally engineering, procurement and construction. The process selection is to be based on a "Decision Analysis" involving many parameters and a final decision taken in favour of a particular process for constructing a commercial plant that can be operated successfully.

According to Canterford [8], all the available 'simple' ore bodies have already been consumed/ exhausted and our current and future feed stocks are far more mineralogically complex and often in more remote locations. So, the cost and availability of energy, water, labour and transport are all becoming increasingly significant in determining project viability. The same author has provided an excellent review of what should be regarded as some very basic ground rules concerning of mineral resource characteristics, a detailed design criteria–process selection–risk analysis, and plant layout – modularation – redundancy – product selection – resource integration. For example, Phelps Dodge undertook extensive technical and economic evaluation of many process options and ultimately selected the medium temperature (150-165^{0}C) option combined with solvent extraction and electrowinning, for medium temperature pressure leaching of copper concentrates.

Concluding Remarks

The particle size distribution and surface area measurements of chalcopyrite and bulk concentrate leach residues gave interesting observations. Those data will be useful for arriving at optimum particle size of starting feed materials and to balance the energy use. The recent literature on process innovations versus comparison of energy consumption for copper extraction technologies indicate that hydrometallurgical processes consume significantly less energy than alternatives such as grinding, flotation, smelting and refining.

Acknowledgement

The author is grateful to Dr. B.K. Mishra, Director, Institute of Minerals and Materials Technology, Bhubaneswar for his permission to present the paper during EPD-2009.

Thanks are due to Dr. R.K. Paramguru, HoD, Hydro- & Electro-metallurgy Department for his encouragement.

References

1. K.R. Rabago, A.B. Lovins and T.E. Feiler. *Energy and sustainable development in the mining and minerals industries.* MMSD project report of International Institute for Environment and Development, and World Business Council for Sustainable Development. Copyright © 2002 IIED and WBCSD.

2. K. Sarveswara Rao, *On promoting energy conservation in hydrometallurgical processing of complex materials,* Proc. of sessions and symposia sponsored by the Extraction and Processing Division of TMS, held at the 2008 TMS Annual Meeting in New Orleans, Louisiana, March 10-Mar 13, 2008, Edited by Stanley M. Howard, ISBN 978-0-87339-715-5.

3. K. Sarveswara Rao and H.S. Ray, A new look at characterization and oxidative ammonia leaching behaviour of multimetal sulphides, *Minerals Engineering*, 1998, Vol. 11, pp. 1011-1124.

4. K. Sarveswara Rao and H.S. Ray, *Leaching Kinetics of Cu-Zn-Pb bulk concentrate – surface area measurements*, Proc. Ninth National Convention of Chemical Engineers and International Symposium on Importance of Biotechnology in the coming Decades. June 5-7, 1993, Visakhapatnam. Tata McGraw-Hill Publishing Company Ltd., Delhi, pp.218-222.

5. B.I. Whittington, R.G. McDonald, J.A. Johnson, D.M. Muir. Pressure acid leaching of arid-region nickel laterite ore, Part I: effect of water quality. *Hydrometallurgy*, 70 (2003):31.

6. John O. Marsden, *Energy efficiency and copper hydrometallurgy,* Hydrometallurgy 2008, Proc. Sixth International Symposium Honoring Robert S. Shoemaker, Society for Mining, Metallurgy, and Exploration, Inc. (SME), ISBN 978-0-87335-266-6, pp.29-42.

7. K. Sarveswara Rao, Changing priorities for processing and utilization of mineral resources – a study, *Everyman's Science*, Vol. 38, No. 5, Dec'03-Jan'04, pp. 292-296.

8. John Canterford, *Some challenges in the design and operation of smart and safe metallurgical plants,* Hydrometallurgy 2008, Proc. Sixth International Symposium Honoring Robert S. Shoemaker, Society for Mining, Metallurgy, and Exploration, Inc. (SME), ISBN 978-0-87335-266-6, pp.14-22.

Energy Technology Perspectives
Edited by: Neale R. Neelameggham, Ramana G. Reddy, Cynthia K. Belt, and Edgar E. Vidal
TMS (The Minerals, Metals & Materials Society), 2009

COPPER ELECTROWINNING USING NOBLE METAL OXIDE COATED TITANIUM - BASED BIPOLAR ELECTRODES

K.Asokan[1 *] and K.Subramanian [1].
[1]Central Electrochemical Research Institute, Karaikudi – 630 006. India.
([*]author for correspondence, E-mail: asokan_tsia@cecri.res.in)

Key words: Copper electrowinning, bipolar electrodes, lead pollution abatement, busbar saving, mass transfer enhancement, energy conservation.

Abstract

Electrowinning of copper by monopolar cells require common anode and cathode busbar and crossbars to connect each electrode to the respective common busbar. On the other hand, the bipolar configuration warrants the end electrodes only to be connected to busbars; there is substantial reduction in copper requirement. In the present work, an electrowinning cell with bipolar electrodes and end electrodes, each having an area of 1000 cm^2 was designed and operated. The bipolar electrode is made of mixed metal oxide coated titanium mesh welded to plain titanium sheet. Coated titanium mesh acts as anode and the plain titanium sheet acts as cathode. Environmentally unacceptable lead anode and the recurring loss of lead are done away with. Closer spacing of the electrodes, paves way for the application of higher current density leading to mass transfer enhancement. Performance of bipolar copper electrowinning cell is reported.

1. Introduction

The normally adopted practice in electrowinning of copper is the conventional monopolar configuration. In this practice, the individual electrodes act either as anode or cathode. Hence there is a need to connect every one of the electrodes to the common anode or cathode busbar and crossbars running across the top of the cell are necessary. The common busbar carries very high current which is divided among the individual crossbars and hence has to be thicker in relation to the current load of the cell. The drawback of the need to use common busbars of thicker cross section and the need to have crossbars are obviated in the bipolar configuration. This configuration warrants the end electrodes only to be connected to busbars. The current through each pair of electrodes is the same as the current through the whole cell; this fact allows the use of thinner busbars for connection to the end electrodes. Hence there is substantial reduction in the requirement of copper busbars in cell construction.

Further, the anode used in monopolar configuration is a slab of lead alloyed with antimony or silver, each weighing 100 to 150 kg. In this, there is sizable initial consumption of lead in addition to recurring loss of 0.9 kg lead per ton of copper produced [1,2]. The drawback of the use of environmentally unacceptable lead is obviated in the present work, by the use of precious metal oxide coated titanium as anode. Also the recurring loss of lead that leads to pollution by the heavy metal is done away with. The cathode current density normally employed is around 200 A m^{-2} in the monopolar art. The mass transport condition near the cathode is not conducive to still increase the said current density for the sake of enhancing the productivity of electrowinning cell [1,3]. In the bipolar configuration, the closer spacing of the electrodes, paves way for the application of higher current density which in turn leads to mass transfer enhancement.

Electrowinning cells of different versions were fabricated and operated. The bipolar electrode is made of mixed metal oxide coated titanium mesh welded to a plain titanium sheet. Coated titanium mesh acts as anode and the plain titanium sheet acts as cathode. Copper sulphate solution containing 30 to 35 gpL copper and 100 to 150 gpL free sulphuric acid is used as the electrolyte. Performance characteristics of these versions of the copper electrowinning cells are reported.

2. Experimental

Phase I: An electrowinning cell suitable for accommodating four bipolar electrodes and two end electrodes, each having an active area of 60 cm^2 was fabricated using a PVC/Acrylic combination. Rubber sheets of appropriate thickness were pasted to the inner walls of the cell and grooves were cut in the rubber sheets so that the end electrodes and bipolar electrodes were tightly fitted with an inter electrode gap of 15 mm. The space between adjacent electrodes forms electrolyte compartments. Each compartment has an electrolyte inlet and outlet.

The electrodes are bipolar in nature, which means that one side of the electrode acts as anode and the other side acts as cathode. The electrode is made up of a titanium sheet. The anode face of the bipolar electrode is given an undercoat of iridium oxide followed by an overcoat consisting of a mixture of oxides of titanium and ruthenium [4], while the cathode face is left bare. The electrodes are placed in a cell with a gap of 1.5 to 2.5 cm. The first and last electrodes in the cell called "end electrodes" are only connected to the direct current source by means of busbars and the intermittent bipolar electrodes are left unconnected. Each of the intermittent electrodes starts behaving in a bipolar way, the moment direct current is applied to the end electrodes and the whole cell becomes operative as a series electrowinning cell.

The electrolyte flow through the cell is accomplished by means of a pump drawing solution from a reservoir. The electrolyte is circulated between the cell and the reservoir and its total volume is 2.5 liters. Though a recirculation procedure is adopted here, conversion to once through mode is also feasible.

The bipolar electrode currently tried is precious metal oxide coated titanium mesh tag-welded to plain titanium sheet. Coated titanium mesh acts as anode, while the plain titanium on the hind side acts as cathode. Copper sulphate solution with 30 -35 gpL copper and 100 to 150 gpL of free sulphuric acid is used as the electrolyte. During the electrowinning operation, the cathodic reaction is copper deposition and the anodic reaction is oxygen evolution. The electrolyte was circulated using a pump.

After passing DC current for a known duration, copper deposited on the cathode was removed using a sharp knife and weighed. Current efficiency for the run was determined from the weight of copper deposited. The cell was operated at different current densities in the range 200 to 600 $A.m^{-2}$. Current efficiency ranged from 85 to 94%. The weight of copper deposited on the cathode side of the different bipolar electrodes varied by ±5%, indicating more or less equal current passage through the bipolar electrodes.

Phase II: The electrowinning cell was scaled up to accommodate bipolar electrodes of area 1000 cm^2. The material of construction and other cell design details were the same as with Phase I. The cell was fitted with 2 bipolar electrodes, one end anode and one end cathode. The electrowinning cell was operated at different current densities in the range of 200 to 400 Am^{-2} for durations ranging from 1 to 6 hours. Current efficiency ranged from 93 to 98 %. The weight of copper deposited on the cathode side of the bipolar electrodes varied by ±3 %, indicating more or less equal current passage through the bipolar electrodes. In one particular run, the cell was operated continuously for 42 hours and the total copper product from this run was 2.83 kg, where in about 0.95 kg was the deposit from each cathode. The thickness of the deposit was more or less uniform which was around 2.0 mm.

Phase III: In order to test the present design of the cell with respect to the containment of by pass current with more number of electrodes, a cell with 14 bipolar electrodes and two end electrodes with an active area of 40 cm^2 each plus other accessories were fabricated. Rubber swells during prolonged usage in aqueous medium and the width of the grooves and inter-electrode gap could not be maintained constant. Grooves were cut on the sides of the PVC cell and the electrodes were fitted with matching PVC frames. This arrangement prevented current leakage through the sidewalls. As grooves were not cut at the cell bottom, current efficiency was only in the range of 55 – 60% indicating current leakage through the cell bottom.

Phase IV : The modified electrowinning cell used in Phase III was scaled up to accommodate six bipolar electrodes and two end electrodes of area 1000 cm^2. Here grooves were cut on all three sides of the cell wall (including the bottom of the cells) and matching PVC frames were provided on all the three sides of the electrodes, coming into contact with the cell walls. The bipolar cell configuration, with details, are shown in Figures 1 to 6. This modified arrangement almost completely prevented leakage current as evidenced by the improvement in current efficiency to around 96.5%. The performance data of the different phases of work is given in Table.1.

3. Results and discussion

Performance data of bipolar cell for copper electrowinning is shown in Table 1.

Table 1

	Parameter	**Phase I**	**Phase II**	**Phase III**	**Phase IV**
1	No of bipolar electrodes	4	2	14	6
2	Active area of electrodes (cm^2)	60	1000	60	1000
3	Current density ($A.m^{-2}$)	200	200	200	200
4	Inter-electrode gap (mm)	15	15	15	15
5	Current passed (A)	1.2	20	1.2	20
6	Current rating of the cell (A)	6	60	18	140
7	Depletion of Copper per run (gpL)	5	5-6	5	5-6
8	Current efficiency (%)	85	93	54	96.5
9	Productivity of the cell ($g\ h^{-1}\ dm^{-3}$)	6.9	5.6	1.3	8.4
10	Energy consumption ($kwh\ kg^{-1}$)	1.38	1.35	2.63	1.32

Advantages of the present development are listed below:

1. Lower power consumption: Specific power consumption is 1.4 kWh kg^{-1} against 2.2 kWh kg^{-1} in the mono polar version, due to low oxygen over voltage and lesser contact drop. Power saving works out to 800 kWh per tonne. A 100 TPD copper plant saves 80,000 kWh per day.
2. Saving in copper busbar for cell construction. In the mono polar version, each electrode has to be connected to the rectifier. In Bipolar version, only the end electrodes are connected to the rectifier. Depending on the number of electrodes in a cell, busbar requirement can be as low as 5%, when compared to the mono polar configuration.
3. Use of lead alloy anode results in a lead loss of ~ 1.0 kg per tonne of copper produced. Lead has been eliminated in this Green Technology.

4. Conclusion

The present improvement in electrowinning of copper from acidic copper sulphate solution by the use of bipolar electrodes obviates the need for cross connecting bars in the cell and the use of busbars of thicker cross section that are needed for the hitherto used monopolar configuration. This development is more eco friendly as the lead anode is substituted with titanium coated with oxides of precious metals such as iridium and ruthenium. It also paves the way for copper electrowinning cells of enhanced productivity. This art can be extended to the electrowinning of zinc and nickel.

References

1. C.L. Mantell, *Industrial Electrochemistry* (Newyork, McGraw-Hill Book Company Inc, 1950)
2. Improvements in or relating to electrowinning of copper. Indian Patent, 48112 (1986).
3. E.C.Potter, *Electrochemistry: Principles and Applications* (London, Macmillan,1961).
4. Coated titanium anodes for electrolysis of aqueous solutions. Indian Patent, 178184 (1990).

Fig.1

Fig.2

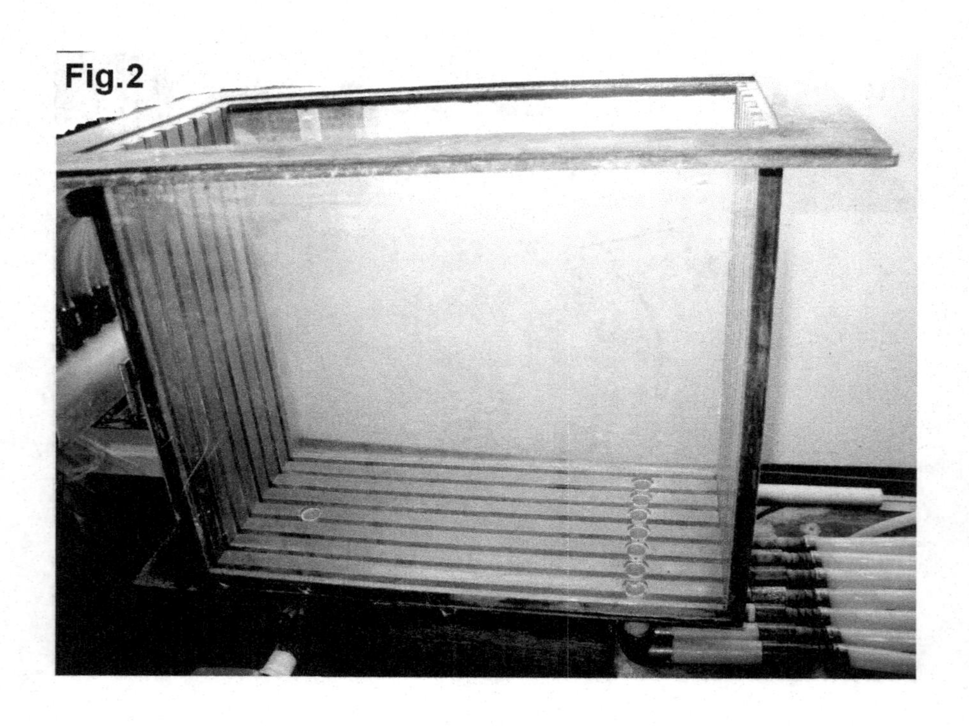

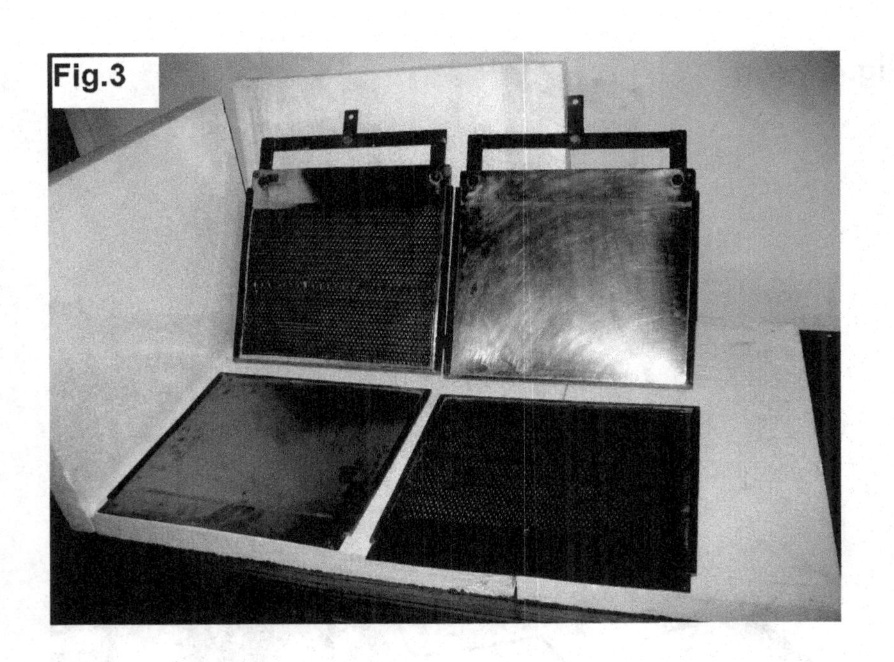

Fig.3

Fig.4

Fig.5

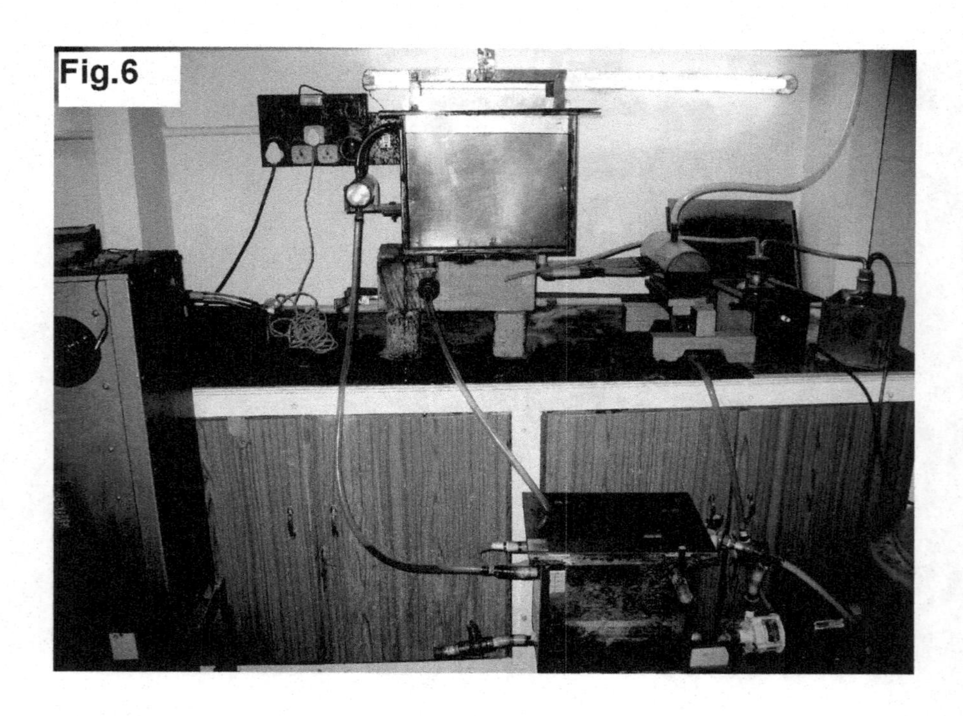

Captions to figures 1 to 6

Fig 1: Three dimensional view the electrowinning cell fitted with cover
Fig 2: A view of the grooves cut in the cell bottom and sides to arrest leakage current
Fig 3: End electrodes and bipolar electrodes
Fig 4: Electrodes fitted to the electrowinning cell
Fig 5: Front view of the electrowinning cell
Fig 6: A view of the bipolar copper electrowinning cell in operation

Energy Technology Perspectives
Edited by: Neale R. Neelameggham, Ramana G. Reddy, Cynthia K. Belt, and Edgar E. Vidal
TMS (The Minerals, Metals & Materials Society), 2009

MICROBIAL REDUCTION OF LATERITIC NICKEL ORE FOR ENHANCED RECOVERY OF NICKEL AND COBALT THROUGH BIO-HYDROMETALLURGICAL ROUTE

L.B. Sukla, B.K. Mishra, N. Pradhan, R.K. Mohapatra, B.K. Mohapatra, B.D. Nayak

Institute of Minerals and Materials Technology
Council of Scientific & Industrial Research (CSIR)
Bhubaneswar-751013, India

Keywords: Nickel ore, Iron (III) Reducing Bacteria (IRB), Goethite, Magnetite and Leaching

Abstract

Present study is about use of a group of Iron[Fe(III)] Reducing Bacteria (IRB) to convert the goethite (α-FeOOH) present in the original lateritic Nickel ore to magnetite under an anaerobic condition and subsequently release the bound Co(III) and Ni(II) through leaching. The lateritic Nickel ore contains 0.8% Ni, 0.049% Co, 1.92% Cr, 0.32% Mn and 50.2% Fe. An anaerobic dissimilatory Fe(III) reducing bacterial consortium capable of using acetate as carbon source (electron donor) and lateritic ore as terminal electron acceptor, changes the initial light brown color of the ore to dark brown. The change in color is due to the conversion of goethite to magnetite, which was confirmed by XRD. When the IRB treated sample was subjected to both bioleaching and acid leaching, it shows a greater recovery of Nickel and Cobalt values as compared to untreated original lateritic Ni ore. Further a magnetic separation method was used to separate the magnetic part of the treated lateritic ore and subjected to acid- and bio-leaching for better recovery of Nickel and Cobalt.

Introduction

It has been known that micro-organisms have the potential to reduce metals but more recent observations have shown that a diversity of specialist bacteria and archaea can use such activities to conserve energy for growth under anaerobic conditions [12]. Natural Fe(III) oxides are high in surface area and are reactive [1]. Fe (III) oxides adsorb a wide range of metal cations and anions by complexation to surface hydroxyl groups [17], and function as the primary redox buffering solid phase in many sediments and subsurface materials [5].

Recently a number of bacteria have been isolated based on their ability to use metal ions, including Fe(III), as electron acceptor under anaerobic conditions. These are capable of producing energy by coupling the oxidation of organic compounds to reduction of Fe(III). Iron[Fe(III)]-Reducing Bacteria (IRB) have been isolated and identified from a wide variety of environments including freshwater and marine aquatic sediments and submerged soil [8]. The biogeochemical cycle of iron is linked to those of the trace metals, as many trace metals co-associate with Fe(III) oxides or reduced iron phases (e.g., magnetite, siderite, or iron sulfide). IRB can utilize Fe(III) oxides as terminal electron acceptors for respiration [18]; thereby reducing all or a fraction of the solid. The reductive portion of the iron biogeochemical cycle in non-phototropic environments (e.g., sediments and subsurface materials) is driven by the direct enzymatic reduction of Fe(III) oxides to Fe(II) by dissimilatory iron reducing bacteria [13,14].

Iron-reducing microbes are found in the genera *Shewanella*, *Geobacter*, *Geovibroi*, *Disulfobulbus propionicus*, etc [2]. Among these metal-reducing bacterias, *Shewanella* is perhaps the most widely studied [2]. Microbial dissimilatory Fe(III) reduction plays an important role in the geochemical cycling of iron and organic matter in anoxic ecology system[13,14]. Metal cations that are co-precipitated or structurally incorporated in FeOOH may be released [9] during bacterial reduction of Fe(III) oxide. Such 'dissimilatory' processes have opened up new and fascinating areas of research with potentially exciting practical applications.

Mechanism of Fe(III) reduction

A wide range of Archaea and Bacteria are able to conserve energy though the reduction of Fe(III) to Fe(II) ion i.e., ferric to ferrous ion. Fe(III) can be the dominant electron acceptor for microbial respiration in many subsurface environments. As such, Fe(III)-reducing communities can be responsible for the majority of the organic matter oxidized in such environments [10, 11]. Fe(III) used as an electron acceptor in many sedimentary environments is present as insoluble Fe(III) oxides because of its low solubility at neutral pH. Three mechanisms of microbial electron transport to insoluble acceptors have been reported in a various processes. First, direct cell-Fe(III) mineral contact is required for reduction, which occurs by cell-oxide adhesion. Second, membrane-bound ferric reductase and cytochromes can make link between Fe(III) oxides and the electron transport system in the cytoplasmic membrane. Third, microbially secreted electron carriers such as extracellular cytochrome and quinon, and electron shuttling compounds may also be involved in Fe(III) reduction [3, 6, 7].

Nickel lateritic ore contains 0.8% Ni, 0.049 % Co, 1.92% Cr, 0.32% Mn and 50.2% Fe. Bacterial leaching of roasted Nickel lateritic ore shows more recovery of Ni & Co as compared to unroasted Ni lateritic ore. Nickel and cobalt are present in adsorbed state with goethite. Roasting of ore resulted in conversion (reduction) of goethite to hematite, which was responsible for releasing nickel and cobalt from $Fe2+$ lattice and resulted better recovery. In the present study attempt has been made to examine the effect of biological reduction of the original ore with help of anaerobic dissimilatory Fe(III) reducing bacterial consortium and subsequent metal leaching to recover Nickel and cobalt values. In this bacterial reduction process, the bioleaching of treated ore can be done in a 9k- (without ferrous sulfate) medium as compared to previously done roasted Ni ore bioleaching in 9k+ (with ferrous sulfate) medium [16]. If bacterial reduction method is followed, cost intensive method of ore roasting process can be eliminated. This would also save the cost of ferrous sulfate which was used in previous bioleaching method. The objective of the present study is to convert the goethite present in the original Ni lateritic ore to hematite/magnetite, which would release Nickel and Cobalt from iron lattice, thereby increase solubilization and result in higher recovery of Nickel and Cobalt values.

Materials and Methods

Lateritic nickel Ore

Lateritic Nickel ore, from Sukinda mines (Orissa) was provided by Orissa Mining Corporation (OMC), India. The sample is a low grade lateritic ore which contains about 0.8% Ni, 0.049% Co, 1.92% Cr, 0.32% Mn, and 50.2% Fe.

Source of organisms

Water sediment samples were taken from three different wetland sites around Bhubaneswar, Orissa, India. After collecting, these were immediately transferred to anaerobic vials and taken back to the laboratory for subsequent enrichment culture in a selective growth media for IRB for isolating Fe(III) reducing bacteria.[15]

Enrichment and isolation of Fe (III)-reducers

One gram of sediment was transferred to vial containing 10 ml of anaerobic saline solution. These vials are thoroughly mixed and then 10 ml of sample was transferred to the 100 ml volumetric flasks containing selective alkaline anaerobic iron reducing medium (Table1) and covered with paraffin oil to prevent further contact with oxygen to its surface.

Table 1: Composition of Media for isolating IRB.

Component	g/l
Fe_2O_3	1.0
K_2HPO_4	3.0
KH_2PO_4	0.8
KCL	0.2
NH_4CL	1.0
$MgCL_2$	0.2
$CaCL_2$	0.1
Yeast Extract	0.05
Na-Acetate ($C_2H_3O_2Na$)*	10 mmol
1% mixture of Vitamins and trace minerals solution	

*(as a carbon source and electron donor)
PH of the media was adjusted to 7.2 with the help of 1N NaOH

Finally the flasks was over flooded by oil and covered with paraffin paper, so that there is no place for air bubble. The flasks were incubated at 35^{0}C for 7 days under dark condition. The enrichment procedure was repeated four times with transfer of 10 ml of culture as inoculum into fresh medium. Finally, enrichment cultures were serially diluted and plated in the anaerobic chamber on to an agar containing iron reducing medium plate with goethite as Fe(III) source and the anaerobic chamber was flushed with CO_2 and kept under dark condition at 35^{0}C for 3 weeks.

Treatment of Lateritic Nickel Ore with enriched IRB consortium

Lateritic Ni ore was treated with well enriched Fe(III) reducing bacterial consortium in the medium (see Table1) in an anaerobic, alkaline medium as mentioned above at 2% pulp density and kept at 35^{0}C for 7 days under dark condition. With increasing incubation period, it was noted that initial light brown colour of ore was changed to dark brown colour. Simultaneously control experiments without any bacterial solution were also run. After treatment, the treated lateritic Ni ore was filtered, dried in air.

Characterization of IRB treated ore

X-ray diffraction (XRD) study was done for original lateritic Ni ore, using an automatic Philips X-ray diffractometer (PW-1710). A part of IRB-treated lateritic ore was found to be magnetic. Hence, the magnetic and non-magnetic part of the IRB-treated ore was separated by the help of a hand magnet, and XRD was taken up for these two separated ores. Electron microscopic study accomplished with compositional mapping was undertaken by JEOL make (JXA-8100)electron probe microanalyser (EPMA).

Chemical and bio leaching of treated ore

The IRB treated ore was subjected to two types of leaching techniques. One was inorganic acid leaching using H_2SO_4 and other using *Acidothiobacillus ferrooxidance* in $9K^+$ and $9K^-$ media. Composition $9K^+$Media was taken as follows in g/lit: $(NH_4)_2SO_4$—3.0; K_2HPO_4,—0.5; $MgSO_4.7H_2O$—0.5; KCl—0.1; $FeSO_4,7H_2O$—44.2; pH—1.8. Without $FeSO_4.7H_2O$ the $9K^+$Media is known as $9K^-$Media. The acid leaching was done by four different concentration of H_2SO_4 (1M, 2M, 4M, 8M) at 2% pulp density by continuous stirring and shaking for different time interval. The bioleaching was also done by *A. ferrooxidans* in two different media at 2% pulp density by continuous shaking at 35^0C and 150 rpm for different time interval. Leaching experiments were performed on original ore, IRB treated ore, magnetic and non magnetic fractions of IRB treated ore. At regular interval of time, samples were collected and subjected to AAS analysis of metals like Ni, Co, and Fe.

Influence of incubation time in IRB-treatment for recovery of metals

Lateritic Ni ore was treated with Iron reducing bacterial consortium culture for different time interval i.e. 0 hr, 5 days, 10 days, 15 days, 20 days, 25 days and 30 days. The reaction medium was supplied with sodium acetate as carbon source after every 5 days. Then the solid residue was filtered, dried and subjected to acid leaching by 8M H_2SO_4 for 10 days to know the relation of IRB-treatment time with the recovery of metals.

Result and Discussion

Lateritic nickel ore from Orissa comprises of goethite/ferrihydrite,gibbsite and surinamite phases and contains 0.8% Ni,0.049% Co,1.92% Cr,0.32%Mn and 50.2% Fe.The goethite/ferrihydrite is a reddish brown mineral of poorly ordered crystal structure.The above ore was subjected to microbial reduction prior to bio-hydrometallurgical leaching and acid leaching with a view to enhance the recovery of Ni and Co. After successive enrichment on selective medium a consortium capable of using acetate as carbon source (electron donor) and lateritic ore containing ferrihydrites as terminal electron acceptor was obtained. With increasing incubation period, it was noted that initial light brown colour of lateritic ore was changed to dark brown colour (Figure 1).

The initial light brown colour of the ore was due to goethite(FeOOH) and its conversion to dark brown was due to formation of hematite(Fe_2O_3) /magnetite(Fe_3O_4). There was no such change in the control experiment. After enrichment, when the enriched culture was spread plated on to an agar containing iron reducing medium plate with goethite as Fe(III) source. The colonies appear brown on the brown background of the medium (see Figure 2) and were visible only on careful observation. Colonies appeared only after 10 days of incubation under anaerobic conditions.

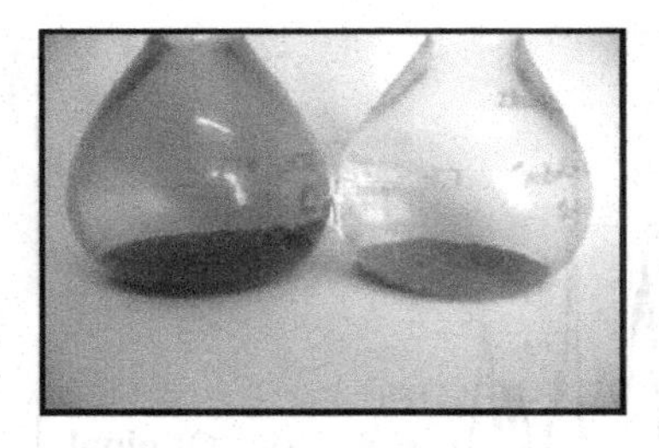

Figure 1: Change of colour of IRB-treated lateritic Ni ore as compared to control.

Figure 2: Isolated colonies appearing iron reducing medium incorporated with laterite ore.

Growth of iron reducing bacteria was studied in a selective medium mentioned above in the presence of lateritic nickel ore. Initially the growth was slow showing the lag phase then the growth increased exponentially which can be observed from Figure 3.

To confirm about the change of goethite present in the ore to magnetite, an experiment was conducted in similar way to generate 20 gm of IRB treated ore. X-ray diffraction (XRD) pattern of the IRB-treated ore and original lateritic ore show that there was about 20 % conversion to magnetite.

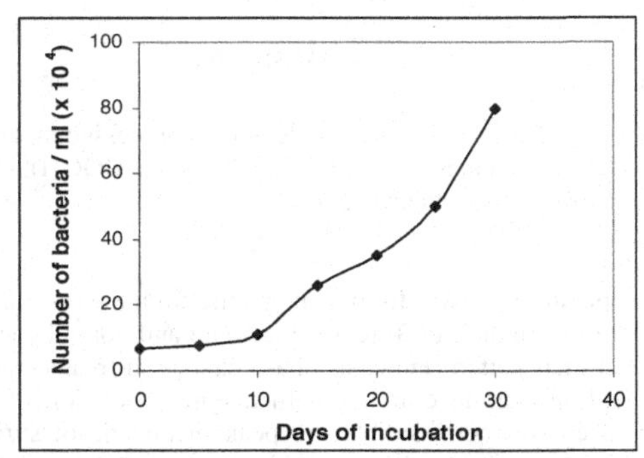

Figure 3: Growth of iron reducing bacteria, with lateritic Ni ore as terminal electron acceptor.

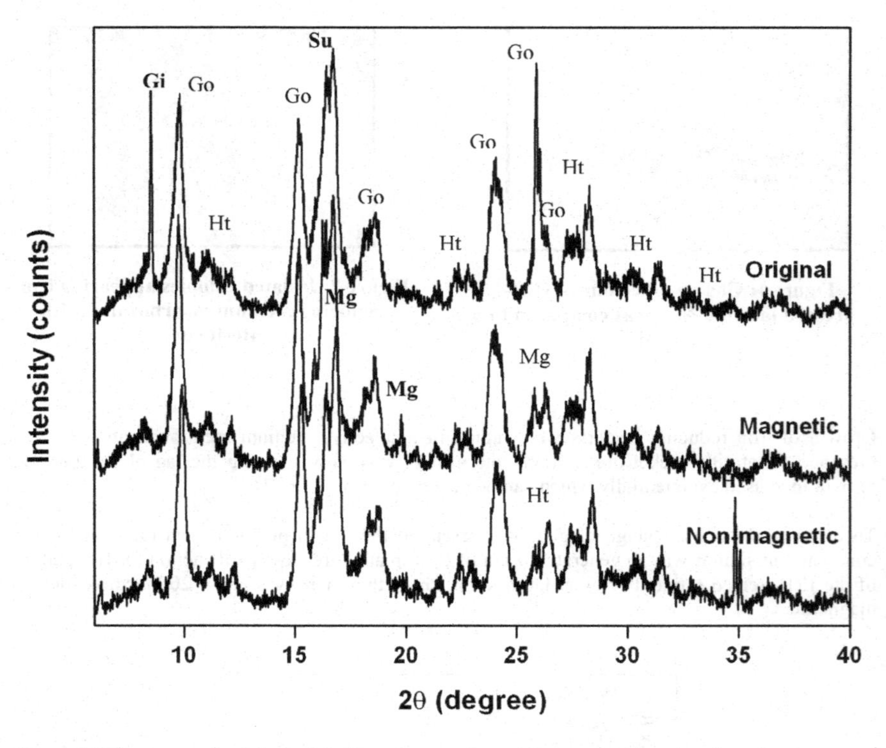

Fig. 4: XRD pattern of original nickel laterite sample and its processed magnetic and nonmagnetic products. Ht: Hematite (Fe_2O_3); Gi: Gibbsite (AlOOH); Go: Goethite (FeOOH); Su: Surinamite; Mg: Magnetite (Fe_3O_4).

It was also possible to separate the newly formed magnetite from the rest of goethite with help of magnet. The XRD pattern of original, IRB treated magnetic and non magnetic fraction is shown in Figure.4. Difference in their pattern indicates phase change in treated sample. In the original ore gibbsite, goethite and surinamite dominates. In magnetic fraction of IRB treated sample gibbsite peak intensity decreases considerably, no peak diagnostic of surinamite occurs, with additional peaks characteristic of magnetite appear at 1.61A° and 2.7A°. Hematite and goethite peaks are also present. In non-magnetic fractions of IRB-treated sample gibbsite peak intensity decreases, hematite peak became more distinct having additional peak at 1.18A°, and with minor amount of goethite and magnetite. With increase in incubation period and other culture conditions, the percentage of conversion of goethite to magnetite can also be increased. So only the magnetic part can be used for leaching and the rest non magnetic part, which was not transformed to the magnetite, can be retreated with IRB solution following previous method.

The compositional maps of original ore sample and magnetic fraction of treated sample brought out through EPMA, display some difference both in their nature and level of pixel concentration (Figure-5). In the original sample Ni and Co are observed to be confined within iron phase. .Average level of Co and Ni is found to be 1 in respect of 5 in Fe level (Figure-5a). In contrast the compositional map in magnetic fraction of treated sample shows more diffused Co and Ni. In this case level of Co and Ni shoot up to 3 in respect of 4 in Fe level (Figure-5b).

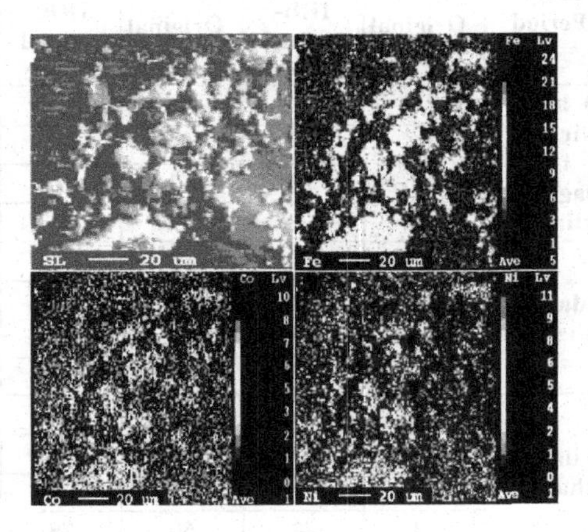

Fig. 5a: Electron micrograph of a selected area(SL) of original nickel laterite sample and X-ray image map of Fe, Ni and Co.

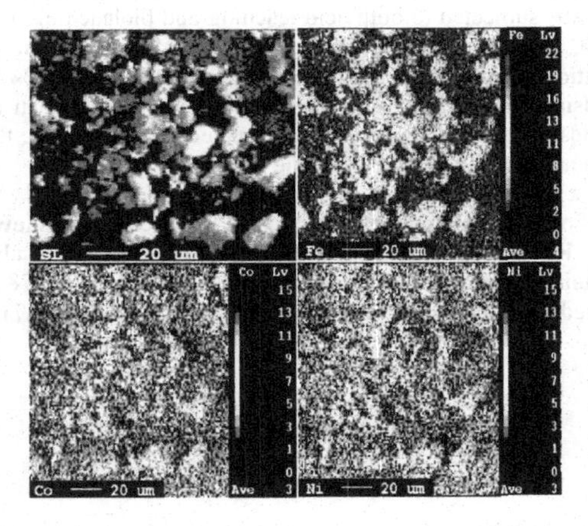

Fig. 5b: Electron micrograph of a selected area(SL) of processed nickel laterite sample (magnetic product) and X-ray image map of Fe, Ni and Co.

Table 2: Comparison of metal recovery by acid leaching of IRB-treated and original ore with H_2SO_4; at 2% pulp density and varying time intervals of leaching.

Sl. No	Conc. Of H_2SO_4 Used (M)	Incubation Period	% of Ni Leached		% of Co Leached		% of Fe Leached	
			Original Ni ore	IRB-treated Ni ore	Original Ni ore	IRB-treated Ni ore	Original Ni ore	IRB-treated Ni ore
1	**1**	**3 hour stirring by magnetic stirrer**	0.6	1.52	2.1	6.1	1.6	3.6
2	**2**		0.8	2.8	4.1	8.2	1.9	5.3
3	**4**		1.1	4.5	6.1	10.2	0.7	5.7
4	**8**		1.3	10.5	8.2	14.3	1.2	13
5	**1**	**5 days in shaker**	0.8	1.8	4.1	14.3	2.0	7
6	**2**		1.0	3.0	6.1	16.3	8.0	36
7	**4**		2.0	6.1	8.2	18.3	22.0	15
8	**8**		7.6	19.5	12.3	24.5	13.8	35
9	**1**	**15 days in shaker**	4.7	5.1	10.2	20.4	13.0	18
10	**2**		8.4	13.7	12.3	26.6	17.9	32
11	**4**		19.8	36.1	22.5	30.6	34.9	50
12	**8**		54	**82**	60	**78**	80.2	**94**

The IRB treated ore was subjected to both acid leaching and bioleaching. It was observed that a maximum of 82% Ni, 78% Co, 94% Fe were recovered in 8M H_2SO_4 in shake flask leaching, under 15 days incubation (Table 2). In case of bioleaching recovery of 12% Ni and 13% Co at 35 days of incubation using *Acidothiobacillus ferrooxidans* was obtained in (Table 3). Thus, the percentage of metal leaching were greater for IRB-treated Ni lateritic ore than that of untreated original ore by both acid leaching and bio-leaching .

Comparative study of original and magnetic fraction of IRB treated ore show clear cut difference in metal leachability. Recovery of metals like Ni, Co, Mn and Fe were always more from the magnetic fraction obtained from the IRB treated ore. Maximum 34% Ni, 78 % Co, 90 % Mn and 35% Fe were recovered from the magnetic part of IRB treated ore (Figure.7).

Table 3:Results of *Acidothiobacillus ferrooxidans* Leaching of Original and IRB treated ore in $9K^{+}$ and $9K^{-}$ medium at 2% pulp density (pH-1.5).

Sl. No.	Incubation (days)	Media	Type of Ore	% Ni Leaching	%Co Leaching
1	2	$9K^{+}$	Original	0.12	3.06
2			IRB Treated	0.64	5.71
3		$9K^{-}$	Original	0.10	2.04
4			IRB Treated	0.63	5.91
5	5	$9K^{+}$	Original	0.18	3.67
6			IRB Treated	0.72	6.32
7		$9K^{-}$	Original	0.12	2.65
8			IRB Treated	0.72	6.73
9	7	$9K^{+}$	Original	0.20	3.87
10			IRB Treated	0.82	6.53
11		$9K^{-}$	Original	0.15	3.06
12			IRB Treated	0.83	7.14
13	14	$9K^{+}$	Original	0.34	4.08
14			IRB Treated	1.15	7.34
15		$9K^{-}$	Original	0.27	3.49
16			IRB Treated	1.26	8.16
17	21	$9K^{+}$	Original	0.37	4.28
18			IRB Treated	1.21	7.55
19		$9K^{-}$	Original	0.29	3.67
20			IRB Treated	2.39	8.36
21	28	$9K^{+}$	Original	0.54	4.89
22			IRB Treated	7.77	9.38
23		$9K^{-}$	Original	1.34	4.08
24			IRB Treated	8.89	10.40
25	35	$9K^{+}$	Original	1.65	5.10
26			IRB Treated	11.89	11.42
27		$9K^{-}$	Original	2.40	4.28
28			IRB Treated	**12.15**	**13.06**

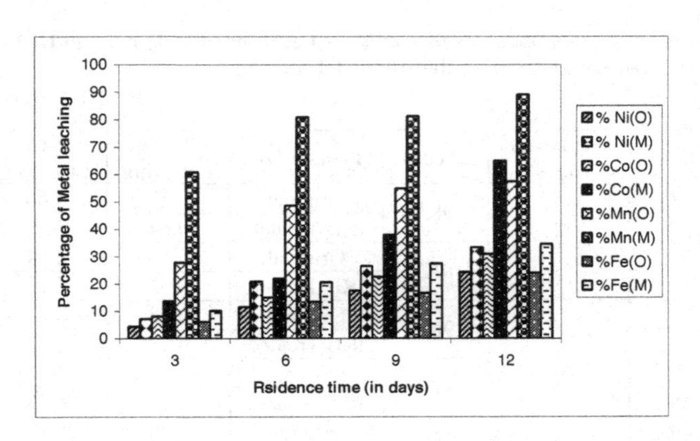

Fig.6: Schematic diagram showing comparative results on acid leaching of Original, IRB-treated magnetic and Non-magnetic of Ni-lateritic ore by 8M H_2SO_4 for 12 days in shaker at 1% pulp density: O=Ni lateritic original ore. M=Magnetic part of IRB treated Ni lateritic ore.

The extent of bioreduction with respect to time was also investigated. It was observed that by increasing the time of treatment of ore with Iron reducing bacteria, the rate of conversion of goethite to magnetite was also increased, which was confirmed by acid leaching of that treated ore. It was also found that longer the incubation time with IRB, greater the conversion and percentage of metal recovery. Illustration in the Fig.7 shows that longer is the treatment time of IRB with lateritic Ni ore, greater is the recovery of Ni, Co, Cr and Fe.

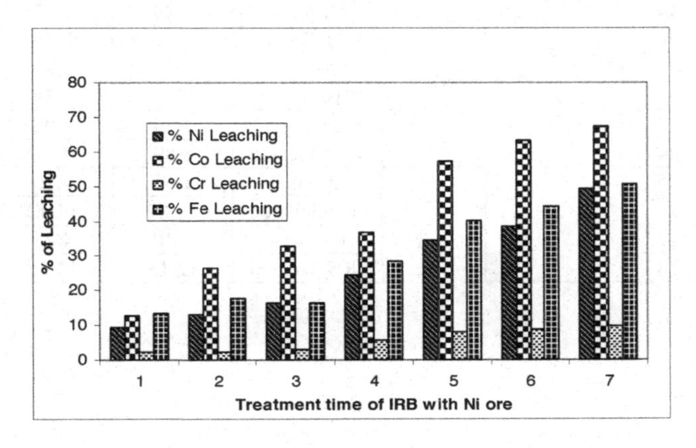

Fig 7: Schematic diagram showing results of acid leaching of varying treatment time of IRB with Ni-lateritic ore by 8M H_2SO_4 for 10 days in shaker at 2% pulp density.

Conclusions

Microorganisms plays important role in the environmental balance of toxic metals and metalloids with physico-chemical, and biological mechanisms effecting the transformations between soluble and insoluble phases. Anaerobic dissimilatory iron (III) reducing bacterial consortia (mostly, the genera of *Shewanella, Geobacter)* isolated from fresh water sediments, capable of using acetate as carbon source (electron donor) and lateritic Ni ore as terminal electron acceptor was developed and utilized for recovery of metals from lateritic ore. The Fe(III) reducing bacterial consortium, when treated with Ni lateritic original ore, converts the goethite present in the ore to hematite/magnetite, by which it solubilizes the Ni, Co and other metals bound to goethite. This way higher percentage of metal recovery is obtained as compared to the original untreated Ni lateritic ore. About 34% Ni, 65 % Co, 90 % Mn and 35% Fe were recovered from the magnetic part of IRB treated ore in 12 days of leaching (Table-4) and 82% Ni 78 % Co, and 94% Fe were recovered from IRB-treated ore in 15 days of leaching (Table-2).12% Ni and 13% Co were recovered with *A.Ferrooxidans* using IRB treated ore.It was almost double to the metals leached from original ore and non magnetic fraction of the IRB treated ore.

Treatment of original Ni lateritic ore with IRB consortium resulted in about 20% of conversion from goethite to magnetite in 2 months of incubation. Increase in the time of incubation, may result in more percentage of conversion. After separating the magnetic part of the treated ore i.e. magnetite, the non-magnetic part can also be retreated with IRB-consortium and this process can be run in cyclic manner.

References

1. Cornell R. M. and Schwertmann U. (1996) *The Fe Oxides: Structure, Properties, Reactions, Occurrences, and Uses.* VCH.
2. Eun Jeong Cho, Andrew D. Ellington, "Optimization of the biological component of a bioelectrochemical cell," Bioelectrochemistry, 70(2007), p.165-172.
3. Eun Younglee, Seung-Rim Noh, Kyung-Suk Cho et al., " Leaching of Mn,Co and Ni from Manganese Nodules using an Anaerobic Bioleaching Method," Journal of Bioscience and Bioengineering,vol.92,issue.4,2001, p.354-359.
4. Geoffrey M.Gadd, "Microbial Metal Transformations," The Journal of Microbiology, vol.39, issue.2, June 2001, p.83-88.
5. Heron G., Crouzet C., Bourg A. C. M., and Christensen T. H. (1994) Speciation of Fe (II) and Fe (III) in contaminated aquifer sediments using chemical extraction techniques. *Environ. Sci. Technol.* **28,** 1698–1705.
6. David J. Scala, Eric L. Hacherl, Robert Cowan, Lily Y. Young, David S. Kosson, "Characterization of Fe(III)-reducing enrichment cultures and isolation of Fe(III)-reducing bacteria from the Savannah River site, South Carolina," Research in Microbiology, 157 (2006), p. 772–783.
7. Doo Hyun Park and Byung Hong Kim, "Growth properties of the Iron-reducing Bacteria, Shewanella putrefaciens IR-1 and MR-1 Coupling to Reduction of Fe(III) to Fe(II)," The Journal of Microbiology,vol.39,issue.4, December 2001, p.273-278.
8. John D. Coates, Elizabeth J.P Philips, Debraj. Lonergan et al., "Isolation of Geobacter Species from Diverse Sedimentary Environments," Applied and Environmental Microbiology, vol.62, issue.5, May 1996, p.1531-1536.
9. John M Zachara, Jim K. Fredrikson, Steven C. Smith and Paul L. Gassman, "Solubilization of Fe(III) oxide-bound trace metals by a dissimilatory Fe(III) reducing bacterium," Geochimica et Cosmochimica Acta,vol.65,issue.1, 2001, p.75-93.

10. Jonathan R. Lloyd, "Microbial reduction of metals and radionuclides" FEMS Microbiology Reviews, 27 (2003), p. 411-425.
11. Karrie A. Weber, Matilde M. Urrutia et al., "Anaerobic redox cycling of iron by freshwater sediment microorganisms," Environmental Microbiology, 8(1), 2006, p.100-113.
12. Kim G.T, M.S. Hyun, I.S. Chang, H.K. Kim, H.S. Park et al., "Dissimilatory Fe(III) reduction by an electrochemically active lactic acid bacterium phylogenetically related to *Enterococcus gallinarum* isolated from submerged soil," Journal of Applied Microbiology,2005,99, p.978-987.
13. Lovley, D.R., "Dissimilatory Fe (III) and Mn (IV) reduction," Microbiol Reviews, 55(1991), p.259-287.
14. Lovley, D.R., "Dissimilatory metal reduction," Annual Rev. Microbiol., 47(1993), p.263-290.
15. Moon-Sik Hyun, Byung-Hong Kim, In-Seop Chang et al., "Isolation and Identification of an Anaerobic Dissimilatory Fe (III)-Reducing Bacterium, *Shewanella putrefaciens* IR-1," The journal of Microbiology, vol.37, issue.4, December 1999, p.206-212.
16. S. Mohapatra , S. Bohidar, N. Pradhan, R.N. Kar, L.B. Sukla, "Microbial extraction of nickel from Sukinda chromite overburden by Acidithiobacillus ferrooxidans and Aspergillus strains," Hydrometallurgy, 85 (2007), p. 1–8.
17. Schindler P. W. and Stumm W. (1987) The surface chemistry of oxides, hydroxides, and oxide minerals. In *Aquatic Surface Chemistry* (ed. W. Stumm), John Wiley & Sons pp. 83–110.
18. Zachara J. M., Fredrickson J. K., Li S.-M., Smith S. C., and Gassman P. L. (1998) Bacterial reduction of crystalline Fe(III) oxides in single phase suspensions and subsurface materials. *Am. Mineral.* 83, 1426– 1443.

Energy Technology Perspectives
Edited by: Neale R. Neelameggham, Ramana G. Reddy, Cynthia K. Belt, and Edgar E. Vidal
TMS (The Minerals, Metals & Materials Society), 2009

Energy Saving Strategies for the Use of Refractory Materials in Molten Material Contact

James G. Hemrick[1], Klaus-Markus Peters[2], and John Damiano[3]

[1]Oak Ridge National Laboratory;
Bethel Valley Road, Oak Ridge, TN 37909, USA
[2]Fireline, TCON, Inc.
300 Andrews Avenue, Youngstown, OH 44505, USA
[3] Minteq® International Inc.
640 North 13th Street, Easton, PA 18042, USA

Keywords: Refractories, Molten Metal, Glass, Lime, Energy Savings

Abstract

Work was performed by Oak Ridge National Laboratory (ORNL), in collaboration with industrial refractory manufacturers, refractory users, and academic institutions, to employ novel refractory systems and techniques to reduce energy consumption of molten material processing vessels found in industries such as aluminum, glass and pulp and paper. The energy savings strategies discussed are achieved through reduction of chemical reactions, elimination of mechanical degradation caused by the service environment, reduction of temperature limitations of materials, and elimination of costly installation and repair needs. Key results of several case studies resulting from US Department of Energy (DOE) funded research programs are discussed with emphasis on applicability of these results to high temperature processing industries.

Introduction

Refractory materials are limited in their application by many factors including chemical reactions between the service environment and the refractory material, mechanical degradation of the refractory material by the service environment, temperature limitations on the use of a particular refractory material, and the inability to install or repair the refractory material in a cost effective manner or while the vessel is in service. All of these limitations reduce the energy efficiency of the process as degraded refractory materials lead to loss of process heat (reduced insulation by refractories) and the need for maintenance through repair or replacement of refractory linings.

This situation is conceptualized in Figure 1 which shows that when the thickness of the refractory wall is reduced by chemical or mechanical degradation, the heat losses through the wall increase exponentially. The subsequent maintenance to repair such conditions often requires cooling of the furnace or refractory lined vessel which entails loss of energy due to cooling and consumption of energy due to reheating. Additionally, production time and capability are sacrificed.

Additionally, in most applications where refractories are used (metal melting furnaces, glass furnaces, steel making furnaces, gasifiers, etc.) wear of the refractory lining is not uniform. Very often, specific areas will wear (become thinner) more rapidly due to exposure to more corrosive or erosive conditions. This concept is illustrated in Figure 2, through the example of excessive metal line corrosion.

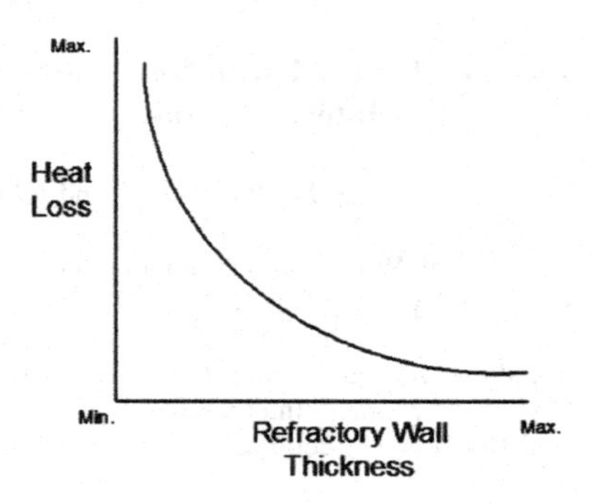

Figure 1. Schematic of hypothetical heat loss in relation to decreasing refractory wall thickness due to chemical or mechanical degradation.

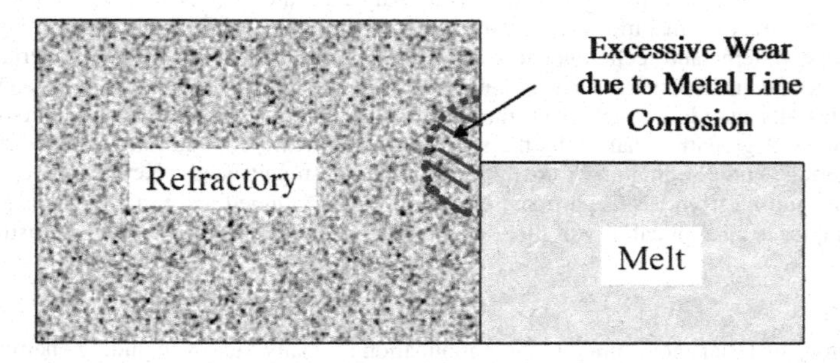

Figure 2. Schematic example of uneven refractory wear as would be seen in the case of metal line corrosion in a molten metal or glass furnace.

This concept is further illustrated through the examination of refractory both above and below the metal line in an aluminum furnace as shown in Figure 3 [1]. For many cases the metal line forms a triple point where solid refractory is in contact with liquid molten metal and the gaseous species above the melt. It is at this location that the most severe reactions are found to occur. Formation of surface agglomerates, such as corundum mushrooms (in the case of molten aluminum) on the wall above the molten metal leads to spalling of the refractory wall as stresses are created by the alumina surface concretion and creation of porosity in the refractory structure through internal corundum growth within the refractory wall. The integrity of the refractory lining below the metal line can also be degraded by metal penetration leading to reduced thickness of the refractory wall and floor. This reduces the thermal efficiency of the refractory lining leading to greater heat losses and less heat retained in the process and can result in catastrophic failure of the molten metal containment vessel. Other similar reactions and degradation mechanism exist for other molten metal/glass-refractory systems.

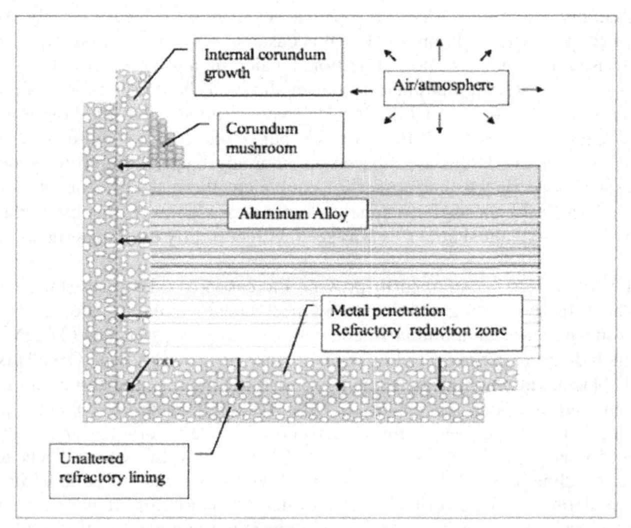

Figure 3. Schematic of corrosion and wear mechanisms associated with aluminum contact refractories [1].

Energy Savings Strategies for Aluminum

Under a project funded by the DOE Energy Efficiency and Renewable Energy/Industrial Technologies Program (EERE/ITP) entitled "Multifunctional Metallic and Refractory Materials for Energy Efficient Handling of Molten Metals"[I], two new refractory materials were developed for aluminum melter applications in collaboration with Missouri University of Science and Technology (MS&T) and industrial partners. The development of these materials was based on understanding of the corrosion and wear mechanisms associated with currently used aluminum contact refractories gained through physical, chemical, and mechanical characterization of salvaged materials, as discussed elsewhere [1]. These new materials were designed to have improved corrosion and wear resistance leading to increased service life, along with improved thermal management through reduced heat losses.

Through physical, chemical, and mechanical characterization of salvaged refractory materials from the aluminum industry it was found that refractory failure was due to three primary sources. The first was poorly performing anti-wetting additives (CaF_2 for example) and degradation of the aggregate chosen for use in the refractory castables. The second was reaction of micro-silica in the refractory matrix with cement binder systems (silica with calcium aluminate). The final source of failure was due to poor furnace maintenance practices leading to mechanical damage of refractory linings.

To address the first two sources of refractory failure (poorly performing additives/aggregates and reaction of micro-silica with calcium aluminate) a new castable refractory material was

I Work supported by the Industrial Technologies Program, U.S. Department of Energy, under Grant No. DE-PS07-031D14425.

developed with industrial partner, Missouri Refractories Company, Inc. of Pevely, MO, based on an alternative form of calcium aluminate. For this castable composition, bonite ($CaO * 6Al_2O_3$) aggregate was used in place of more traditional calcium aluminate ($CaO * 2Al_2O_3$). This mineralogical phase of calcium aluminate has been shown to be more resistant to reaction with molten aluminum due to lower wettability [2,3]. Testing showed that the bonite aggregate had similar thermal characteristics to tabular alumina, but required either more water or advanced flow aids to get similar flow to castables based on other aggregates. Therefore, a more corrosion resistant material was produced that possessed equivalent thermal efficiency to currently used materials, but for which the thermal efficiency was not degraded over time by degradation of the refractory lining due to chemical attack as was seen with currently used materials.

A second material was also developed to provide increased corrosion resistance (resistance to first two sources of failure), along with increased mechanical strength to address the third source of failure. This material was an alumina/silicon carbide composite material (TCON®) which was developed with industrial partner Fireline TCON, Inc. of Youngstown, OH. This material is produced by displacement reactions between silica, carbide phases, and aluminum metal to form alumina, silicon carbide, and silicon metal. The resulting material contains a continuous microscopic network of interpenetrating microscopic scaled ceramic and metallic phases (approximately 53 wt. % SiC, 35 wt. % Al_2O_3, 12 wt. % Al/Si) which leads to significant improvement in toughness and damage tolerance, while the presence of ceramic phases leads to high hardness and improved performance at elevated temperatures. It was expected that this material would exhibit corrosion resistance to molten metal contact due to its low porosity, its high alumina/low silica content, and the formation of an in-situ passivating layer during interaction with molten aluminum leading to non-wetting behavior. Abrasion/wear resistance to damage caused by mechanical dross removal was also expected due to the presence of the secondary silicon carbide phase.

The resistance to degradation (both chemical and mechanical) of both materials was validated through commercial trials by the manufacturers and through the exposure of material samples for over 2000 hours in an industrial setting at Energy Industries of Ohio (EIO). In the EIO trial, TCON plates backed by bonite castable were subjected to contact with Mg-rich molten aluminum. From the results of this test, it was estimated that a total energy savings of over 0.55 trillion Btu per year could result from use of these improved materials in the aluminum industry. Currently, these materials are involved in a full industrial trial as a lining system in an aluminum rotary degasser.

Energy Savings Strategies for Glass

Under a project funded by the DOE Energy Efficiency and Renewable Energy/Industrial Technologies Program (EERE/ITP) entitled "Compressive Creep and Thermophysical Performance of Refractory Materials"[II], new materials for use as crowns in oxy-fuel fired glass furnaces were investigated [4]. These materials were of interest since tighter environmental standards and cost constraints had led many industrial glass plants to consider either converting older, conventional air-fuel fired furnaces to oxy-fuel firing, or building new oxy-fuel fired furnaces instead of new air-fuel fired furnaces. This not only decreases furnace emissions, but

---

II Work supported by the Industrial Technologies Program, U.S. Department of Energy, under Grant No. DE-AC05-000R22725.

leads to higher melting temperatures resulting in greater thermal efficiency and higher glass throughput (more glass per unit energy and time).

However, the internal furnace environment produced by oxy-fuel firing creates many problems with refractory life not previously encountered with conventional firing [5]. Furnaces employing oxy-fuel firing have increased potential for corrosion and rat holing along with deterioration of traditionally-used conventional silica crown refractories. Also, due to a significant increase in the concentration of volatiles in the furnace, more intense alkali attack occurs as water vapor reacts with alkali from the glass melt and batch to form hydroxide vapors. Therefore, alternative refractories to silica were considered for superstructure applications and data were obtained to predict the performance of these refractories in service. Furthermore, the cost of the alternate refractories must be balanced with their effects on furnace life and glass quality while also considering the comparative performance of the other refractories in the furnace and overall furnace life.

Creep testing and analysis of refractories of interest to glass manufacturers were performed at representative service temperatures. Under the study, six types of conventional silica, ten types of mullite, two types of fusion-cast alumina, and one type of fusion-cast spinel refractory were measured with the data generated for the silica refractories serving as a baseline. The mullite refractories were examined because they are used in borosilicate glass furnace crowns and superstructures and in sidewall applications. Additionally, despite their high cost they are less expensive than other refractories such as chrome alumina or fusion-cast alumina, which are used as replacements for traditional silica refractories in harsh oxy-fuel environments. Fusion-cast alumina refractories, which are a popular choice for crowns and sidewalls in oxy-fueled glass furnaces, were selected for this study because of their homogeneity, low porosity, high refractoriness, low levels of glassy phase, and good spall resistance. Although more expensive, fusion-cast spinel refractories were evaluated since they offer improved creep and corrosion resistances.

Compressive creep testing was performed at 1300-1650°C at static stresses between 0.2-0.6 MPa on a pneumatically controlled compressive creep frame in an electrically heated furnace. The phase content, microstructure, and secondary phase composition of the refractories were characterized before and after their creep testing in an attempt to correlate the creep responses with any observed microstructural changes. Corrosion resistance of each of the refractories was also examined.

In the case of silica refractories, the amount of compressive creep was negligible for all brands tested at 1550°C and for five of the six brands tested at 1600 and 1650°C. This would be expected since this is the traditional material used in glass tank crowns. The corrosion resistances of the silica refractories tested were found to be high when they were exposed to sodium carbonate at 1400°C for 24 hours, and the amount of recession was found to increase linearly with temperature.

For the mullite refractories, creep rates ranged from low (10^{-11} s^{-1} – 10^{-10} s^{-1}) to significant (10^{-9} s^{-1} – 10^{-8} s^{-1}) depending on composition. The dominant mechanism of creep rate deformation was also found to vary from creep controlled by diffusion to grain boundary creep. Poor creep performance was observed in materials with high levels of matrix porosity and/or fine grain structure. Materials with lower porosity and broader grain structure performed better. Although changes in chemistry or in microstructure were seen in many samples due to the creep-testing conditions, only minor amounts of recession (much less than that seen for the silica refractories) was evident in these materials due to exposure to sodium carbonate (1400°C, 24 hours). These corrosion results are thought to be due to the presence of Al_2O_3 which incorporates Na_2O into its structure forming β-alumina that results in refractory swelling, but not recession.

The fusion-cast alumina refractories studied were found to be highly creep resistant, exhibiting creep rate values on the order of 10^{-10} s^{-1}. These materials showed minimal recession (less than that seen for the mullite refractories studied) when subjected to sodium carbonate at 1400°C for 24 hours. The high creep and corrosion resistance exhibited by these materials is expected to be due to a combination of their chemistry, as seen with the presence of alumina in the mullite refractories, and their low porosity (<2%).

The fusion-cast spinel ($MgO * Al_2O_3$) refractory studied was Mg-rich in stoichiometry and was found to also be highly creep resistant (creep rate values on the order of 10^{-10} s^{-1}) and even more corrosion resistant than the fusion-cast alumina refractories (when subjected to sodium carbonate at 1400°C for 24 hours). Again, the creep resistance and corrosion resistance is expected to be due to a combination of chemistry (presence of Al_2O_3) and low porosity (<2%), with added corrosion resistance due to the presence of Mg in the refractory structure which inhibits β-alumina formation by occupying sites Na wants to occupy in the Al_2O_3 structure.

Through this study, several observations were made concerning candidate refractories for use in oxy-fuel fired crowns. For greatest creep resistance, a high density material is required. High density also aids in corrosion resistance, but chemistry is also important. Silica refractories will readily react with Na and other alkali species. The presence of alumina will help stabilize the silica in the presence of alkali, but formation of the β-alumina phase can lead to deterioration of the refractory over the long run. Therefore, incorporation of other cations into the refractory structure, such as Mg to form spinel, can aid in further increasing the corrosion resistance to alkali salts.

Energy Savings Strategies for Pulp and Paper

Under a project funded by the DOE Energy Efficiency and Renewable Energy/Industrial Technologies Program (EERE/ITP) entitled "Novel Refractory Materials for High-Temperature, High-Alkali Environments"[III], alternative refractory materials (such as spinel-based materials like those studied previously) and installation techniques (such as shotcreting as opposed to traditional brick linings) are being investigated to increase thermal efficiency through longer lining life.

To achieve this, ORNL has teamed with Minteq® International Inc. of Easton, PA and Missouri University of Science and Technology (MS&T) to develop a family of $MgO\text{-}Al_2O_3$, $MgAl_2O_4$, or other similar spinel structured unshaped refractory compositions utilizing new aggregate materials, bond systems, protective coatings, and phase formation techniques. The newly developed materials are expected to offer alternative material choices for high-temperature, high-alkali environments that may be capable of operating at higher temperatures (goal of increasing operating temperature by 100-200°C depending on process) or for longer periods of time (goal of twice the life span of current materials or next process determined service increment). This will

III Work supported by the Industrial Technologies Program, U.S. Department of Energy, under Award Number CPS Agreement #14954.

lead to less process down time, greater energy efficiency for associated manufacturing processes (more heat kept in process), and materials that can be installed/repaired in a more efficient manner. The overall goal of the project is a 5% improvement in energy efficiency (brought about through a 20% improvement in thermal efficiency) resulting in a savings of 3.7 TBtu/yr (7.2 billion ft^3 natural gas) by the year 2030 through implementation of this technology in lime kilns and other industrial applications such as black liquor gasifiers, aluminum furnaces, and coal gasifiers.

This project is still on-going, but currently an in-situ spinel forming shotcrete material has been developed and laboratory tested. Corrosion testing through static cup testing has shown it to be highly resistant to attack by lime mud at processing temperatures characteristic of those provided by industrial partners with no adherence of the mud to the refractory cup. Additionally, preliminary energy analyses have been performed that substantiate significant projected energy and economic savings (as compared to use of current state of the art materials) which could be realized by using this material in conjunction with an insulating refractory back-up material.

Applicability of Strategies to Other High-Temperature Processes

The strategies used to achieve energy savings in aluminum furnaces are applicable to any application where refractory linings are degraded due to corrosion or mechanical damage. Although the two developed materials are specifically tailored for aluminum contact, similar characterization of salvaged materials from other processes can lead to understanding of failure mechanisms specific to each individual process and to materials which address relevant issues. Further, the use of multi-component linings and composite materials is applicable to any situation where multiple properties are needed such as mechanical strength, corrosion resistance, and thermal insulation.

Although the results cited for the creep and corrosion study discussed above are specific to refractories for glass applications, the techniques and approaches developed through that project are applicable to refractories used in all industrial applications and have been used in studies performed for black liquor gasification and molten metals other than aluminum [6,7]. Mechanical properties must be considered any time alternative refractory materials are being sought since many times they will function in a structural as well as thermal insulation role. Additionally, chemical reactions and phase changes can lead to swelling of materials resulting in spalling or other deterioration of refractory linings.

Alternative material selection is a common strategy applied across all three applications. Specifically for the lime kiln, materials which form favorable phases in-situ were sought. This approach allows for materials that improve in performance with time, instead of degrading in properties. In addition, alternative installation techniques were considered. The use of shotcreting technology, instead of traditional brick and mortar installation, can lead to substantial energy, time, and economic savings. This also allows for seamless linings since mortar joints (a common source of weakness) are removed from the construction. Similar to the aluminum application, a multi-component lining system is also used for the lime kiln. In this instance, the primary lining provides corrosion and wear resistance, while the back-up lining provides improved thermal efficiency.

Summary

Work funded by the DOE Energy Efficiency and Renewable Energy/Industrial Technologies Program (EERE/ITP has led to investigation of refractory issues related to energy efficiency of industrial processes affected by refractory materials. It has been found that through characterization of salvaged materials to gain insight into failure mechanisms; development of materials targeted at improved corrosion resistance, wear resistance, and thermal performance; utilization of multi-component linings with optimized performance of layers; understanding of material properties in-situ; development of materials that improve through interaction with the service environment; and utilization of alternative installation techniques improvements in energy efficiency of these industrial processes can be gained. Application of these same techniques and approaches can also lead to energy efficiency improvements in other high-temperature industrial applications.

References

1. J.G. Hemrick, W.L. Headrick, and K.M. Peters, "Development and Application of Refractory Materials for Molten Aluminum Applications", International Journal of Applied Ceramic Technology, Vol. 5, No. 3, (2008).

2. G. Buchel, A. Buhr, D. Gierisch, and R.P. Racher, "Bonite – A New Raw Material Alternative for Refractory Innovations", UNITECR '05 – Proceedings of the Unified International Technical Conference on Refractories: 9th Biennial Worldwide Congress on Refractories, (2006).

3. V.A. Perepelitsyn, V.I. Sizov, V.M. Rytvin, and V.G. Ignatenko, "Wear-Resistant Refractories in Aluminum Pyrometallurgy", Refractories and Industrial Ceramics, Vol. 48, No. 4, (2007).

4. M.K. Ferber, A.A. Wereszcazk, and J.G. Hemrick, "Final Technical Report: Compressive Creep and Thermophysical Performance of Refractory Materials", ORNL Technical Report, ORNL/TM-2005/134, (2006).

5. J.G. Hemrick, "Creep Behavior and Physical Characterization of Fusion-Cast Alumina Refractories", Doctoral Dissertation, University of Missouri – Rolla, (2001).

6. J.R. Keiser, J.G. Hemrick, R.A. Peascoe, C.R. Hubbard, and J.P. Gorog, "Studies and Selection of Containment Materials for High-Temperature Black Liquor Gasification", Proceedings of 2006 TAPPI Engineering, Pulping & Environmental Conference, Atlanta, GA, November (2006).

7. X. Liu, V. Sikka, J.G. Hemrick, W.L. Headrick, "Development of Next Generation of Metallic and Refractory Materials for Molten Metals Handling", Proceedings of AIST 2005 Iron & Steel Technology Conference and Exposition, Charlotte, NC, (2005).

Energy Technology Perspectives
Edited by: Neale R. Neelameggham, Ramana G. Reddy, Cynthia K. Belt, and Edgar E. Vidal
TMS (The Minerals, Metals & Materials Society), 2009

ENERGY SAVINGS THROUGH PHOSPHATE-BONDED REFRACTORY MATERIALS

Jens Decker

Stellar Materials Incorporated
7777 Glades Blvd, Suite 200, Boca Raton, FL, 33434, USA

Keywords: Liquid Phosphate Bonded Refractories, Non-Wetting Properties

Abstract

In consideration of energy savings the ideal refractory furnace lining should possess the following features:

- Lowest thermal conductivity possible in order to avoid heat loss.
- Single component lining in order to allow freeze plane changes caused by wear, infiltration and temperature changes of the furnace.
- Resistance against mechanical wear from cleaning tools and stirring in order to allow maximum output without equipment downtime due to maintenance.
- Resistance against thermo-chemical attack from aluminum and alloying elements.

However, refractory materials with a low thermal conductivity typically possess a higher porosity and this leads to lower strengths and resistance against chemical attack and wear. Hence, as a compromise, multi layer linings are required in order to meet energy saving standards.

In this paper we present the features of chemically phosphate bonded dense and light weight refractories and how such refractories can contribute to energy savings.

Introduction

In several reports, the U.S Department of Energy published theoretical energy requirement limits for aluminum production as a future target, with recommendations on how to reduce energy losses. The reports provide a basic description of the processes and equipment involved, their interrelationship, and their effects on energy consumed. The report uses onsite process measurements as benchmarks that industry uses to compare performance between facilities and companies [1]. The energy consumptions used in this paper are primarily based on numbers published by the Department of Energy. However, their reports do not consider the potentially significant energy losses from process interruptions due to a lack of preventive maintenance and material deterioration over time. Further, the Department of Energy reports do not consider equipment costs and availability. Energy input to metal output depends on factors beyond energy losses through the furnace wall. For instance, a focus on energy savings should consider fixed output costs due to depreciation of the equipment and the plant.

What is the Potential for Refractory Materials to help Provide Energy Savings?

Refractory walls are engineered with consideration given to the freeze plane of molten aluminum and the thermal profile necessary to entrain the required heat in the furnace. The primary design objective is to keep furnace conditions stable over time in order to avoid interruptions of the production process. Interruptions such as additional door openings, furnace maintenance downtime, or changes in the thermal profile change the heat containment conditions and increase total energy consumption.

Heat loss in aluminum melting and holding furnaces through the refractory lining can be distinguished between wall losses at steady state operating conditions and heat storage losses during transient conditions.

Transient conditions occur during furnace cleaning, skimming, fluxing, charging and heat up / dry out procedures after a shut down. Almost 40% of the total heat input is flue gas losses and only an estimated 10% of the remaining available heat is heat losses through the refractory wall due to steady state operating conditions [2]. The Sankey diagram in Figure 1 shows that the majority of the heat loss of the remaining available heat is due to opening losses and heat storage loss of the refractory and is estimated to be 20%-30% depending on the furnace type and operating conditions. For instance, significant radiation losses through open doors or other openings in the furnace enclosure can add up to 49,000 BTU/hr/ft^2 based on a furnace with a production rate of 15,000 lbs/hour. Furnace shut downs can lead to process downtime and additional heat losses of other connected units like launders, holding or melting furnaces and filter/degassing systems. In this case, the production capacity loss far outweighs the energy losses.

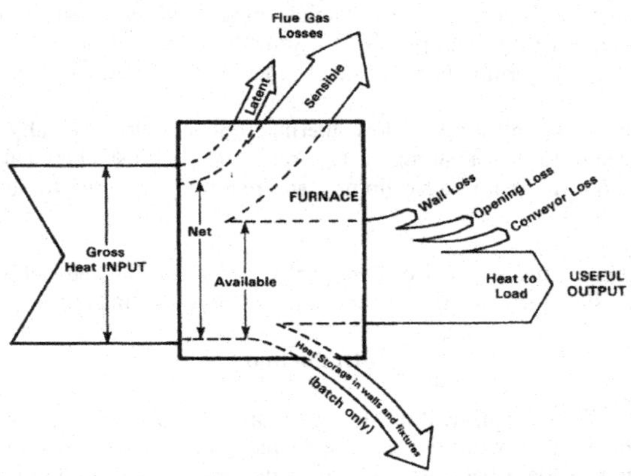

Figure 1. This Sankey Diagram depicts the magnitude and points of energy loss in an aluminum melting furnace. Only an estimated 10% of available heat is lost through the refractory wall during steady state operating conditions.

Aspects of the refractory lining in aluminum furnaces and the potential for energy savings will now be discussed, and in particular, how phosphate-bonded monolithic materials can contribute significantly to overall energy savings.

Steady State Conditions of the Refractory Lining

In consideration of energy savings, the ideal refractory furnace lining under steady state conditions should possess the following features:

- Lowest thermal conductivity possible in order to avoid heat loss.
- Single component lining in order to allow freeze plane changes caused by wear, infiltration and temperature changes of the furnace.

- Resistance against mechanical wear from cleaning tools and stirring in order to allow maximum output without equipment downtime due to maintenance.
- Resistance against thermo-chemical attack from aluminum and alloying elements.

One of the major problems in aluminum melting and holding furnaces is the corrosiveness of molten aluminum and its alloying elements. In addition, the refractory lining is exposed to low viscosity fluxes that penetrate the lining, and to mechanical wear from charging, cleaning and skimming. Often, these conditions do not allow the use of energy efficient insulating refractory materials as a hot face lining due to the lower chemical and mechanical resistance of high porosity light weight refractory products. Less energy efficient dense refractory materials must then be used.

The higher thermal conductivity of dense refractory materials, however, requires a multi layer furnace lining even though a single component lining is preferred to keep the aluminum freeze plane in the refractory hot face lining after the hot face has changed in operation. Changing of the refractory hot face lining during furnace operation commonly leads to the freeze plane moving into the insulating material, and that can lead to aluminum infiltrations deep into the porous insulating lining if the hot face layer shows cracking. Thus, it is essential to install hot face refractory materials that are resistant to interactions with the metal and atmosphere in order to maintain stable furnace conditions over a long period of time.

Figure 2 shows the thermal profile of a typical aluminum furnace side wall with a freeze plane (590°C to 660°C depending on alloy) inside a dense 88% alumina lining. An additional 75mm fireclay safety lining protects the insulating lining and steel shell from potential leakages. This type of lining can be considered "conservative" with regards to safety. The heat loss is 1100 W/ m^2 (350 Btu/sq.ft/h).

Figure 3 shows a lining that uses a microporous insulation board with only 0.03W/m^2°K thermal conductivity and a resulting 800 W/m^2 (254 Btu/sq.ft/h) heat loss. This would lead to more than 25% heat loss savings but the freeze plane of the aluminum can move into the microporous board leading to a higher risk of aluminum leakage.

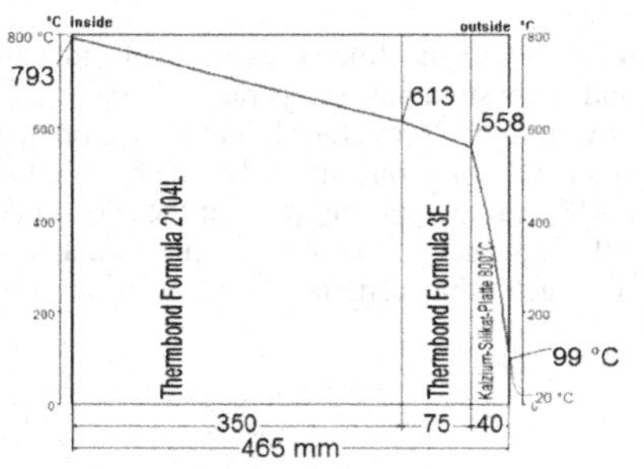

Figure 2. Thermal profile of a typical aluminum furnace side wall with a freeze plane

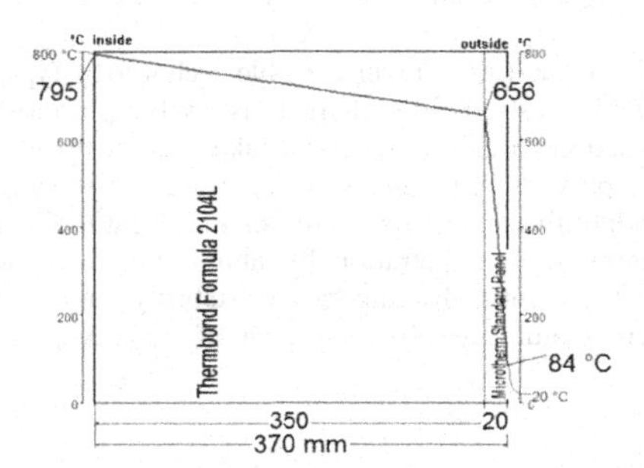

Figure 3. Thermal profile of a lining that uses a microporous insulation board.

This example shows the limits between potential energy savings and safety risks. Refractory linings below the metal line should allow freeze plane changes without risking aluminum leakage.

How to Reduce Corrosion and Wear

Because of the dynamics of refractory corrosion and wear, the temperature profile of a furnace lining can change within a short period of time depending on the operating conditions and furnace design. As a consequence, the refractory lining needs to be monitored because the initial calculated thermal profile of the lining under ideal conditions is only temporary. For instance, the thermal profile of the lining changes with the following factors:

- Corundum build-up on the lining.
- Flux salt infiltration of the lining.
- Aluminum infiltration in pores and cracks.
- Reduction of the wall thickness due to mechanical wear.

Corundum Build-Up

There are two types of corundum build up, internal and external corundum growth.

External corundum growth is a function of oxygen partial pressure in the furnace in the presence of aluminum and alloying elements like silicon and magnesium. The corundum build-up occurs at the metal line, or belly band, of the furnace at the interface of the atmosphere, aluminum bath, and the refractory lining. The capillary system inside the corundum build-up sucks up metal which oxidizes on the surface and promotes further growth. As a consequence the effective furnace volume shrinks, reducing the energy input to metal output ratio.

In contrast, internal corundum formation is the result of a reduction process of refractory components by aluminum metal and alloying elements. This reaction takes place in the pore system of the refractory material and leads to a decomposition of the matrix. With an increase in temperature the reduction process develops faster [3]. Typical chemical reactions are:

$$3\ SiO2\ \text{(refractory)} + 4\ Al\ \text{(metal bath)} = 2\ Al2O3\ \text{(corundum)} + 3\ Si \quad (1)$$
$$3\ Mg\ \text{(alloying metal)} + 4\ Al_2O_3\ \text{(refractory)} = 3\ MgAl_2O_4\ \text{(spinel)} + 2\ Al \quad (2)$$

In order to diminish molten metal reactions, low silica content in the refractory is favorable.

However, low cement castable technology typically relies on ultra fine silica fume in order to obtain desired flow characteristics for placement and high strengths in operation. Since any reaction with silica can only take place if the metal is wetting the refractory lining, one approach to prevent reaction is to employ non-wetting additives. In conventional cement bonded monolithic materials, additives like $BaSO_4$, CaF_2 or AlF_2 can temporarily prevent the refractory lining from penetration by aluminum metal and alloying agents. However, these additives possess limited temperature stability and tend to react with certain fluxes even at low temperatures thereby losing their effectiveness [3,4].

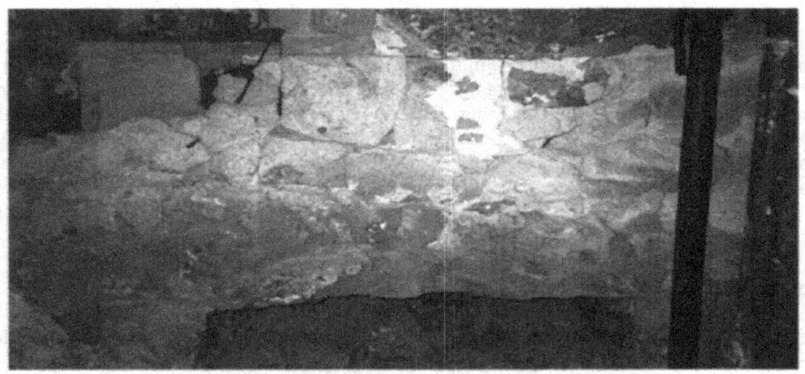

Figure 4. Corundum build-up on a hot wall of a side well furnace.

It is evident that corundum build up is one of the major factors impacting energy consumption because it:

- Changes the thermal conductivity of the wall.
- Reduces the furnace capacity due to corundum growth in the hearth area.
- Reduces the lifetime of the refractory lining and leads to downtime costs.
- Requires intense cleaning of the lining with cleaning fluxes and mechanical tools which relates to wear on the hot face lining, heat loss in the furnace and production downtime.

Flux Salt Infiltrations of the Lining

Flux salts penetrate refractory linings whenever the temperature of the furnace wall is above the melting point of the flux. Evaluations of refractory samples taken from a furnace lining have shown that the wall was deeply penetrated with flux salts after one year in operation. The wall was originally installed with an 85% alumina refractory castable with 18% porosity. After 1 year the wall showed infiltrations of 3% sodium and 3% potassium. The flux used in this furnace contained 47.5% KCl, 47.5% NaCl and 5% fluoride salt with a melting point of 1130 F. The flux infiltration not only causes a higher thermal conductivity of the wall but also a reduction in hot strengths due to glass phase penetration. The mechanical abuse due to cleaning and skimming typically causes wear on the interface metal, flux cover, and atmosphere of the so called belly band. Besides the use of cover fluxes, it is also a common practice to apply exothermic cleaning fluxes which can lead to serious damages of the furnace wall. The purpose of using cleaning fluxes is to loosen and disperse corundum build up on the wall in order to maintain the furnace capacity. However, components of the refractory lining can also get attacked by aggressive fluoride salts such as Na_2SiF_6. This practice requires process downtime because cleaning flux is applied on the hot walls of a drained furnace. After coating the walls with 3-6mm of flux, the reaction will be facilitated with burners turned on high for at least 10-15 minutes [5].

Aluminum Infiltration in Pores and Cracks

Refractory linings are prone to cracking due to shrinkage and chemical attack. In brick linings the weak link is the brick joint. In monolithic linings the cold joints between cast panels can be the weak link. Aluminum infiltration, and consequently corundum formation, leads to pressure build up from crystallization in the pores and cracks and eventually spalling of the refractory material [6].

Transient conditions of the refractory lining

Transient conditions occur during the initial dry out of a cast refractory lining and the heat up after a shut down or during door openings. The lining is often exposed to thermal shock during skimming, cleaning, fluxing and charging. This all leads to energy losses and as a consequence to stresses in the lining that can cause damage and eventually a reduction of the furnace refractory lifetime. To maximize energy savings, the ideal refractory furnace lining under transient conditions should possess the following features:

- Excellent thermal shock properties.
- Low heat capacity in order to reduce storage heat loss.

Thermal Shock properties

Thermal shock resistance of refractory materials in aluminum furnaces is required in the door area of the furnace due to door openings during charging, skimming and cleaning. In particular sills, jambs, and lintels and the floor of aluminum furnaces are prone to thermal shock.
An ideal thermal shock resistant material should possess the following thermo-mechanical properties:

- Low thermal expansion.
- High thermal conductivity.
- High strengths under consideration of the stress/strain relationship. (A high σ/ E ratio leads to good thermal shock properties) [7].

In order to avoid thermal shock, mullite containing alumina silica refractories or silicon carbide materials are the preferred materials in aluminum melting furnaces. However, the silica content of these choices is a source for potential chemical attack and corundum growth which can lead to limited furnace lifetime.

The energy absorption rate of a refractory lining during heat up to operating temperature after a shut-down is not dependant on the heat up time. However, a 140 hour heat up process leads to a production loss of 2.1 million lbs of aluminum (15,000 lbs/h production rate). It is evident that a faster heat up would lead to large production and energy savings particularly if other process units connected with the repaired furnace are on hold under temperature waiting for the furnace to be put into operation and production to commence.

How Can Phosphate-Bonded Refractory Materials Contribute to Energy Savings?

Based on the described factors which have an impact on refractory lifetime and the stability of the aluminum manufacturing process, it can be concluded that refractory materials which could eliminate or reduce the discussed problems could also contribute to energy savings.

Phosphate bonded refractory castable materials have long been used in the aluminum industry because of favorable properties in contact with molten aluminum. Phosphate bonded refractory materials support almost every installation method including casting, gunning, patching and ramming. The major characteristics of these materials are as follows:

Non-Wetting to Molten Aluminum and Aluminum Alloys.

Due to the non wetting properties of liquid phosphate bonded refractories, primary corundum build-up can be totally avoided. It also reduces the necessity and intensity of mechanical cleaning of the walls and the need for cleaning fluxes. The result is that phosphate bonded materials do not show corundum build up and therefore eliminate related problems like production capacity losses and wall deterioration. Figure 5 shows the same furnace wall shown in Figure 4 after the reline with a liquid phosphate bonded high alumina castable after one year. The wall stayed clean and the corundum build up problem was solved. Figure 6 shows a cup test conducted with an 85% alumina phosphate bonded castable tested with 7075 alloy enriched with 5% Mg. The test temperature was 1800°F and the holding time 120 hours. The result shows that phosphate bonded materials are resistant to high Mg-alloys which potentially cause spinel formation and consequently cracking of the refractory lining.

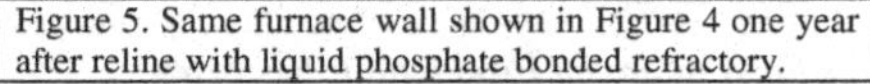

Figure 5. Same furnace wall shown in Figure 4 one year after reline with liquid phosphate bonded refractory.

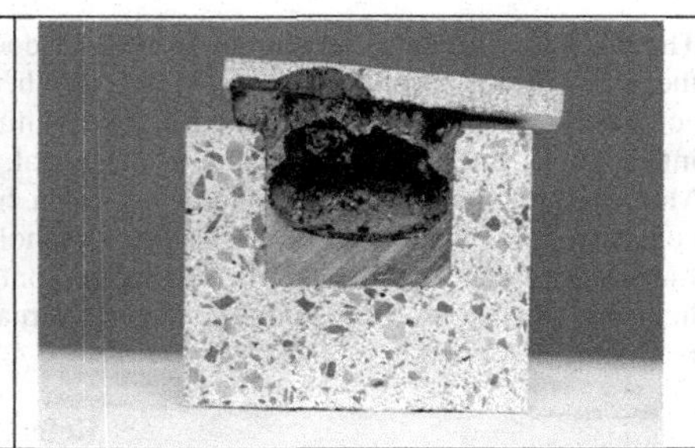

Figure 6. Castable refractory cup test shows phosphate bonded materials are resistant to high Mg-alloys.

Moreover, light weight phosphate bonded materials are also non-wetting, reducing the risk of deep penetration and leakage should the hot face lining be compromised.

Chemical Bond to Existing Refractory

Liquid phosphate bonded materials allow veneer repairs that have monolithic properties. This reduces maintenance downtime and stops corundum build up on old conventional monolithic materials after the repair. The outcome is less refractory waste because up to 80% of existing linings can be maintained indefinitely rather than torn out and disposed of repeatedly. Due to the use of a two component phosphate bonded material, phosphoric acid penetrates into the pores of the existing lining and forms, with the alkali and alumina in the existing refractory, a strong chemical bond with up to 2000 PSI tensile strengths.

Fast Firing Properties

During setting, liquid phosphate bonded products develop a bonding system that contains meta- and polyphosphates. These are ring and chain structures with polymer like behavior that allow extremely fast heat up rates of the material because this flexible binder system compensates for occurring thermal stresses. Hence this property leads to faster maintenance turnaround with reduced downtime. Linear heat up rates of 200°F per hour or greater, without holding time, are possible for upper walls [8].

Extremely Thermal Shock Resistant

Due to the very flexible polymer-like bonding system, even high alumina materials can withstand direct flame impact without pre-drying. This means big block shapes can be installed without pre-dry-out. Burner throats can be rammed in place without heat up restrictions and furnace heat up can be conducted without costly external equipment and manpower.

High Strengths and Abrasion Resistance

Depending on the liquid content, phosphate bonded materials show abrasion resistance of less than 3cc, making them suitable for full thickness linings in high impact floor, sill and ramp applications. These materials contain little or no silica. and are non-wetting against aluminum. It is also possible to install these high performance materials as a veneer over worn out refractory in impact areas.

Summary

The refractory lining of an aluminum furnace does more than determine the heat losses through the walls during steady state conditions and heat losses due to storage heat during transient conditions. To a greater extent, the refractory lining has an impact on energy losses from process interruptions due to interactions with the metal, mechanical wear, and equipment availability. Moreover regular downtime maintenance of a furnace can lead to additional energy losses in connected units like launders, degassing units, holding furnaces, and other heated equipment.
Liquid phosphate bonded refractories can contribute to energy savings because production interruptions can be reduced due to cleaner furnace walls and a refractory lining that is easy to maintain.

References

1. W.T. Chaote, J.A.S Green: *Energy Requirements for Aluminium Production: Historical Perspective,Theoretical Limits and New Opportunities*(Department of Energy 2003) 8

2. D. Whipple, *Basics of Combustion* (TMS, Furnace Systems and Technology, March 2008 Seminar)

3. Siljan et al, *Refractories for molten aluminium contact, part 1* (UNITECR'01, Cancun, Mexico, 2003) 13

4. M. Allahevrdi et al., *Additives and the corrosion resistance of aluminosilicate refractories in molten Al-5Mg* (JOM, February 1998) 31-34

5. T.A.Utigard et al., *The properties and uses of fluxes in molten aluminum processing* (JOM, November 1998) 38 - 40

6. Siljan et al, *Effect of Mineralogy and pore size on aluminum resistance and mechanical wear of refractories, part 1*(UNITECR'01, Cancun, Mexico, 2003) 9

7. D.P.H Hasselmann., *Thermal Stress Resistance parameters for brittle refractory ceramics* (Bulletin of the Amer. Ceram. Soc. 49. 1970) 1033 - 1037

8. Plibrico, *Technoclogy of Monolithic Refractories* (Plibrico Japan 199) 239

Energy Technology Perspectives
Edited by: Neale R. Neelameggham, Ramana G. Reddy, Cynthia K. Belt, and Edgar E. Vidal
TMS (The Minerals, Metals & Materials Society), 2009

ADVANCED CERAMIC COMPOSITES FOR IMPROVED THERMAL MANAGEMENT IN MOLTEN ALUMINUM APPLICATIONS

Klaus-Markus Peters[1], Robert M. Cravens[2], James G. Hemrick[3]

[1]Fireline TCON, Inc.; 300 Andrews Avenue, Youngstown, OH 44505, USA
[2]Rex Materials Group; 5600 East Grand River, Fowlerville, MI 48836, USA
[3]Oak Ridge National Laboratory; 1 Bethel Valley Road, Oak Ridge, TN 37831, USA

Keywords: Refractories, Molten Metal, Aluminum, Energy Savings

Abstract

Degradation of refractories in molten aluminum applications leads to energy inefficiencies, both in terms of increased energy consumption during use as well as due to frequent and premature production shutdowns. Therefore, the ability to enhance and extend the performance of refractory systems will improve the energy efficiency through out the service life. TCON® ceramic composite materials are being produced via a collaboration between Fireline TCON, Inc. and Rex Materials Group. These materials were found to be extremely resistant to erosion and corrosion by molten aluminum alloys during an evaluation funded by the U.S. Department of Energy and it was concluded that they positively impact the performance of refractory systems. These findings were subsequently verified by field tests. Data will be presented on how TCON shapes are used to significantly improve the thermal management of molten aluminum contact applications and extend the performance of such refractory systems.

Introduction

The production of molten aluminum creates aggressive environments for processing equipment, thereby requiring refractory materials to contain the aluminum during melting, transfer, treatment, and casting operations. Degradation of these refractories not only causes issues with reduced product quality and production yields, but also increases heat losses as the insulative properties of the refractories are compromised. Furthermore, in addition to disrupting production output, refractory failures lead to further impacts upon energy efficiency as large amounts of energy are lost during the cooling and subsequent reheating of the equipment as the refractory linings are repaired or replaced.

The need for improved refractory materials is recognized throughout the industry. A joint U.S. Department of Energy and industrial workshop held in 2000 on applications for advanced ceramics in aluminum production [1] noted that refractory materials used in melt processing equipment have many limitations. The workshop report noted that these issues may be addressed by using advanced ceramics, but new materials must be cost effective when compared to current refractory materials. Also, in 2005 the Canadian aluminum industry produced a technology roadmap [2] having the goal of identifying and bridging gaps between current technological resources in the industry and future requirements. One of the areas identified as being crucial to the future of the industry was the need for improved molten aluminum-resistant materials.

Refractory deterioration and failures can arise from several mechanisms [3,4] such as: chemical reactions between the molten aluminum and the refractory material (corrosion); mechanical degradation of the material by the process environment (erosion); and thermal stresses, leading to failure by fatigue or shock. All of these limitations reduce the energy efficiency of the process, as degraded equipment linings cause a loss of process heat (through reduced insulation) and the need for maintenance through repair or replacement of the linings.

TCON ceramic composites are newly developed materials that appear to address the issues raised by the aluminum industry workshops; they have been found to be extremely resistant to erosion, corrosion, and thermal stresses caused by molten aluminum alloys in a variety of applications. These materials are now being produced and distributed via a collaboration between Fireline TCON, Inc. (Fireline) and Rex Materials Group (RMG), with RMG marketing them under the RX-TCON trademark. As will be discussed below, these materials offer unique, cost-effective solutions to the problems that compromise the performance of refractory systems in molten aluminum processing equipment. Not only can TCON composites extend the performance life of the equipment, they can be integrated into refractory systems that outperform conventional refractories by significantly reducing the heat losses from the equipment.

TCON Composite Materials

In a classical composite, one material (the matrix or binder) surrounds and binds together a collection of particles or fibers of a different material (the aggregate or reinforcement). The different materials work together to give the composite unique properties, but within the composite the different materials can be easily discerned as discrete, dispersed, and isolated phases embedded in an otherwise homogeneous matrix material [5]. Given the well-known performance benefits of conventional composites, it can be expected that a material in which both phases form continuous networks may exhibit even better performance characteristics [5-7]. Such materials with continuous networks are called interpenetrating phase composites (IPCs) [5].

An especially interesting IPC family are co-continuous ceramic/metallic composites, as they often exhibit the attractive properties of metals such as high strength and tolerance towards thermal shock and those of ceramics such as corrosion resistance and tolerance of high temperatures without necessarily also sharing their respective undesired properties. Particularly attractive methods for their production are displacement reactions in which sacrificial oxides are reduced by a metal [8-13]. There are numerous reactions that are thermodynamically, kinetically, and mechanistically favorable [8]. Of these IPC materials produced via displacement reactions, the alumina/aluminum (Al_2O_3/Al) composite is the one that has been most investigated [9-13], which is produced by the reaction shown in equation 1:

$$3\ SiO_2 + 4Al = 2Al_2O_3 + 3Si_{[Al]} \qquad (1)$$

This is carried out by reacting the silica (SiO_2) precursor with molten aluminum at elevated temperatures, yielding a continuous network of alumina interlaced with a continuous network of aluminum as seen in Figure 1. The aluminum network forms due to volume contractions as the silica is converted to alumina, and the silicon by-product dissolves into the aluminum. The alumina and aluminum networks typically have cross-sectional thicknesses ranging from one to a few micrometers across [8,10,11] and fine nanometer-sized grains [13]. Also, the interfaces between the alumina and aluminum phases have unique nano-scale features [14-16] that are believed to directly impact the macro-scale properties.

Inert additives that do not participate in the displacement reaction, such as silicon carbide particles, can be added to the silica precursor, yielding a multiphase composite where the inert ceramic particles are bonded together by the IPC matrix, as shown in Figure 2. These inert additives can have a significant effect on the final composite properties, such as improving the thermal shock characteristics of the material [17].

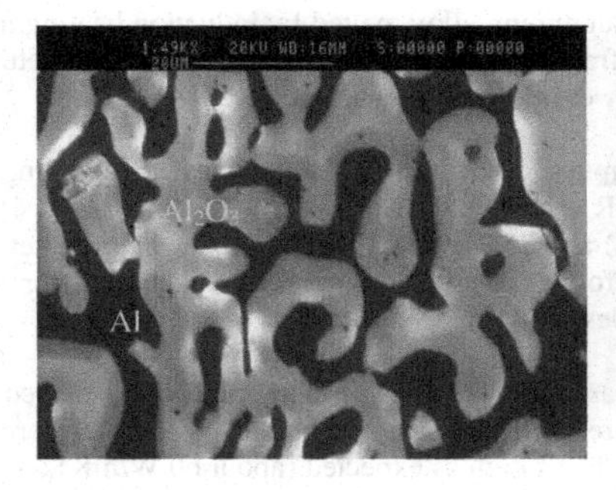

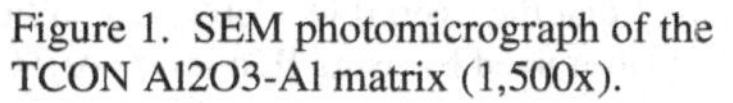

Figure 1. SEM photomicrograph of the TCON Al2O3-Al matrix (1,500x).

Figure 2. Optical photomicrograph of the TCON grade TC2 (50x).

The TCON process involves two basic steps: 1) Fabricating preforms of various sizes and shapes via conventional processing techniques and from readily-available precursor and inert additive materials; 2) Carrying out the displacement reaction by fully immersing the preforms in molten aluminum metal under specific processing conditions and for certain periods of time, resulting in the complete reaction of the ceramic materials and transforming them into the ceramic-metallic interpenetrating phase composites. During development of the TCON process, the following key characteristics were identified: 1) The properties of TCON materials can be tailored by varying the preform materials and process parameters. For example, the room temperature strength can be increased by a factor of up to 10 times. 2) The process is capable of producing shapes of various sizes and complexities, ranging from a few millimeters to over a meter in a single dimension. This makes the TCON process a very cost-effective method to produce net shapes with enhanced material properties.

Evaluation of TCON Materials in Molten Aluminum Applications

Laboratory Testing

A U.S. Department of Energy project entitled, "Multifunctional Metallic and Refractory Materials for Energy Efficient Handling of Molten Metals" was initiated in 2004 and one of the stated goals was to extend the life of molten-metal containment refractories by an order of magnitude through the development of new materials by the industrial collaborators. The project was supported by the Industrial Technologies Program, U.S. Department of Energy, under Grant No. DE-PS07-031D14425, and led by West Virginia University with research participation by the University of Missouri – Rolla (UMR, now called the Missouri University of Science and Technology) and Oak Ridge National Laboratory (ORNL). Fireline TCON, Inc. participated in this project as an industrial collaborator. Extensive testing by project researchers found excellent performance by the TCON materials as summarized below:

Static and Dynamic Corrosion Tests. Bars of TCON materials were immersed into molten aluminum alloys at 700°C for durations between 500 and 1,000 hours. Cross sectional samples examined by the Oak Ridge researchers using both optical and electron microscopy, revealed only minimal interaction of the TCON material with the molten aluminum [18].

Sessile Drop Testing. Static and dynamic sessile drop methods were employed for studying the wetability of TCON materials by liquid aluminum. The dynamic test more closely replicates application conditions, with a small quantity of aluminum alloy melted by induction heating in a reducing atmosphere and dropped onto the refractory substrate. Results showed non-wetting behavior on a macroscopic scale and only slight wetting on a microscopic level [18,19].

Room and Elevated Temperature Modulus of Rupture. MOR testing was performed according to ASTM C133. Four-point bending tests gave MOR values on the order of 50-75 MPa (7-11 ksi) at room temperature (depending upon composition) and 25 MPa (3.5 ksi) at 700°C. Hemrick et al. noted that this is an extremely good strength for a refractory material, good enough for the material to be used in a structural application if desired [18].

Thermal Conductivity. Thermal conductivity was evaluated using a new method developed at ORNL, utilizing a unique High Intensity Infrared (HDIR) lamp technique [20]. The thermal conductivity of the TCON material was found to be as high as expected (about 60 W/mK) due to the presence of SiC. However, the researchers concluded that the overall insulating capability of a refractory system containing TCON material could be optimized by using thin layers of TCON backed by a highly insulating refractory [18].

Thermal Shock/Thermal Cycling. Thermal shock and thermal cycling evaluation was performed according to ASTM C1171. Strength of originally tested compositions was found to decrease due to thermal cycling/thermal shock. Testing of an optimized TCON grade (TC2), reformulated based on results from initial testing, yielded a material that showed no measurable drop off in strength (original RT MOR = 49 ±2 MPa, RT MOR after thermal cycling = 48 ±3 MPa) due to thermal cycling/shock.

The ORNL and UMR researchers considered the lab scale testing of TCON to be a success [18], and initiated an industrial scale test to further validate the material. As shown in Figure 3, six plates of the TCON refractory were inserted into the wall of a 325-lb melting furnace.

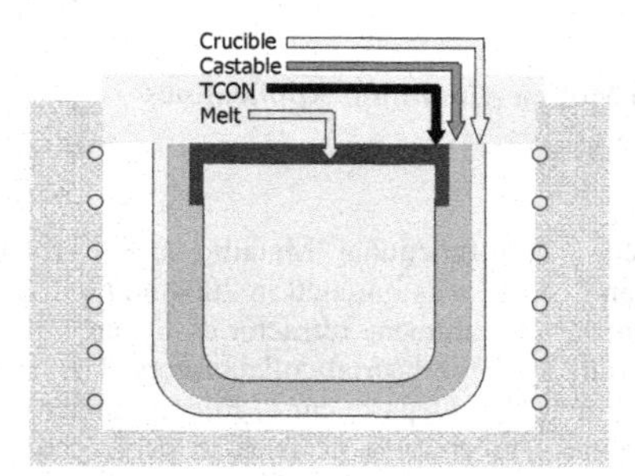

Figure 3a. Industrial scale evaluation of TCON materials - schematic

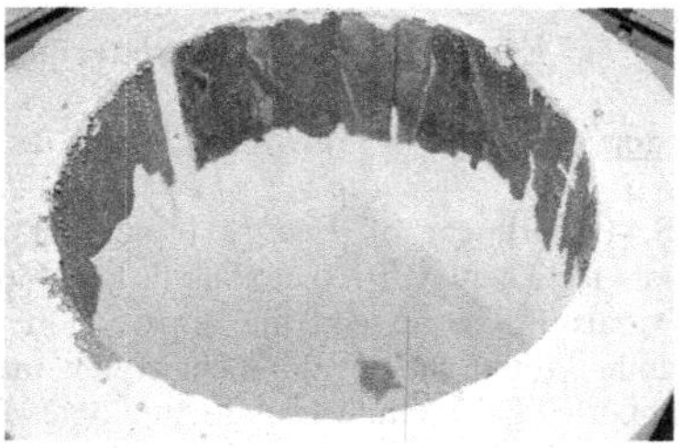

Figure 3b. Industrial scale evaluation of TCON materials - actual installation.

The melt line of the 5083 aluminum alloy was located in the center of the six inch TCON plates (i.e., three inches from the top of the furnace). The alloy was kept molten for 2,000 hours (83.3 days) and the testing included periodic (every one to three days) physical scraping off of the dross and corundum from the TCON plates to simulate refractory cleaning in industrial applications. Mg content was replenished using solid magnesium metal and the metal line was kept constant by adding additional solid 5083 alloy. Dross removal was performed in a fashion to simulate the harsh standard practices of industry described above by scraping with a metal rod. Following the test, examination of the TCON plates led the ORNL researchers to conclude that the test was successful, as it was found that solidified aluminum present on the TCON plates was easily pealed away from the refractory and optical and electron microscopy revealed no detectable signs of aluminum penetration [18]. In addition, the researchers could find no evidence of mechanical degradation from the surface cleaning, demonstrating the excellent wear resistance of the TCON material. In summary, the ORNL and UMR researchers concluded that their testing had been successful, showing that the TCON refractory material exhibited improved corrosion and wear resistance for use with molten aluminum [18].

Field Testing

Fireline and Rex Materials Group have successfully trialed SiC-Al_2O_3-Al TCON composite materials (grades TC1, TC2, and RX1) in several different applications, as summarized below:

Hooks. The first test was in an automated production cell for casting aluminum alloy diesel pistons, where iron rings are preconditioned in molten aluminum prior to being embedded into pistons during casting. Hooks made out of TCON TC1 were used to hold iron rings during the preconditioning process, and were subsequently subjected to corrosion by molten aluminum above 700°C and erosion due to rotation of the hooks in the bath. The TCON hooks were found to last up to ten weeks, as compared with competitive hooks that typically lasted one to three days. Additionally, the TCON hooks were found to fail in a predictable and controlled manner, whereas the competitive hooks failed catastrophically and unpredictably.

Inductively Heated Channels for Holding Furnaces. Another successful evaluation was in an inductively heated channel located in the bottom of a melt holding furnace. This channel was comprised of two round, upright tubes leading into a square, horizontal tube, and an induction coil was wrapped around the square tube upon installation in the furnace. During furnace operation, cold molten aluminum traveled down into channel and was heated up by the induction field, subsequently traveling back up into the holding chamber. This tube was traditionally formed out of a castable refractory but over time there were significant issues with deposits and build up, requiring the operators to "rod" the channels and clean them out every shift. Channels made out of TCON RX-1 were evaluated and at it was found that these furnaces remained free of build up, subsequently eliminating the need for "rodding". Life of this component went from a few weeks to about one year, saving the end user substantial dollars in maintenance costs and down time. This customer has subsequently equipped all of their furnaces with TCON channels.

Ladle Impact Pads. The second application was in a 500 lb. ladle for the transfer of molten aluminum alloy from a melting furnace to a holding furnace within a foundry. TCON TC2 was used as a melt impact pad (located in the bottom of the ladle) to reduce erosion of the refractory lining. Conventional refractory materials located in the bottom of the ladle were found to erode as the aluminum melt was poured into the ladle. This erosion led to the need to remove the ladle from service for repair as the erosion became severe. Previous maintenance practices require patching of the ladle every two to three weeks and replacement of ladles every 18-24 months adding cost and the need to have multiple ladles on hand. TCON plates were installed in the bottoms of two ladles while they were being relined. After 17 and 24 weeks of service, respectively, the pads in both ladles showed no observable corrosion or erosion. Further, during inspection the aluminum skin adhering to the TCON surface was easily peeled away indicating a lack of wetting. Based on the performance of these two ladles, the customer has added TCON plates to all of its transfer ladles.

Tap-Out Trough. In another application, a tap-out trough at a major secondary aluminum processor was retrofitted with a TCON TC2 impact pad by chiseling out a portion of the existing refractory and mortaring in a 12 x 12 x 0.75" TCON plate. This trough was connected to a reverb furnace that was tapped two to three times a day, with approximately 250,000 lbs. of aluminum alloy being drained into the trough with each tapping. The plate lasted for over five months with no discernable effect on the TCON pad. (The trial concluded when the refractory material surrounding the TCON plate eroded away.) As a result of this positive trial, this and other molten aluminum processors will be installing TCON impact pads into tap-out troughs.

Energy Efficient Refractory Systems Utilizing TCON Composite Materials

The ability of TCON materials to mitigate erosion/corrosion issues and extend the service life of molten aluminum processing equipment indirectly results in energy savings by reducing the frequency of refractory lining repairs and replacements. However, the frequency of these intermittent energy savings is highly dependant upon the type of equipment and the severity of the environment, with months or years possibly transpiring between events. More importantly, significant and immediate energy efficiency gains can be realized by incorporating TCON shapes into refractory systems that utilize highly insulative materials. In spite of the fact that TCON materials have high thermal conductivity values, due to the significant amount of silicon carbide present, the ability to produce a variety of useful TCON shapes allows for the assembly of refractory systems that can outperform conventional refractories in terms of both longevity and energy efficiency. This is accomplished by placing relatively thin TCON shapes where needed most, namely at the hot face where it is exposed to the molten aluminum. Given the proven ability of TCON to combat erosion/corrosion, this type of refractory system allows for the use of back-up linings that are optimized for their insulative properties and less so for their resistance to molten aluminum, thereby maximizing the energy efficiency of such a system.

The benefits of this type of refractory system are illustrated in Figure 5. Cold face temperatures were calculated for the following types of refractory systems: 1) A conventional 70% alumina castable refractory; 2) 0.75" of TCON backed by Rex Material Group's FUSIO$_2$N^{TM} SS fused silica; 3) A conventional 70% alumina castable refractory backed by ceramic paper (0.125" thick); 4) 0.75" of TCON backed by Rex Material Group's Fusio$_2$n SS fused silica plus ceramic paper; and 5) 0.75" of TCON backed by Rex Material Group's Pyrolite® 3500 ceramic fiber. Figure 5 shows how the cold face temperature is reduced as the overall thickness of each refractory system is increased. (The thickness of TCON was held at 0.75", while the thickness of the back-up materials was incrementally increased.)

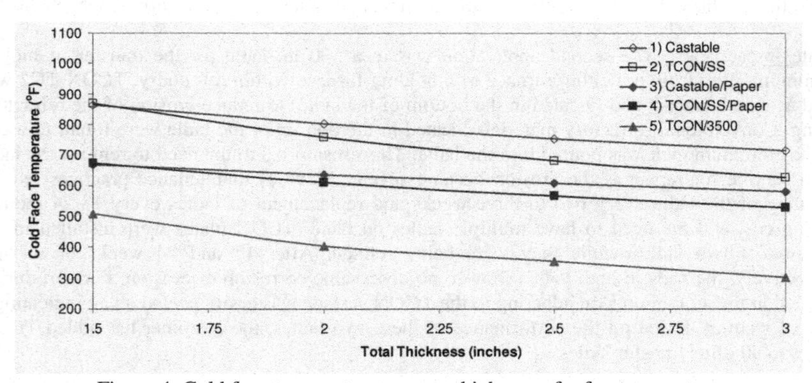

Figure 4. Cold face temperatures versus thickness of refractory system

These calculations show how a refractory system incorporating TCON at the hot face backed by an insulating refractory layer can be very effective in reducing heat losses and improving energy efficiency. At a total thickness of 3 inches, TCON backed by $FUSIO_2N$ SS fused silica results in cold-face temperatures that are 9 to 12% lower than that achieved by a conventional 70% alumina castable refractory material (with and without ceramic paper). More significantly, 3 inches of TCON plus Pyrolite 3500 ceramic fiber achieves a cold face temperature that is 48% lower than an equivalent amount of conventional 70% alumina castable refractory material backed with ceramic paper. Therefore, the use of such a TCON/Pyrolite refractory system can achieve significant energy savings. Conversely, in the case where an aluminum cast house or foundry needs to maximize the internal dimensions of its molten aluminum processing/transfer equipment, the use of a TCON faced refractory system would be very beneficial. In such a case a 1.5 inch thick TCON/Pyrolite refractory system will be just as effective as 3 inches of conventional 70% alumina castable refractory material backed with ceramic paper.

Conclusions

TCON materials are unique refractory materials for use in molten aluminum applications and have been proven to be extremely resistant to the erosion, corrosion and thermal stresses encountered in these processing environments. Just as important to increasing the up time of processing equipment, utilizing TCON materials in the hot face of refractory systems can achieve significant energy savings throughout the operational lifetime of the equipment.

References

1. *Applications for Advanced Ceramics in Aluminum Production: Needs and Opportunities*, Proceedings of a workshop sponsored by the United States Advanced Ceramics Association, the US Department of Energy & the Aluminum Association, February 2001.

2. *Canadian Aluminum Industry Technology Roadmap*, proceedings published by Industry Canada, May 2005

3. T.W. Dunsing, "Tips on Good Refractory Practice for Aluminum-Melting Furnaces", *Industrial Heating*, April 2002, pp 27-29.

4. P. Bonadia, M.A.L. Braulia, J.B. Gallo & V.C. Pandolfelli, "Refractory Selection for Long-Distance Molten –Aluminum Delivery", *American Ceramic Society Bulletin*, Vol. 85, No. 8, p9301.

5. D.R. Clarke, "Interpenetrating Phase Composites", *J. Am. Ceram. Soc.*, 75 [4] 739-59 (1992).

6. H.X. Peng, Z. Fan, J.R.G. Evans, "Bi-Continuous Metal Matrix Composites", *Mater. Sci. Eng. A*, 303, 37-45 (2001).

7. O. Sigmund, "Composites With Extremal Thermal Expansion Coefficients", *Mech. Mater.*, Vol. 20, No. 4, 351-368 (1995).

8. W. Liu and U. Koster, "Criteria for Formation of Interpenetrating Oxide/Metal- Composites by Immersing Sacrificial Oxide Preforms in Molten Metals", *Scripta Materialia*, Vol. 35, No. 1, pp 35-40 (1996).

9. M.C. Breslin et al., "Processing, microstructure, and properties of co-continuous alumina aluminum composites", *Mater. Sci. Eng. A*, 195 (1995) 113-119.

10. W.G. Fahrenholtz et al., "Synthesis and Processing of Al2O3/Al Composites by In Situ Reaction of Aluminum and Mullite", *In-Situ Reactions for Synthesis of Composites, Ceramics, and Intermetallics*, pp. 99-109, ed. by E.V. Barrera et. al., The Minerals, Metals, and Materials Society, Warrendale, PA, 1995.

11. R.E. Loehman and K. Ewsuk, "Synthesis of Al2O3-Al Composites by Reactive Metal Penetration", *J. Am. Ceram. Soc.*, 79 [1] 27-32 (1996).

12. E. Saiz and A.P. Tomsia, "Kinetics of Metal–Ceramic Composite Formation by Reactive Penetration of Silicates with Molten Aluminum", *J. Am. Ceram. Soc.*, 81 [9] 2381–93 (1998).

13. N. Yoshikawa, A. Kikuchi, and S. Taniguchi, "Anomalous Temperature Dependence of the Growth Rate of the Reaction Layer between Silica and Molten Aluminum", *J. Am. Ceram. Soc.*, 85 [7] 1827–34 (2002).

14. J. Ringald et. al, "Scanning and transmission Electron Microscopy on Composite Materials prepared by SMP and In-Situ Displacive Reactions", *Inst. Phys. Conf. Ser. No 147*: Section 13, 1995.

15. G.S. Daehn et. al, "Elastic and Plastic Behavior of a Co-Continuous Alumina/Aluminum Composite", *Acta Materialia*, Vol. 44, No. 1, pp. 249-261(1996).

16. X.F. Zhang, G. Harley, L.C. De Jonghe, "Co-continuous Metal-Ceramic Nanocomposites", *Nano Letters*, Vol. 5, No. 6,, 1035-1037 (2005).

17. G.S. Daehn, M.C. Breslin, *JOM*, 58, 87-91 (2006).

18. J.G. Hemrick, W.L. Headrick & K-M Peters, "Development and Application of Refractory materials for Molten Aluminum", *Int. J. Appl. Ceram. Technol.*, Vol. 5, No. 3, 2008, pp 265-277

19. J. Xu, X. Liu, E. Barbero, J.G. Hemrick & K-M Peters, "Wetting Reaction Characteristics of Al2O3/SiC Composite Refractories by Molten Aluminum and Aluminum Alloys", *Int. J. Appl. Ceram. Technol.*, Vol. 4, No. 6, 2007, pp 514-523

20. J.G. Hemrick, R.B. Dinwiddie, and E.R. Loveland, "Technique Development for Large Sample Thermal Conductivity Measurement of Refractory Ceramics", *Proceedings of the 43rd Symposium on Refractories*, St. Louis, Missouri, March (2007).

Energy Technology Perspectives
Edited by: Neale R. Neelameggham, Ramana G. Reddy, Cynthia K. Belt, and Edgar E. Vidal
TMS (The Minerals, Metals & Materials Society), 2009

Energy Efficient, Non-Wetting, Microporous Refractory Material for Molten Aluminum Contact Applications

Kenneth McGowan

Westmoreland Advanced Materials, LLC
110 Riverview Drive, Monessen, PA 15062

Keywords: Aluminum, Energy, Efficient, Corundum, Refractory, Non-wetting

Abstract

The aluminum industry is the 5th largest consumer of energy and the largest consumer of energy on a per-weight basis. Process heating accounts for 25% of the energy consumed to manufacture aluminum. Consequently, the generation of this energy results in the production of a significant amount of CO_2 (over 3.75 million metric tons in 2003)[1]. As a result, the Aluminum Industry Technology Roadmap of 2003 listed the development of improved refractory materials as a goal to reduce energy use and decrease greenhouse gas creation. The result of a R&D effort addressing this desire is the development of patented microporous refractory materials for metal contact applications in all furnace areas including the belly band. It has been demonstrated that this material can reduce energy consumption up to 38% in furnaces and significantly reduce heat loss in launder and trough systems. Furthermore the material developed is non-wetting and remains un-penetrated by metal throughout its lifetime. The material does not contribute to the formation of corundum. As a result, the energy efficiency of an older lining remains intact compared to standard refractory materials which may have an initial high insulating value (such as lightweights and board) or standard dense refractory (even with penetration inhibitors) both of which show a rapid increase in thermal conductivity as the refractory is penetrated with aluminum and formed corundum

Introduction

The aluminum industry is the second largest producer of metal in the United States, behind the iron and steel industry. It is the 5th largest consumer of energy and the largest consumer of energy on a per-weight basis. Process heating is required to melt, hold, purify, alloy, and heat treat the metal and accounts for 25% of the energy consumed to manufacture aluminum. As such, process heating is the second largest energy consuming operation. Smelting, which utilizes an electrolytic reduction process is the largest energy consuming operation[1].

Most process heating is accomplished through the use of fossil fuels, either directly or indirectly. The choice of fuel is dependent on a plant's location and fossil fuel availability. Table I shows the processes impacted by refractories and the calculated BTU's generated by usage of the respective fuels as well as the Metric Tons (MT) of metal passed through each process in 2000[1].

It is useful to convert the BTU values into a dollar value in order to gain an understanding of the cost of these processes. Table II was generated using the data listed in Table I and reported fuel prices in 2000. Over half a billion dollars was spent on fuel alone for these processes in that year. Even a modest 5% improvement in efficiency would have accounted for over $25,000,000 in savings. Compared to the current price of energy; the potential savings is much more significant.

Table I. Energy Utilized as a Function of Fuel and Aluminum Processed[2,3,4,5]

Process	Fuel Oil (BTU/MT)	Natural Gas (BTU/MT)	Propane (BTU/MT)	Al Processed (MT)
Primary Casting	814,554	1,878,693	19,196	3,668,000
Secondary Casting	1,981,346	4,569,793	46,692	3,450,000
Rolling	0	1,211,358	818	2,749,000
Extrusion	109	3,735,624	107,285	1,719,000
Shape Casting	0	8,704,367	0	2,513,000

Table II. Estimated Cost of Fuel Consumed

Process	Fuel Cost / MT Aluminum processed	Total Cost of Fuel Consumed
Primary Casting	$21.12	$77,460,000
Secondary Casting	$51.37	$177,219,000
Rolling	$9.58	$26,323,000
Extrusion	$30.26	$52,023,000
Shape Casting	$68.76	$172,805,000
Total		$505,830,000

In addition to the actual production costs it is also important to consider the decrease in greenhouse emissions that will result from burning less of these fossil fuels. Table III shows the CO_2 equivalent of BTU's produced as a result of burning fuel oil, natural gas, and propane in the processes impacted by refractories. The same modest 5% improvements in efficiency would have resulted in a reduction of more than 193,000 MT of CO_2 produced.

Table III. Estimated CO_2 Generation[1]

Process	Total Kg of CO_2 per Metric Ton of Aluminum Produced	Total MT of CO_2 Produced per Process
Primary Casting	165	605,220
Secondary Casting	402	1,386,900
Rolling	115	316,135
Extrusion	211	362,709
Shape Casting	475	1,193,675
Total MT CO_2 Produced		3,864,639

In February of 2003 the aluminum industry released their technology roadmap[6]. This document highlights the process areas where the need for technological improvement is recognized and listed by the aluminum industry. One of the biggest recognized shortfalls listed as a priority problem is "Low fuel efficiency in melting and holding furnaces; furnaces are not optimized for scrap heating and waste heat recovery." Six R&D needs were identified. One of these is Energy-Efficient Technologies. Within this need, the top priority listed includes the development and design of a furnace that improves cost effectiveness and improves fuel/energy efficiency[6]. The furnace lining, and hence the refractory, plays a very large role in defining these parameters.

In addition to causing desired chemical reactions to occur, process heating also causes undesirable reactions. The most common and detrimental of these reactions affect the vessel lining, in particular the refractory. On a small scale, most of these reactions would not be considered a problem. However, the methods by which the reactions propagate, how the reaction products are removed, and the effects over time these reactions have on the properties of the refractory (e.g. high temperature strength, thermal conductivity, thermal shock and impact resistance) often cause premature refractory failure. The following are the major detrimental reactions which have a negative impact on the process: corundum formation, spinel formation, and the formation of organic phosphates and nitrates.

Corundum is a naturally occurring form of aluminum oxide (Al_2O_3) also known as alpha alumina or α-Al_2O_3. This material can form as a result of oxidation of the liquid aluminum metal in a vessel. Manufacturers attempt to control the atmosphere of the vessel to limit the free oxygen available for this reaction. Because of constantly opening these vessels for charging, and poor sealing around doors and ports, it is impossible to completely eliminate available oxygen. The corundum that forms not only forms on the top of the liquid aluminum metal; but forms in and on the surface of the refractory just above the metal line. Liquid aluminum, which has a viscosity lower than water, is able to penetrate the porosity of typical refractory ceramic. Inside the porosity, capillary action allows the metal to move upward above the metal line. Exposed to the atmosphere and available oxygen the aluminum reacts causing the formation and subsequent growth of corundum crystals. The area at the metal line is commonly referred to as the 'belly band'. This is due to the formation of a band of corundum at and above the metal. It is also recognized as the highest refractory wear area and is often patched and repaired to keep a furnace in operation. A second source of oxygen is attributed to the reduction of lower energy oxides (compared to Al_2O_3) particularly SiO_2[7] by aluminum metal at temperature. These are often components of the refractory material. The corundum that forms is not detrimental to the quality of the aluminum except if present as inclusions. Once the crystals reach a certain size and volume, they interfere with the physical operations of the furnace. They can impede the charging process, cause doors and ports to close improperly or incompletely, deflect burner flame, and insulate the metal from desired heat transfer. Manufacturers will remove the corundum by physically scraping the walls of the vessel with a mechanical arm or other device. In this process, the refractory is damaged in several ways. First, the corundum is strongly bonded to the refractory surface[8]. Removing the corundum typically removes the surface of the refractory as well. Metal in the refractory, previously denied oxygen contact due to surface corundum growth, now oxidizes to start the process over. Physical damage to the refractory in the form of cracks and the reduction of the refractory thickness allow aluminum to penetrate further into the lining. Because this penetration increases the thermal conductivity of the refractory, more waste heat is produced by the process. Metal penetration into, and corundum formation within, the refractory cause it to become brittle. This causes the refractory to become susceptible to differential thermal expansion between the refractory and the corundum/metal matrix. The shear stress of the differential expansion causes the refractory to fracture. These fractures exacerbate the problem by allowing further penetration. Repetition of this process will cause the refractory lining to wear to the point of having to be replaced.

A magnesia spinel mineral phase is formed by the reaction of aluminum and magnesium in the presence of oxygen (O_2) or of alumina (Al_2O_3) with periclase (MgO). It occurs naturally as $MgAl_2O_4$, (MgO-Al_2O_3). Certain alloys of aluminum, such as 7075, contain magnesium metal. When these alloys are exposed to free oxygen, spinel can form. This crystal expands on formation and if this reaction occurs in the porosity of a refractory, the ceramic can fracture. It is more commonly present near door seals and ports where free oxygen is available due to atmospheric leakage. The crystals can be excessive and if not removed regularly, can cause jamming of door mechanisms or other interference with operations. The problems resulting from removing these crystals are similar to those described with corundum removal. Manufacturers have recently been running their combustion processes lean (additional oxygen) for the purpose of decreasing the amount of incomplete combustion products such as hydrocarbons, sulfur and carbon monoxide. If the fuel/oxygen mixture is too lean, it has the side effect of causing additional spinel formation with certain alloys.

The formation of organic phosphates and nitrates are of concern because their inherent instability at temperature is a safety hazard. Under certain conditions their presence can cause explosions. In order for an explosion to occur three conditions must be met. There must be oxygen, fuel and an ignition source. Meeting these conditions is a rare occurrence in aluminum manufacturing. However, the potential results of such an occurrence are worth guarding against. The source of organic nitrates is primarily a result of fuel composition and/or inhomogeneous burning conditions within a furnace. Nitrogen is also available via aluminum nitrate and nitride present in dross (an aluminum rich slag sometimes used as feed in secondary production). In this case the aluminum nitride can also decompose into ammonia and alumina in the presence of water present as a combustion by-product. Incomplete combustion of fuel, especially fuels containing polyaromatic hydrocarbon (PAH), in the presence of nitrogen or certain nitrogen compounds can result in the formation of explosive organic nitrates. There are two primary sources of phosphates in aluminum production. One source is the phosphoric acid which is sometimes used to clean aluminum scrap. The other is through phosphate bonded refractory. Aluminum phosphate, calcium phosphate, and magnesium phosphate are commonly used to bond refractory brick, ramming mixes, some two component castables and plastics. The typical phosphate bonded refractory will contain both phosphoric acid and reactive hydrate of alumina, magnesia or calcium oxide. Most castable or gunite refractories contain organic fiber, which can form organic phosphates with phosphate sources under reducing conditions, high temperature and/or pressure.

Technical Objectives

The development of a new type of refractory ceramic was proposed. This refractory material will have the following properties and characteristic in order to minimize or solve the stated problems.

A. Reduced thermal conductivity to gain a minimum of 10% improvement in heat loss compared to current lining configurations.
    1. Reduced BTU required/MT Al produced.
    2. Reduced Kg CO_2 generated/MT Al produced.
    3. Reduced cost to produce aluminum metal (refractory must provide savings in fuel, be less costly than current products, or both).
B. Suitable for direct contact with liquid aluminum metal (i.e. working lining).
    1. Resistant to liquid aluminum penetration
    2. Inhibit or reduce growth of corundum and/or spinel
    3. Used in all areas of sidewalls and roofs of furnace.
    4. Resistant to thermal cycling/shock.
    5. Have strengths equivalent to current products.
C. Phosphate and nitrate free.

Research efforts targeted an aggregate based on the calcium aluminate mineral system as the choice to manufacture a refractory product. Calcium aluminates, like their barium counterparts, are naturally non wetting to liquid aluminum. Figure 1 shows the phase diagram for the CaO-Al_2O_3 system as a function of temperature[9]. The use of calcium aluminate as an aggregate for aluminum application is not new. What is new is the specific form and mineralogy we are proposing for this application. For this study we are interested in phases where the weight percent of Al_2O_3 is greater or equal to about 64%(wt.) favoring the formation of CA and higher alumina phases (CA is an industry convention, where C=CaO and A=Al_2O_3). This is due to the

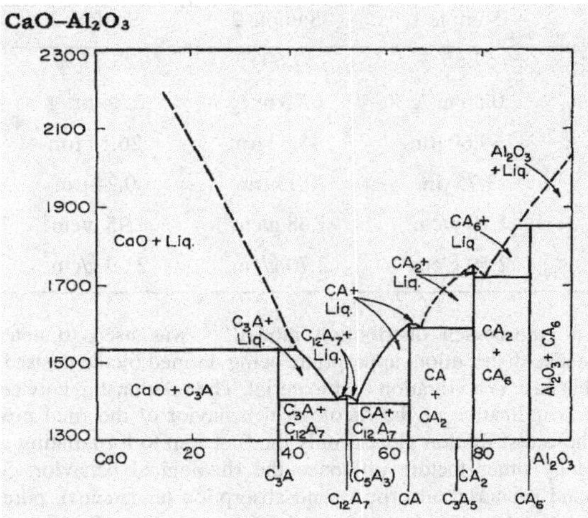

Figure 1. Phase diagram of the CaO-Al2O3 System[9]

reported melting points being 1608°C or greater upon increased Al_2O_3 content above 64% (wt.). Aluminum manufacturing temperatures can reach 1500°C in the furnace atmosphere. Figure 1 also shows that at concentrations of Al_2O_3 less than 64% (wt.), the formation of C12A7 is prevalent and represents an unsuitable low melting phase. Calcium aluminate manufacturers have offered aggregates which have a concentration of Al_2O_3 around 52% (wt.). This aggregate is claimed to be aluminum resistant[10] but is suitable for only low temperature conditions and does not offer a thermal conductivity advantage. This study limited initial investigations to aggregate containing primarily CA and CA2.

Aggregate is produced by a sintering reaction between alumina and limestone. The sintering process causes decarburization of the calcite in limestone. This introduces porosity to the calcium aluminate aggregate. Process parameters such as temperature and residence time also play a role in the amount and size of porosity. Preliminary studies indicate an average apparent porosity of 50% with an average pore size of 44μm or less could be expected. It was hypothesized that optimizing these properties, in an aggregate used to make a refractory, would result in a refractory product with aluminum penetration resistance due to the small pore size and low thermal conductivity as a result of the large total pore volume.

Experimental

Process variables were altered to create four samples of a calcium aluminate material with chemistries around a 70% (wt.) target of Al_2O_3. XRD analysis confirmed the target mineralogy of CA and CA2 as major components and C12A7 either not detected or possibly present as a trace component. Table IV shows the porosity and density characteristics of each sample.

Aggregate of various size distributions were created by crushing, grinding, and screening the material in the

Table IV. Porosity and Density Values

	Sample 1	Sample 2	Sample 3	Sample 4
Total Porosity	57.65 %	37.69 %	27.37 %	42.55 %
Total Pore Area	0.56 m^2/g	6.73 m^2/g	2.34 m^2/g	0.29 m^2/g
Med. Pore Vol.	69.60 μm	45.93 μm	26.14 μm	19.49 μm
Avg. Pore Dia.	3.75 μm	0.13 μm	0.24 μm	3.51 μm
Bulk Density	1.10 g/cm^3	1.68 g/cm^3	1.95 g/cm^3	1.66 g/cm^3
App. Density	2.59 g/cm^3	2.70 g/cm^3	2.69 g/cm^3	2.89 g/cm^3

normal fashion. An Andreessen distribution model[11,12] was used to determine appropriate packing and particle size distribution; appropriate being defined by the desired nature of the final product, which in this case is a vibration cast material. The relationship between particle size and cumulative volume is indicative of the rheological behavior of the final product. It should be noted that the use these distribution plots is only the first step to formulating a mix. It represents a starting point. Many other factors influence the rheological behavior. Some of these are density, water demand (material adsorption and absorption tendencies), pore size, and particle shape. After several iterations an optimum vibration cast formulation was determined and subjected to a variety of standard tests. The results of these tests are shown in Table V.

Table V. Physical Properties of Test Formulation

Test		Test	
Density after 110 °C (g/cm^3)	2.08	MOR after 110 °C (MPa)	8.34
Density after 815 °C (g/cm^3)	1.89	MOR after 815 °C (MPa)	5.03
Density after 1093 °C (g/cm^3)	1.84	MOR after 1093 °C (MPa)	5.58
Density after 1371 °C (g/cm^3)	1.89	MOR after 1371 °C (MPa)	7.24
Linear reheat change after 815 °C (%)	0.0	HMOR at 815 °C (MPa)	6.69
Linear reheat change after 1093 °C (%)	-0.1	HMOR at 1093 °C (MPa)	4.48
Linear reheat change after 1371 °C (%)	-1.4	Thermal Cond. 17°C (W/m-°C)	1.43
CCS after 110 °C (MPa)	38.6	Thermal Cond. 400°C (W/m-°C)	0.76
CCS after 815 °C (MPa)	22.9	Thermal Cond. 800°C (W/m-°C)	0.69
CCS after 1093 °C (MPa)	21.7	Thermal Cond. 1201°C (W/m-°C)	0.76
CCS after 1371 oC (MPa)	26.3		

Aluminum contact testing was performed to evaluate the material for potential field service trials. A 72 hour cup test was performed as well as a 72 hour immersion test. The alloy used for both studies was 7075. Figure 2 shows the result of the cup test. With no adherence, the calcium aluminate was rated excellent. Figure 3 shows the immersion test results. As a basis of comparison, the top photo is a sample of a standard 60% alumina castable with added aluminum penetration inhibitors. The microporous calcium aluminate material is shown on the bottom. It can be seen how the aluminum has penetrated the 60% alumina castable while the calcium aluminate based castable is not penetrated. With such a small average pore size the molten aluminum could not penetrate the castable.

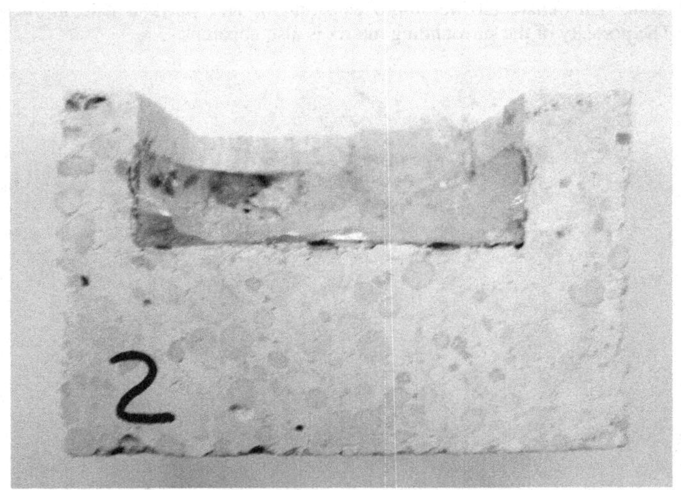

Figure 2. Cup Test Results

Figure 3. Immersion Test Results

Figure 4 shows an SEM micrograph of the cast material. A relatively porous grain was chosen for this photo to show the pore structure of the original aggregate now surrounded by matrix material. Figure 5 shows how the aggregate plays a role in the bonding mechanism. It can be seen how the original surface of the aggregate has reacted and been bonded to the surrounding matrix material. This characteristic improves strength, and plays a role in thermal shock resistance. The porosity of the surrounding matrix is also apparent.

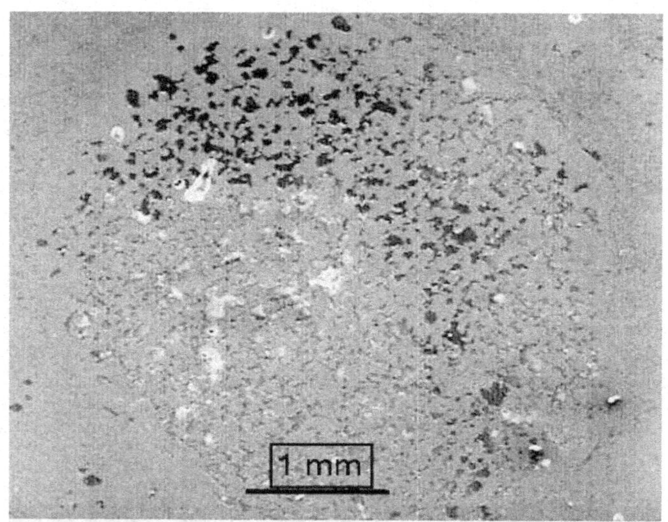

Figure 4. Cross Section of Individual Aggregate Surrounded by Matrix

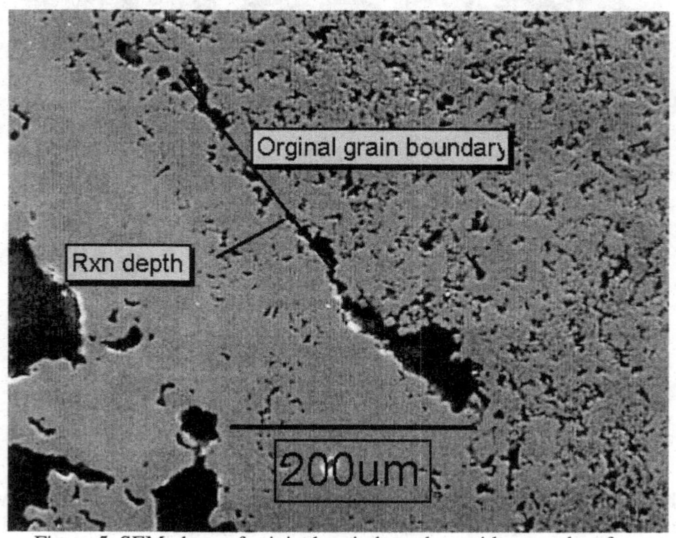

Figure 5. SEM photo of original grain boundary with reacted surface

Field Service Trials / Results and Discussion

Field service trials have been run at several different customers. Some of these are completed and some, involving new customers, are ongoing. The results of several of these are shared. The first field trial involved the direct comparison of two aluminum holding furnaces at a die cast shop. These furnaces were identical with the only difference being the choice of hot face material. The first furnace was lined with the product of this research effort, now branded WAM® AL II. The other furnace was lined with an identical thickness of a common phosphate bonded 60% alumina castable material historically used at this shop. Thermocouples were placed at various depths in the lining and on the shell to evaluate thermal performance. After several weeks in service, the two furnaces were evaluated. The furnaces were re-evaluated several times during their lifetime. Table VI shows the results for the initial evaluation and the last evaluation before the furnace with the phosphate bonded castable had to be taken out of service because it was no longer able to hold temperature. The original WAM® AL II furnace is still in service as of this writing. Figures 6 shows the furnace with the phosphate bonded castable lining drained, cold cleaned and scaled after one year of service. Figure 7 shows the furnace with WAM® AL II after the same one year of service. The effect of aluminum penetration and subsequent corundum formation is apparent on the phosphate bonded castable lining. The WAM® AL II lining shows only mechanically attached aluminum adhering to surface imperfections of the face and no corundum formation on or in the lining.

Table VI. Holding Furnace Trial

Data Collection Time	November 2004		December 2007	
Furnace Lining	WAM® AL II	Phos. Bonded	WAM® AL II	Phos. Bonded
Metal Temperature	682 °C	682 °C	686 °C	686 °C
Energy Usage	2920kWh (7d)	3460 kWh (7d)	485 kWh (27h)	516 kWh (27h)
Delta Energy Usage	540 kWh (7days)		31kWh (27hours)	
Metal Processed	equivalent	equivalent	3081 Kg	1604 Kg
Energy/Unit Metal	n/a	n/a	0.157 kWh/Kg	0.322 kWh/Kg
Energy Savings	15.6%		51.0%	

A second field trial involved the use of WAM® AL II in place of a tabular alumina based lining in a special high temperature aluminum furnace used in a proprietary production process. Table VII shows the results of a comparison shell temperatures and energy usage of the two lining materials. Noteworthy is the average lifetime of the tabular alumina based castable at six months. The first WAM® AL II lining was installed in the spring of 2005 and is still in service. Therefore, no comparative data is available after six months. However, the measured energy usage has remained constant for the WAM® AL II lining over the 3+ years it has been in service.

Table VII. Shell Temperature and Energy Usage Comparison

Furnace Lining Material		Steady State Energy Consumption	
WAM® AL II		29.7 kWh	
Tabular Alumina Castable		35.6 kWh	
Lining Material	Front of Furnace	Left Side Furnace	Right Side Furnace
WAM® AL II	88 °C	82 °C	88 °C
Tab Alumina Castable	121 °C	109 °C	118 °C
Temp Diff	33 °C	27 °C	30 °C
Improvement	27.3 %	24.7 %	25.4 %

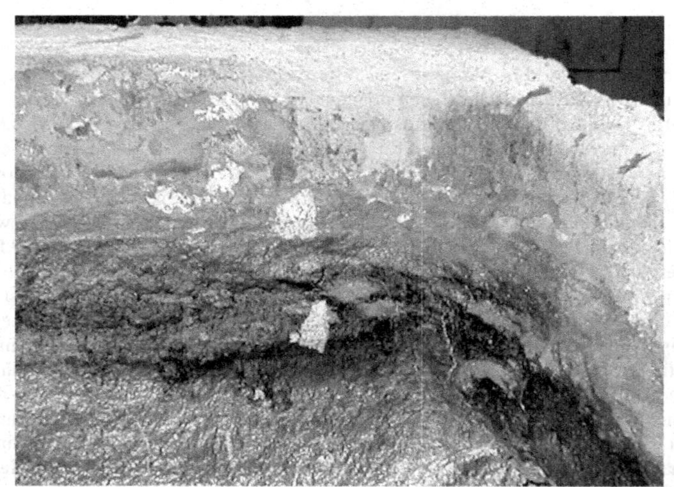

Figure 6. Lining of Phosphate Bonded Castable at 1 Year Service

Figure 7. WAM® AL II Furnace Lining at 1 Year Service

Another field trial involved the installation of several furnaces in the casting shop of a U.S. based automotive manufacturer. At this facility, engine blocks are cast. The maintenance personnel have monitored the energy usage of the WAM® AL II lined furnaces as well as furnaces lined with other standard castable materials. Their findings to date are shown in Table VIII.

Table VIII. Calculated Energy Savings with WAM® AL II

	2 months	14 months	4 years
Holder A	20%		
Holder B		46%	
Launder			25%
Cleaning Frequency Improvement	n/a	400%	+100%

In addition to the obvious savings in energy; the cold cleaning, scaling and patching frequency requirements for the vessels have decreased from every six months to once every two years. Furthermore, the time required to cold clean, scale and patch for a crew of 3 workmen went from 5 days to one day if no patching was required or two days if patching was required. The savings in maintenance costs has not yet been calculated but it would seem significant.

Conclusions

Based on the technical objectives listed the following can be concluded:

- The target of 10% energy savings has not only been met but exceeded. The least favorable savings seen at start up has been a 15% improvement over the previous lining material. Furthermore, during the service life of the vessel the standard refractory linings degrade resulting is less efficient performance while the WAM® AL II linings retain their efficiency. This has resulted in measured energy savings from 26% to over 50% compared to the standard linings.
- Decrease in the BTU or kWh requirements to produce the product has been demonstrated. If the BTU or kWh is generated using fossil fuel there is a corresponding reduced production of CO_2.
- WAM® AL II is similar in cost to some high quality phosphate bonded castable systems. It can be 3X the cost of lower quality 70% alumina based castable systems. However, as an example of cost effectiveness, using the initial kWh savings in November (Table VI) and using the $0.0463/kWh that customer paid in 2004, the energy saving in dollar per year was $1,300 for that furnace. Because of the reduced density of WAM® AL II it requires about 25% less material compared to a standard dense lining like a 70% alumina low cement castable. For this customer who used the phosphate bonded castable, the increased lining expense was paid for in 3 months. The overall payoff is considerably higher (and is still increasing) as the WAM® AL II lining is still operating while the other lining had to be replaced.
- WAM® AL II has been shown to be suitable for use in metal contact via immersion testing and long term field trials.
- The material has demonstrated resistance to corundum attachment and does not promote the formation of either corundum or spinel via oxidation/reduction reactions with material such as SiO_2.
- It has been used in all areas of metal contact, including the belly band areas of reverberatory melting furnaces. It should not be used in hearth areas where solid sows or aluminum block will be pushed or dropped on the surface of the refractory.
- Due to the large amount of porosity in the product the material is very thermal shock resistant. The porosity arrests crack propagation occurring as a result of thermal shock.

- Also due to the large amount of porosity WAM® AL II does not have the modulus or crushing strengths of standard dense castables. The claim that dense, strong material is needed in the belly band is made irrelevant because no corundum forms that is bonded to the lining. Cleaning of the belly band area becomes a gentle and easy process. Furthermore corundum that forms on standard refractory materials far exceeds the strength of available refractories and therefore removal of the corundum always damages these refractories[8].
- The material is phosphate and nitrate free and therefore will not contribute to the formation of organic phosphates or organic nitrates.

Products resulting from this work are covered under U.S. Patent Application Publication No. U.S.2008/0061465 A1, U.S. Patent Application Serial No.11/899,128; and U.S. Patent No. 7,368, 010 B2, granted May 6, 2008, as well as various international patent applications.

Westmoreland Advanced Materials™ would like to acknowledge and extend our appreciation to the follow: U.S. DOE for support of the development of these products under the SBIR program. Honda Manufacturing of Alabama, L & P Aluminum Company and Airo Die Casting, and Fireline TCON, Inc. for their initial and ongoing support.

References

1. W.T. Choate, J.A.S. Green, "U.S. Energy Requirements for Aluminum Production: Historical Perspective, Theoretical Limits and New Opportunities" (Report V1.1, U.S. DOE, Energy Efficiency and Renewable Energy, Industrial Technologies Program, Washington D.C., 2003).
2. Petroleum Marketing Monthly, "Distillate Prices by Sales Type, PAD District, and Selected States" (Energy Information Administration, Table 39: No 2, November 2003).
3. Natural Gas Navigator, "Industrial Gas Prices per Area per Mcf" (Energy Information Administration, November 2003)
4. Heating Oil and Propane Update, "U.S. Heating Oil and Propane Prices" (Energy Information Administration, November 17, 2003).
5. NAFA / Energy Equivalents, "Gasoline Gallon Equivalent (GGE) Table" (National Association of Fleet Administrators, November 2003).
6. P. Angelini, et al. "The Aluminum Industry Technology Roadmap" (The Aluminum Association, Washington D.C. February 2003).
7. Ellingham-Richardson Diagram www.industrialheating.com.
8. K.A. McGowan, P.A. Beaulieu, "Comparative Strength of Refractories verses Corundum in Furnace Linings," *Die Casting Engineering* 2008, v52 n6.
9. M.K. Reser, ed., *Phase Diagrams for Ceramists* (Columbus, OH: American Ceramic Society/NIST, 1964), 231-232.
10. Product Flyer for "r50" (Lafarge Calcium Aluminate (Kronos, Inc.) December 2003).
11. A.H.M. Andreassen, J. Andersen, *Kolloid Z,* 50 (1930), 217-228.
12. Myhre, A.M. Hundere, "The Use of Particle Size Distribution in Development of Refractory Castables" (Paper presented at XXV ALAFAR, December, 1996).

AUTHOR INDEX
Energy Technology Perspectives

A
Ablitzer, D. 65
Asokan, K. 201

B
Bauer, P. 169
Bell, N. 87
Birat, J. 65

C
Choi, M. 93
Cox, A. 77
Cravens, R. 241

D
Damiano, J. 225
Decker, J. 233

E
Eckert, C. 149

F
Fray, D. 77

G
Gamweger, K. 169
Goel, M. 53
Guo, Z. 139

H
Hemrick, J. 225, 241

J
Jaiswal, A. 173
Jha, A. 115, 127

K
Keiser, J. 225

L
Lahiri, A. 115, 127
Lekakh, S. 103
Li, C. 139
Lin, Z. 139
Losif, A. 65
Lu, H. 39

M
McGowan, K. 249
Mirgaux, O. 65
Mishra, B. 213
Mohapatra, B. 213
Mohapatra, R. 213

N
Nayak, B. 213
Niehoff, T. 161

O
Ogura, K. 25
Osborne, M. 149
Overfield, K. 181

P
Peaslee, K. 103
Peters, K. 225, 241
Peterson, R. 149
Pradhan, D. 17
Pradhan, N. 213

R

Ramos, J. 93
Rao, K. 195
Rawlins, C. 103
Reddy, R. 17
Richards, V. 103
Rogov, S. 49

S

Shabala, S. 5
Sohn, H. 93
Stewart, D. 181
Stoffel, D. 161
Subramanian, K. 201
Sukla, L. 213

W

Wang, P. 39
Wikoff, D. 187
Williams, E. 181
Williams, S. 187

Y

Yang, X. 127

Z

Zavodinsky, V. 49
Zhang, X. 39
Zhang, Y. 93

SUBJECT INDEX
Energy Technology Perspectives

1

1-Ethyl-3-methyl-imidazolium Chloride 17
1-n-butyl-3-methylimidazolium Hexafluorophosphate 39

A

Aluminum 149, 241, 249
Aluminum Slugs 169
Anhydrase Enzyme 103
Application 139
Aqueous Processing 195
Artificial Photosynthesis 5
Aspen PlusTM Software 65

B

Bipolar Electrodes 201
Busbar Saving 201

C

Calcium Leaching 103
Carbon Capture 53
Carbon Dioxide 39, 49
Carbon Dioxide Assimilation 5
Carbon Dioxide Emission 93
Carbon Dioxide Reduction 87
Carbon Dioxide Sequestration 103
Carbonic 103
Catalytic Mechanism 139
Cations 115
CeO2 139
CO2 Emissions 65
Combustion 139
Combustion Systems 161
Complex Ores 195
Copper Electrowinning 201
Corundum 249

D

Dendrite 17

E

EAF 173
Efficient 249
Electric 173
Electro-deoxidation 77
Electrochemical Reduction 39
Electrodeposition 17
Energy 161, 181, 187, 249
Energy Conservation 201
Energy Efficiency 149
Energy Savings 87, 195, 225, 241

F

Ferric Oxide 77
Foamy Slag 173
Furnaces 161

G

Gas Purging 169
Glass 225
Goethite 213

H

Hot DRI 173
Hot Metal 173
Hydrogen Reduction 93

I

Ionic Liquid 17, 39
Iron (III) Reducing Bacteria (IRB)213
Iron Oxide Concentrate 93
Isothermal Melting 149

L

LCA 65
Lead Pollution Abatement 201
Light 5
Lime.. 225
Liquid Phosphate Bonded
Refractories.............................. 233

M

Magnetite and Leaching.............. 213
Mass Transfer Enhancement........ 201
Materials 53
Melting.............................. 149, 181
Melting Rate 169
Metal Cations.............................. 5
Molten Metal 225, 241

N

Nickel Ore................................ 213
Non-wetting 249
Non-Wetting Properties 233

O

Oxy-fuel............................. 87, 161
Oxy-fuel Burners 173

P

Particle Size Distribution 195
Photosynthetic Light Conversion..... 5
Porous Plugs 169
Power Consumption.................... 173
Process Modifications................. 161
Pseudo-potential Calculations........ 49
Pulverized Coal.......................... 139

R

R&D 53
Rare Earths 115
Rate Analysis 93
Reduction 77
Refractories 225, 241
Refractory................................ 249
Reliability................................. 187
Roasting 115

S

Scrap Preheating........................ 173
Secondary Aluminum.................. 149
Sequestration 87
Silicon Dioxide 49
Sodium Hydroxide 77
Solid State 49
Steelmaking................................ 65
Steelmaking Slag........................ 103

W

Water Salinity 195